Anton Pannekoek: Ways of Viewing Science and Society

# Anton Pannekoek: Ways of Viewing Science and Society

*Edited by*
*Chaokang Tai, Bart van der Steen, and Jeroen van Dongen*

Routledge
Taylor & Francis Group
LONDON AND NEW YORK

First published in 2019 by Amsterdam University Press Ltd.

Published 2025 by Routledge
4 Park Square, Milton Park, Abingdon, Oxon OX14 4RN
605 Third Avenue, New York, NY 10158

*Routledge is an imprint of the Taylor & Francis Group, an informa business*



ISBN: 9789462984349 (hbk)
ISBN: 9781041175643 (pbk)
ISBN: 9781003690924 (ebk)

NUR 686

Cover design: Coördesign, Leiden

DOI 10.5117/9789462984349

Cover illustration: (Background) Fisheye lens photo of the Zeiss Planetarium Projector of Artis Amsterdam Royal Zoo in action. (Foreground) Fisheye lens photo of a portrait of Anton Pannekoek displayed in the common room of the Anton Pannekoek Institute for Astronomy. Source: Jeronimo Voss

For Product Safety Concerns and Information please contact our EU representative:
GPSR@taylorandfrancis.com
Taylor & Francis Verlag GmbH, Kaufingerstraße 24, 80331 München, Germany

# Table of Contents

## List of Figures

# 1 Anton Pannekoek: Ways of Viewing Science and Society

*Chaokang Tai, Bart van der Steen, and Jeroen van Dongen*

Astronomer and Marxist Anton Pannekoek was a remarkable figure. As an astronomer, he pioneered quantitative astrophysics and founded the renowned Astronomical Institute in Amsterdam that now carries his name. Before World War I, however, he was employed as a Marxist theorist by the Social Democratic Party of Germany, making him one of the leading intellectuals of international socialism. Because of his significant contributions to such diverse subjects as astronomy and socialism, Pannekoek's life and work uniquely capture the fascinating connections between conceptions of nature, society, and their representations in the early decades of the twentieth century. This book aims to study these connections through the prism of Pannekoek's biography. In doing so, it sets out to explain Pannekoek's particular epistemic, aesthetic, and political choices, while placing them in the broader context of the early twentieth century.

Pannekoek tried to keep connections between his political and academic life hidden from view. He had pragmatic reasons to do so. His academic career had suffered from his controversial political reputation on more than one occasion, most dramatically in 1919 when his appointment to deputy director of the Leiden Observatory was obstructed by the Dutch government.[1] From the mid-1910s onwards, he kept his socialist efforts at a distance from his career in astronomy, and even ended up writing two separate autobiographies: one focusing on his career in the labour movement, while the other discussed his astronomical research.[2]

Remarkably, this separation has been carried over into scholarship on his life and work. This either discusses Pannekoek's role in the labour

1 This episode is discussed in detail in: Baneke 2004; and his 'Pannekoek's One Revolution', in this volume, 87-108.
2 Pannekoek 1982.

Tai, Chaokang, Bart van der Steen, and Jeroen van Dongen (eds), *Anton Pannekoek: Ways of Viewing Science and Society*. Taylor & Francis Group, 2019
DOI: 10.5117/9789462984349_CH01

movement, or in astronomy – but rarely their possible relations.[3] This book, on the other hand, seeks to identify and elucidate the relations between Pannekoek's various contributions to science and political theory. This offers the opportunity to gauge the unity and singularity of Pannekoek's work on the one hand, while providing more insight into the wider relations between academia, politics, and ways of viewing the world on the other. Finally, to address the last aspect, the book will also engage the visual arts, both historically and in its attempts to capture the social and natural world.

Multiple chapters in this volume draw attention to the visual aspect of Pannekoek's work – in particular to his engagement with photography and his drawings of the Milky Way. This focus on aesthetics and the visual offers insight into Pannekoek and his time, as well as in current relations between the arts and sciences. Throughout the modern era, both have influenced each other in crucial ways. This was especially true in the early modern period, but even after their institutional separation began to emerge in the nineteenth century, their mutual influence never fully disappeared.[4] This may be exemplified by the way in which Pannekoek's drawings of the Milky Way have captured not only the scientific, but also the artist's imagination, as revealed by the work of contemporary artist Jeronimo Voss, who engaged with Pannekoek's images to find novel representations of both the cosmos and the ideals of communism.[5] These were presented in the installation *Inverted Night Sky*, which was exhibited at the Stedelijk Museum Bureau Amsterdam in May and June of 2016.[6] At the same time, the conference 'Anton Pannekoek. Ways of Viewing Science and Society' was held at the Royal Netherlands Academy of Arts and Sciences in Amsterdam, of which this book is the result.

By engaging with the aesthetics of Pannekoek's drawings, we can learn more about the relation between science and art as they persist into the present. Moreover, a focus on the visual aspect of Pannekoek's work elucidates key elements of his scientific methodology. For a long time, historians of

3 When Pannekoek's autobiography was published in 1982, for example, it contained two introductions that separately discussed his socialism and his astronomy. The lack of any attempt to combine the two was already criticized by Klaas van Berkel (1984). For recent scholarship that does attempt to arrive at a unified understanding of Pannekoek, see Tai and van Dongen 2016; Tai 2017.

4 On the mutual development of art and science in the early modern period, see, e.g. Bennett 1982; Edgerton 1991; Kwa 2005; Smith 2006; Long 2011; for their separation in the nineteenth century, see Daston 1998; Jones and Galison 1998; for examples of their mutual influence in the late modern period, see Henderson 1983; Galison 1990; Wilder 2009; Kojevnikov 2016.

5 For more on Pannekoek's influence on artists, see Lütticken 2018.

6 Voss 2016; For descriptions and images of the exposition, see SMBA 2016; Voss 2017.

science ignored images as mere tools, intended only to illustrate knowledge that was mainly conveyed in words and equations. Yet, scientific images are objects worth studying in their own right when trying to understand how science is practised.[7] In particular, the aesthetic and technical choices scientists make in producing and reproducing images do not just reveal aspects of the knowledge that they wish to convey; they also reflect how scientists believe nature should be observed – indeed, what skills and virtues are required to do these observations.[8] Thus, by looking at how Pannekoek decided to represent the Milky Way, it becomes possible to explore what he believed proper scientific practice was and how he believed scientific knowledge should be constructed.

Pannekoek's life is a rich source of information on the relations between visual culture, scientific scholarship, and leftist politics in the early twentieth century. Of course, he was not the only left-wing radical who moved among these various domains.[9] Similar connections can be found in the lives of socialist physicists like Friedrich Adler, Léon Rosenfeld, and Yakov Frenkel, to name only three examples. Adler was trained as a physicist and at one time had been in close contact with Albert Einstein. He is perhaps best known, however, for his assassination of the Austrian Prime Minister Karl von Stürgkh in 1916, which he hoped would start a socialist revolution in Austria. While imprisoned, he struck up a correspondence with Einstein on the foundations of relativity theory. He tried to reconcile Einstein's relativity principle with the classic concept of a privileged reference frame, much like he attempted to reconcile the revolutionary ideals of the Bolsheviks with his support of the social-democratic Second International.[10] In the case of Belgian quantum theorist Rosenfeld, the connection can be found in his vehement defence of the principle of complementarity in quantum mechanics – mainly against criticism by Soviet physicists, who considered it idealist and subjective; Rosenfeld argued that the principle was the result

7 For an overview of how historians started to research scientific images, see Pang 1997; Jones and Galison 1998; Kusukawa 2016.

8 See, e.g. Daston and Galison 1992; 2007; Winkler and van Helden 1992; Schaffer 1998; Nasim 2013.

9 It is worth mentioning that two of Pannekoek's closest socialist companions, Herman Gorter and Henriette Roland Holst, were also key members of the innovative and progressive impressionist 'Tachtigers' movement in Dutch literature; thus, they moved between literary and socialist circles, as Pannekoek crossed the boundaries between the sciences and socialism. In their case, the subject matter and purpose of their poetry was drenched in socialist themes. For Herman Gorter, see de Liagre Böhl 1996; Zwart 2019; for Henriette Roland Holst, see Etty 1996.

10 Galison 2008.

of a practical application of the dialectic method.[11] Soviet condensed matter physicist Frenkel, finally, used the social concept of collectivism, as it was understood by early twentieth-century radicals, as a metaphor to explain the collective behaviour of electrons in metals, crystals, and plasmas.[12] These examples raise the question whether a similar close connection between political thought and scientific work can be revealed in the case of Pannekoek. Can we get a better understanding of both Pannekoek's astronomy and his Marxism if we investigate how they might relate to one another?

The role of aesthetics at the crossroads of scholarship and political activism is particularly pertinent in the case of Otto Neurath and Rudolf Carnap. Neurath was a socialist philosopher and political economist who famously visualized statistical data through the invention of pictorial 'isotypes', in collaboration with modernist artist Gerd Arntz. By displaying statistical information visually, he strove to enable the masses to access and interpret it.[13] Carnap was the author of *Der logische Aufbau der Welt* (1928), in which he attempted to develop a framework that reduced all empirical knowledge to direct sensory experience. His philosophy brought him in close contact with the architects of the Bauhaus art school, who shared his left-technocratic vision of the world as built up from simplest elements; at the invitation of Walter Gropius, Carnap gave several lectures at the Bauhaus in Dessau.[14] Both Neurath and Carnap were prominent members of the Vienna Circle, the group of philosophers who shared the explicit goal of making philosophy 'scientific' by stripping it of its metaphysical content. This desire was shared by Pannekoek and it is therefore no coincidence that he published an article on the 'essence of natural laws' in *Erkenntnis*, the journal co-edited by Carnap and affiliated with the Vienna Circle.[15]

A scientist's (presumed) close relations with radical politics and avant-garde art could be cause for suspicion for both the authorities and the public at large. This is not only illustrated by Pannekoek's thwarted Observatory position, but also by the delayed appointment of Albert Einstein to a visiting professorship at Leiden University in 1920. In that year, politically charged debates on the truth and significance of relativity theory reached their apex. Einstein was first appropriated by the Dada art movement in a collage by artist Hannah Höch, while reactionary critics of relativity accused

11 Jacobsen 2007.

12 Kojevnikov 1999.

13 Cartwright et al. 1996; Leonard 1999; Mattick 2016.

14 Galison 1990; 1996. For a discussion of Neurath's connections with the Dessau Bauhaus, see Potochnik and Yap 2006.

15 Pannekoek 1932.

him of being a political revolutionary and giving a false representation of nature, which they identified as 'scientific Dadaism'. Einstein himself had a somewhat traditional taste in art and was not a communist but rather a democrat and pacifist – but this did not stop others from labelling him as a radical and accusing his scientific theories of being politically subversive.[16] Influenced by these debates in Germany, the Dutch government, in turn, confused Albert Einstein with the German art critic Carl Einstein, who was in fact a true far-left revolutionary. Carl Einstein, an early promotor of cubism and African tribal art, had been a leading member of the German soldiers' council that had mutinied during the retreat from Brussels in 1918. As a result of this confusion, Albert Einstein's appointment as visiting professor to the Leiden physics department was held up for nearly a year in 1920. Dutch officials wished to be absolutely certain about his political persuasions, as they wished to avoid a repetition of the botched appointment of Pannekoek at the Leiden Observatory a year earlier.[17]

In 1934, there was yet another incident in which a leftist scientist was barred from a Dutch university. This time, Marxist mathematician Dirk Jan Struik was withdrawn as candidate for a guest professorship at Delft University of Technology following objections from the Dutch government.[18] Although Struik was more than two decades younger than Pannekoek, their lives and careers show remarkable similarities. Struik, too, strove to keep socialism and science separate domains of his activity. After deciding to become a socialist professional, rather than a professional socialist, Struik's Marxist beliefs were mainly reserved for his historical writings. He founded the interdisciplinary Marxist journal *Science and Society* and pioneered a dialectic-materialistic approach to the historiography of science in his monograph *Yankee Science in the Making*.[19]

As the above examples show, studying Pannekoek and others at similar junctures in the early twentieth century offers us not only biographical insights, but it also promises to elucidate the ways in which Pannekoek and his contemporaries balanced scientific and political ambitions. Furthermore, it will show us how contemporaries reflected on how progressive, 'revolutionary' science and politics interacted, and the role that the era's innovations in visual culture played in this. These scholars all advanced extraordinary intellectual innovation, while sharing the tumultuous rhetoric of revolution

16 Goenner 2005; van Dongen 2007.

17 van Dongen 2012.

18 Alberts 1994, 281.

19 Alberts 1994.

– for which they were considered a vanguard by some yet abhorred by others. This book focuses exactly on these themes: on how understanding the links between science and society informed representations of nature as well as scientific and political choices in the revolutionary cultures of the early twentieth century. Clearly, Pannekoek offers a uniquely rich starting point for such an endeavour.

Although Pannekoek worked as an astronomer for most of his professional life, it is his political career that has received the largest share of attention from historians and biographers so far. Interest in his political work was revived in late 1960s as the New Left began to pay attention to Pannekoek due to his opposition to both moderate social democrats such as Karl Kautsky and to dogmatic Marxists such as Vladimir Lenin. Former collaborators of Pannekoek subsequently republished his work, and provided a synthesis of his ideas in an effort to rekindle the council communist programme.[20] In the following decades, scholarly reconstructions of Pannekoek's political development were produced, while activist interest in council communism mostly subsided.[21] Contemporary historiography is of course less interested in reconstructing the council communist programme but rather aims to understand the council communist movement in its proper historical context.[22]

Pannekoek's astronomical career has received less attention. Although some of his contributions have found their way into more general surveys in the history of astronomy,[23] these do not offer more than a superficial indication of his research and methodology. Only recently historians of science have attempted more thorough investigations of Pannekoek's astronomical research in an effort to understand and contextualize his scientific research.[24] Yet, there is still much left to be explored before a comprehensive overview of all of his major contributions to astronomy can be provided. This volume indeed aspires to deepen our understanding of Pannekoek's scientific contributions, and to do so by engaging equally his contributions to epistemology and socialist theory. Only then can we begin to unravel their intricate relations.

Astronomer Edward P.J. van den Heuvel, former director of the Anton Pannekoek Institute for Astronomy, first offers a biographical overview of

20 See Brendel 1970; Kloosterman 1972; Smart 1978; Bricianer 1978; Sijes 1982.

21 See, in particular, Boekelman 1980; Gerber 1989; Malandrino 1987; Bock 1992; 1993.

22 See, e.g. van der Linden 2004; van der Steen 2006.

23 See, e.g. Hearnshaw 2014; Baneke 2015.

24 Houziaux 2001; Tai and van Dongen 2016; Tai 2017.

Pannekoek's life based on many conversations with former colleagues and students of Pannekoek. Van den Heuvel shows how Pannekoek's rejection for the Leiden position turned out to be a blessing in disguise as he subsequently was offered a position at the University of Amsterdam, where he had the opportunity to shape his own research agenda independently (in Leiden, he would have had to work under Willem de Sitter). Because his newly founded Astronomical Institute lacked an observatory, Pannekoek became involved in the emerging field of theoretical astrophysics, which developed in tandem with highly innovative studies in atomic physics.

Focusing more on Pannekoek's political career, Gerrit Voerman asks why Pannekoek ended up in numerous heated conflicts with socialist leaders such as Pieter Jelles Troelstra, Karl Kautsky, and Lenin – conflicts that effectively marginalized him as a socialist activist. Voerman points to the principled nature of Pannekoek's character and his preference for theoretical analysis over practical considerations. This meant that he rejected any form of compromise and would become frustrated when the outcomes of his analyses were not acceded to. He was willing to accept the personal consequences of his steadfastness and break off relations with close collaborators if consistency of his political positions dictated such a course of action.

Klaas van Berkel searches for commonalities between Pannekoek as astronomer and Pannekoek as socialist by reflecting on why he made this distinction in his biography in the first place. Van Berkel finds that the distinction is a historical construct that had been created by Pannekoek because of the incidents in which his socialist activism had hindered his astronomical career – not just in 1919 when his Leiden appointment was blocked, but also in 1903 when he was reprimanded by Dutch prime minister Abraham Kuyper for his outspoken support of a general labour strike that year. According to Van Berkel, the most fundamental element that was shared between Pannekoek's approaches to astronomy and socialism was an emotional commitment: a utopian longing for wholeness and purity in both nature and society.

The historical context of Pannekoek's astronomy is the focus of the following contributions. David Baneke provides an overview of Pannekoek's influence on the astronomy community in the Netherlands. After a detailed analysis of Pannekoek's role in the reorganization of the Leiden Observatory and his rejection as assistant director there, he discusses Pannekoek's close relations with Utrecht's Marcel Minnaert, another communist astronomer. Together they established the Dutch school of astrophysics, which first focused on the properties of stellar atmospheres. Pannekoek further contributed to Dutch astronomy by supporting the creation of a Dutch astronomical society and journal. Baneke contends that Pannekoek's actual 'revolution' is found not in the political realm but in the modernization of Dutch astronomy, both institutionally and academically.

Robert W. Smith situates Pannekoek in the wider development of astronomy in the early twentieth century, during which it underwent rapid changes. Not only were ideas about the shape and size of the galaxy in flux, the notion of what astronomy should study, and how and where this should be studied changed as well. Pannekoek, as Smith argues, was at the forefront of many of these developments: his methods and concerns were both influential and representative of the era. Not only was he one of the first astronomers to provide supporting evidence for Harlow Shapley's new model of the galaxy, he was also one of the earliest practitioners of the new quantitative astrophysics that applied the latest developments in atomic physics and quantum mechanics to the stars. Smith concludes that, as an astronomer, 'Pannekoek [...] was both very much of, as well as a maker of, his time.'

Pannekoek considered Marxism to be a science in its own right. This position was shared by many of his socialist contemporaries and predecessors, including, as Bart van der Steen explains, Karl Marx and Friedrich Engels. They had introduced the term 'scientific socialism' for their own approach to socialism. Engels had contended that their approach was preferable because, rather than simply imagining better societies, they used a scientific method to analyse how socialism would evolve out of the contradicting tendencies inherent to capitalism. Even so, the exact content and method of scientific socialism remained strongly contested. In his contribution, Van der Steen reconstructs Pannekoek's understanding of what scientific socialism implied. He finds that three distinct but closely related definitions of scientific socialism can be found in Pannekoek's writing. Socialism was 'scientific' because 1) it made predictions about the future (which entailed that the socialist revolution was imminent); 2) it provided a method for analysing past and present social developments; and 3) because it argued for a worldview that strove for truth through scientific research. This final position offered Pannekoek the opportunity to align his socialism with his astronomical research.

Pannekoek's understanding of scientific socialism deviated from that of many of his contemporaries, as Annemarie Rullens shows. Pannekoek considered scientific socialism a method for analysing human behaviour, which had to be developed further by the working classes. Thus they would gain the consciousness that would enable them to establish a socialist society. Rullens contrasts this view with that of Pannekoek's contemporary Willem Bonger, a prominent Dutch socialist and professor of criminology at the

University of Amsterdam. According to Bonger, society had to be transformed by using the latest insights offered by statistics, social science, economics, and even biology. As member of the Social Democratic Workers' Party (SDAP), he advocated for policies aimed at this goal. For Bonger, scientific socialism was not a philosophical stance, as it was for Pannekoek. Instead, it contained a practical imperative. This position was shared by many of the generation of Dutch leftist ideologues that came after Pannekoek, and of which Bonger can be seen as a representative.

One of Pannekoek's struggles was to make his astronomical research socially relevant. Jennifer Tucker argues that Pannekoek found a way to achieve this by engaging the public and broadening its understanding of science. Pannekoek outlined a method for amateur astronomers to observe and record the Milky Way in his earlier life, for example. He later wrote several popular histories of astronomy. These emphasized the socio-economical context in which astronomy had developed and the progressive values it promoted – in line with the work of other Marxist historians like Boris Hessen, Edgar Zilsel, and J.D. Bernal. In these studies, Pannekoek highlighted the collaborative and elaborate practical effort involved in astronomical research, and he discussed at length the struggles and errors involved in the scientific process. As such, he intended to show the scientific worker 'in overalls'.

Pannekoek's historical studies are also the subject of Bart Karstens' contribution. He addresses how Pannekoek's research should be positioned within contemporary developments in historical sociology of science. As Karstens indicates, Pannekoek's historical research has been appropriated by members of the so-called 'strong programme' in the sociology of scientific knowledge, like Stevin Shapin and Barry Barnes: they saw in Pannekoek an early example of their preferred type of analysis. According to the strong programme, both the development and the content of scientific knowledge is strongly determined by social factors. After analysing Pannekoek's discussion of the discovery of the planet Neptune, Karstens argues that this appropriation of Pannekoek was misguided: far from an early example of the strong programme, Pannekoek's approach most closely resembles that of contemporary sociologist Robert Merton. In his case, too, social factors may influence the direction and pace of scientific research but not its content.

Pannekoek's Milky Way drawings provide an excellent opportunity to establish the deeper epistemic links between his astronomy and Marxism. An analysis of these drawings is provided by Chaokang Tai, who argues that Pannekoek's methods of investigating and depicting the Milky Way reflected his Marxist understanding of how the mind processes information.

According to Pannekoek, the mind instinctively and intuitively synthesizes valuable information about the world from the continuous flow of disparate human observations. To forego such insights would leave a scientist without a well of knowledge, which was the reason Pannekoek held that drawings of the Milky Way could display insights that photographic images could not. When Pannekoek did employ photography, he used a method that allowed the photographic plate to mimic the properties of the human eye, effectively mechanizing human observation – but even then, the end result had to be displayed through drawings.

Omar W. Nasim also searches for the connection between Pannekoek's astronomy and socialism in his Milky Way research. But rather than discussing the role of the mind in Pannekoek's research, Nasim focuses on the role of the hand. He points out that both in his *Marxism and Darwinism* and in *Anthropogenesis*, Pannekoek assigned great significance to the role of manual labour in the development of mankind. According to Pannekoek, the use of increasingly sophisticated tools led to the development of speech and abstract thought. Nasim shows that this emphasis on the value of manual labour was reflected in Pannekoek's Milky Way research, in which hand drawn images of the Milky Way were to be trusted over mechanically produced photographs: it is by the hand that we know. Like Tai, Nasim recognizes that Pannekoek indeed employed photography in his Milky Way studies, but that in the end, his methods were really grounded in laborious handwork.

In combination with his socialist writings, Pannekoek's Milky Way drawings also provided a crucial inspiration for Jeronimo Voss' work *Inverted Night Sky*. In his conversation with cultural theorist Johan Hartle, included in this volume, Voss reflects on his exhibition and the inspiration that Pannekoek's life and drawings of the Milky Way offered. It leads Voss to explore the historical ties between imaginations of the cosmos and communism, and reflect upon how these can enrich both contemporary art and social criticism. Voss is used to transgressing boundaries and aspires to, in his own words, 'a universalist perspective that goes beyond [...] the traditionally separated domains of visual art, documentarism, science, politics, and every-day life' – just as Pannekoek did, one may add. Thus, Voss has created dome structures with projections of Pannekoek's Milky Way drawings that collapse various techniques and that offer both a unique inverted perspective on our nearby cosmos, while they are blended and framed with social commentary.

Alena J. Williams offers a 'close reading' of Voss's art, which she relates to how revolutionaries from Pannekoek's time to today have used images and conceptions of the cosmos to imagine both revolutionary ideals and their catastrophes. A case in point was Louis Auguste Blanqui, a revolutionary who played a leading role in the Paris Commune of 1871: he took to astronomical musings to process the dramatic defeat of the Commune and to rekindle his hopes for a revolutionary future. Williams shows how Voss is inspired by Blanqui's hypothesis 'that all possible variations of our own past, present, and future are real material facts located within infinite space' as it promotes 'a worldview that conceptualizes history as a product of collective decisions rather than as an independent stream of time'. Voss investigates Pannekoek's life, politics, and especially his visualisations of the Milky Way from this perspective, according to Williams. Voss's work and Williams's contemplations on them give greater urgency to Pannekoek's aesthetic choices and their possible political implications.

The articles in this volume reaffirm that Pannekoek's contributions to astronomy and socialism cannot be considered as independent from each other. By investigating his work in both science and political theory, along with his broader epistemology, a multifaceted view emerges that not only reveals the many connections and similarities between his socialist and scientific career, but also clearly shows that they are deeply interconnected in Pannekoek's approach, methods, and goals. Moreover, Pannekoek's case uniquely illustrates the arrival of modernity, and its upheavals: as new ways of being were introduced, new ways of viewing were required – as has famously been documented in the arts, reflected in the sciences, and expressed in the social revolutions that spread across Europe. Pannekoek stood at the epicentre of these developments and contributed to them at least as much as he reflected them: innovation in perspective was often translated into the language of revolution, and Pannekoek was a revolutionary in spirit at least as much as he was an intellectual in temperament.

## Acknowledgements

We thank the Royal Netherlands Academy of Arts and Sciences (KNAW), the Amsterdam University Fund (AUF), the Netherlands Research School for Astronomy (NOVA), the Leids Kerkhoven-Bosscha Fonds (LKBF), and Stichting Pieter Zeeman Fonds for their support of the 2016 conference 'Anton Pannekoek (1873-1960): Ways of Viewing Science and Society' of which this book is the result. We are also grateful to the Vossius Fund for the History of the University of Amsterdam, the Anton Pannekoek Institute for Astronomy, and the Jan van Paradijs Fund for their generous financial support to make the open access publication of this book possible.

## Bibliography


Alberts, Gerard. 1994. 'On Connecting Socialism and Mathematics: Dirk Struik, Jan Burgers, and Jan Tinbergen'. *Historia Mathematica* 21(3): 280-305.

Baneke, David. 2004. '"Hij kan toch moeilijk de sterren in de war schoppen". De afwijzing van Anton Pannekoek als adjunct-directeur van de Leidse Sterrewacht in 1919'. *Gewina* 27 (1): 1-13.

—. 2015. *De ontdekkers van de hemel. De Nederlandse sterrenkunde in de twintigste eeuw*. Amsterdam: Prometheus Bert Bakker.

Bennett, Jim A. 1982. *The Mathematical Science of Christopher Wren*. Cambridge: Cambridge University Press.

Berkel, Klaas van. 1984. Review of *Herinneringen. Herinneringen uit de arbeidersbeweging. Sterrekundige herinneringen*, by Anton Pannekoek. *BMGN – Low Countries Historical Review* 99(3): 465-467.

Bock, Hans Manfred. 1992. 'Die Marx-Dietzgen-Synthese Anton Pannekoeks und seines Kreises'. In *Die Rezeption marxschen Theorien in den Niederlanden*, ed. by Marcel van der Linden. Trier: Karl-Marx Haus, 106-123.

—. 1993. *Syndikalismus und Linkskommunismus von 1918 bis 1923. Ein Beitrag zur Sozial- und Ideengeschichte der frühen Weimarer Republik*. Darmstadt: Wissenschaftliche Buchgesellschaft.

Boekelman, Marinus A.M. 1980. 'The Development of the Social and Political Thought of Anton Pannekoek, 1873-1960: From Social Democracy to Council Communism'. PhD Thesis, Toronto: University of Toronto.

Brendel, Cajo. 1970. *Anton Pannekoek, theoretikus van het socialisme*. Nijmegen: Socialistische Uitgeverij Nijmegen.

Bricianer, Serge. 1978. *Pannekoek and the Worker's Councils*. Saint Louis: Telos Press.

Cartwright, Nancy, Jordi Cat, Lola Fleck, and Thomas E. Uebel. 1996. *Otto Neurath: Philosophy between Science and Politics*. Cambridge: Cambridge University Press.

Daston, Lorraine. 1998. 'Fear and Loathing of the Imagination in Science'. *Daedalus* 127(1): 73-95.

Daston, Lorraine, and Peter Galison. 1992. 'The Image of Objectivity'. *Representations* 40: 81-128.

—. 2007. *Objectivity*. New York: Zone Books.

Dongen, Jeroen van. 2007. 'Reactionaries and Einstein's Fame: "German Scientists for the Preservation of Pure Science", Relativity, and the Bad Nauheim Meeting'. *Physics in Perspective* 9(2): 212-230.

—. 2012. 'Mistaken Identity and Mirror Images: Albert and Carl Einstein, Leiden and Berlin, Relativity and Revolution'. *Physics in Perspective* 14(2): 126-177.

Edgerton, Samuel Y. 1991. *Heritage of Giotto's Geometry: Art and Science on the Eve of the Scientific Revolution*. Ithaca, NY: Cornell University Press.

Etty, Elsbeth. 1996. *Liefde is heel het leven niet: Henriette Roland Holst, 1869-1952*. Amsterdam: Balans.

Galison, Peter. 1990. 'Aufbau/Bauhaus: Logical Positivism and Architectural Modernism'. *Critical Inquiry* 16(4): 709-752.

—. 1996. 'Constructing Modernism: The Cultural Location of Aufbau'. In *Origins of Logical Empiricism*, ed. by Ronald N. Giere and Alan W. Richardson. Minnesota Studies in the Philosophy of Science 16. Minneapolis: University of Minnesota Press, 17-44.

—. 2008. 'The Assassin of Relativity'. In *Einstein for the 21st Century: His Legacy in Science, Art, and Modern Culture*, ed. by Peter Galison, Gerald Holton, and Silvan S. Schweber. Princeton: Princeton University Press, 185-204.

Gerber, John P. 1989. *Anton Pannekoek and the Socialism of Workers' Self-Emancipation, 1873-1960*. Dordrecht: Kluwer.

Goenner, Hubert. 2005. *Einstein in Berlin 1914-1933*. Munich: C.H. Beck.

Hearnshaw, John B. 2014. *The Analysis of Starlight: Two Centuries of Astronomical Spectroscopy*. New York: Cambridge University Press.

Henderson, Linda Dalrymple. 1983. *The Fourth Dimension and Non-Euclidean Geometry in Modern Art*. Princeton, NJ: Princeton University Press.

Houziaux, Léo. 2001. 'Schuster-Schwarzschild's and Milne-Eddington's Viewpoints on the Physical Structure of Stellar Atmospheres'. In *Optics and Astronomy*, ed. by Gérard Simon and Suzanne Débarbat. Proceedings of the XXth International Congress of History of Science (Liège, 20-26 July 1997) 12. Turnhout: Brepols, 217-228.

Jacobsen, Anja Skaar. 2007. 'Léon Rosenfeld's Marxist Defense of Complementarity'. *Historical Studies in the Physical and Biological Sciences* 37 (suppl.): 3-34.

Jones, Caroline A., and Peter Galison, eds. 1998. 'Introduction'. In *Picturing Science, Producing Art*. New York: Routledge, 1-23.

Kloosterman, Jaap, ed. 1972. *Anton Pannekoek: Partij, raden, revolutie*. Amsterdam: Van Gennep.

Kojevnikov, Alexei. 1999. 'Freedom, Collectivism, and Quasiparticles: Social Metaphors in Quantum Physics'. *Historical Studies in the Physical and Biological Sciences* 29(2): 295-331.

—. 2016. 'Universe in Upheaval: Relativistic Cosmology in the Context of the Russian Revolution', lecture at 'Anton Pannekoek (1873-1960): Ways of Viewing Science and Society', Royal Netherlands Academy of Arts and Science, Amsterdam.

Kusukawa, Sachiko. 2016. 'Classics from This Journal: Martin Rudwick's "The Emergence of a Visual Language for Geological Science 1760-1840"'. *History of Science* 54(1): 98-104.

Kwa, Chunglin. 2005. 'Alexander von Humboldt's Invention of the Natural Landscape'. *The European Legacy* 10(2): 149-162.

Leonard, Robert J. 1999. '"Seeing Is Believing": Otto Neurath, Graphic Art, and the Social Order'. *History of Political Economy* 31 (suppl.): 452-478.

Liagre Böhl, Herman de. 1996. *Met al mijn bloed heb ik voor U geleefd. Herman Gorter 1864-1927*. Amsterdam: Balans.

Linden, Marcel van der. 2004. 'On Council Communism'. *Historical Materialism* 12(4): 27-50.

Long, Pamela O. 2011. *Artisan/Practitioners and the Rise of the New Sciences, 1400-1600*. Corvallis, OR: Oregon State University Press.

Lütticken, Sven. 2018. 'Council Aestheticism? Pannekoek, the Avant-Garde and Contemporary Art'. *Studi di estetica* 46(12): 173-197.

Malandrino, Corrado. 1987. *Scienza e socialismo. Anton Pannekoek (1873-1960)*. Milan: Franco Angeli.

Mattick, Paul. 2016. 'Socialism, Art, and Science in the Work of Otto Neurath', lecture at 'Anton Pannekoek (1873-1960): Ways of Viewing Science and Society', Royal Netherlands Academy of Arts and Science, Amsterdam.

Nasim, Omar W. 2013. *Observing by Hand: Sketching the Nebulae in the Nineteenth Century*. Chicago, IL: University of Chicago Press.

Pang, Alex Soojung-Kim. 1997. 'Visual Representation and Post-Constructivist History of Science'. *Historical Studies in the Physical and Biological Sciences* 28(1): 139-171.

Pannekoek, Anton. 1932. 'Das Wesen des Naturgesetzes'. *Erkenntnis* 3(1): 389-400.

—. 1982. *Herinneringen: herinneringen uit de arbeidersbeweging; sterrenkundige herinneringen*. Eds. Ben A. Sijes, Johanna M. Welcker, and J.R. van der Leeuw. Amsterdam: Van Gennep.

Potochnik, Angela, and Audrey Yap. 2006. 'Revisiting Galison's "Aufbau/Bauhaus" in Light of Neurath's Philosophical Projects'. *Studies in History and Philosophy of Science Part A* 37(3): 469-488.

Schaffer, Simon. 1998. 'On Astronomical Drawing'. In *Picturing Science, Producing Art*, ed. by Caroline A. Jones and Peter Galison. New York: Routledge, 441-474.

Sijes, Ben A. 1982. 'Anton Pannekoek 1873-1960'. In *Herinneringen: herinneringen uit de arbeidersbeweging; sterrenkundige herinneringen*, by Anton Pannekoek, ed. by Ben A. Sijes, Johanna M. Welcker, and J.R. van der Leeuw. Amsterdam: Van Gennep, 11-69.

Smart, D.A., ed. 1978. *Pannekoek and Gorter's Marxism*. London: Pluto Press.

Smith, Pamela H. 2006. 'Art, Science, and Visual Culture in Early Modern Europe'. *Isis* 97(1): 83-100.

SMBA. 2016. *Jeronimo Voss – Inverted Night Sky*. SMBA Newsletter 146.

Steen, Bart van der. 2006. 'Anton Pannekoek en het orthodoxe marxisme'. *Vlaams Marxistisch Tijdschrift* 40(2): 73-82.

Tai, Chaokang. 2017. 'Left Radicalism and the Milky Way: Connecting the Scientific and Socialist Virtues of Anton Pannekoek'. *Historical Studies in the Natural Sciences* 47(2): 200-254.

Tai, Chaokang, and Jeroen van Dongen. 2016. 'Anton Pannekoek's Epistemic Virtues in Astronomy and Socialism: Personae and the Practice of Science'. *BMGN – Low Countries Historical Review* 131 (4): 55-70.

Voss, Jeronimo. 2016. Inverted Night Sky. Installation at Stedelijk Museum Bureau Amsterdam.

—. 2017. *Inverted Night Sky*. Dortmund: Verlag Kettler.

Wilder, Kelley E. 2009. *Photography and Science*. London: Reaktion.

Winkler, Mary G., and Albert van Helden. 1992. 'Representing the Heavens: Galileo and Visual Astronomy'. *Isis* 83(2): 195-217.

Zwart, Hub. 2019. 'Poetry, Science and Revolution: The Enigma of Herman Gorter's Pan'. *Journal of Dutch Literature* 10(1): 24-49.

## About the Authors

**Chaokang Tai** is a PhD researcher in history of science at the University of Amsterdam. He is a member of the Vossius Center for History of Humanities and Sciences, University of Amsterdam, and the Descartes Centre for the History and Philosophy of the Sciences and the Humanities, Utrecht University. His current research focuses on the astronomical research of Anton Pannekoek in historical and societal context.

**Bart van der Steen** is University Lecturer at Leiden University's Institute of History. His research focuses on interwar labour movements and New Social Movements from 1968 to the present. His most recent publications include: *In Leiden moet het anders. Geschiedenis van een SP-afdeling, 1970-1982* (2019), and *A European Youth Revolt. European Perspectives on Youth Protest and Social Movements in the 1980s* (2016), edited with K. Andresen.

**Jeroen van Dongen** is Professor of History of Science at the University of Amsterdam, where he also obtained his PhD in 2002. He has taught at Utrecht University and has held research positions at the Max Planck Institute for the History of Science and the Einstein Papers Project at the California Institute of Technology. He currently serves as co-director of the Vossius Center for the History of Humanities and Sciences at the University of Amsterdam.

# 2 Anton Pannekoek's Astronomy in Relation to his Political Activities, and the Founding of the Astronomical Institute of the University of Amsterdam

*Edward P.J. van den Heuvel*

**Abstract**

An overview is given of Anton Pannekoek's life as an astronomer, from his first steps at age twelve as an amateur until the end of his life. Particular attention is given to the interactions between his political activities and his astronomical career. Although Pannekoek was employed full-time in the service of the German Social Democratic Party in the years 1906-1914 and had intensive contacts with Lenin in the period 1913-1920, he remained involved in astronomy in his spare time, which enabled him to return to an astronomical career in 1919. Here we describe the events which led to his appointment at the University of Amsterdam, the founding of its Astronomical Institute in 1921, his Milky Way research, and his work as founder of astrophysical research in the Netherlands.

**Keywords:** Anton Pannekoek, Milky Way research, stellar astrophysics, council communism, University of Amsterdam, history of astronomy

## Introduction

In online bookstores one can still find several of Pannekoek's books in print, for example, *Workers Councils* (Dutch original: 1946), with a preface by Noam Chomsky, *Lenin as Philosopher* (German original: 1938), translated in at least

Tai, Chaokang, Bart van der Steen, and Jeroen van Dongen (eds), *Anton Pannekoek: Ways of Viewing Science and Society*. Taylor & Francis Group, 2019
DOI: 10.5117/9789462984349_CH02

eighteen languages, and *A History of Astronomy* (Dutch original: 1951). Both his political and his scholarly work continue to resonate into the present.

In the final winter of the German occupation of the Netherlands (1944-1945), also known in Dutch as *de hongerwinter* ('the hunger winter'), Pannekoek wrote two separate autobiographies, one about his political life, 147 printed pages, and one about his life as an astronomer, 45 printed pages. Written 'by candlelight' as there was no electricity, they were originally meant only for his family. With their consent, the two autobiographies were published in 1982 in a single volume.[1] The difference in size might suggest that astronomy played a less important role in his life than politics, but I hope to convince the reader that astronomy certainly was the *love of his life*, from his early youth until his last days, and that his contributions to astronomy are of lasting value. However, from 1900 onwards, Pannekoek's scientific career was regularly influenced by his political work. Because the latter played an important role in the developments leading up to his appointment at the University of Amsterdam, I will also discuss parts of his political life.

In this paper, I will focus on Pannekoek's life as an astronomer, building on his published papers and autobiographies, but also on recollections shared with me by family members, colleagues, and students of Pannekoek. Especially the conversations with Dr David Koelbloed, who worked with Pannekoek from 1921 until the latter's retirement (Koelbloed himself retired in 1974), form a rich source of information about Pannekoek as a person and as a scientist.[2]

## Youth and Early Astronomical Career

Anton Pannekoek was born on 2 January 1873 in Vaassen, a small village in the eastern part of the Netherlands, where his father was the manager of a small metal foundry and his mother a midwife. Their two daughters and two sons all received secondary education. Like his elder brother, Adolf, Anton went to the Hoogere Burger School (HBS) in nearby Apeldoorn. This type of

I am very grateful to Chaokang Tai and to the International Institute of Social History (IISG) of the Royal Netherlands Academy of Arts and Sciences for important background information about Pannekoek's astronomical and political work, and for helping with illustrations, and to Dr Karel A. van der Hucht for several illustrations about Pannekoek in Indonesia.

1 Pannekoek 1982.

2 Other notable sources include Pannekoek's colleagues, Dr Elsa van Dien, Prof. Jan de Boer, Prof. Frank Muller, Prof. Bart Bok, and his children, Prof. Antonie Johannes Pannekoek and Dr Anneke ten Houten-Pannekoek.

secondary school was created by the Dutch government in 1863 to provide a solid education for middle-class children to prepare them for a career in the practical higher professions, such as engineering. It focused strongly on the exact and natural sciences, including biology and astronomy. The introduction of the HBS had the – unintended – effect that it provided its pupils with a solid preparation for a university study in the exact sciences. The majority of the Dutch Nobel laureates came from this type of school, as well as almost all prominent Dutch astronomers.[3]

Astronomy was part of the HBS curriculum because the Dutch colonial empire in the East Indies – now Indonesia – required a large Dutch merchant fleet. At the outbreak of the World War II the Dutch merchant fleet was one of the largest in the world. The officers of this fleet needed astronomy for navigation and it was therefore logical to include astronomy in the curriculum of schools that prepared for a professional career. A side effect of this was that interest in astronomy could be stimulated from an early age and this may be one of the reasons why gifted HBS students regularly continued their education as students of astronomy. This was precisely what happened with Pannekoek.

Pannekoek was a fast learner and at the age of fifteen he had already completed the HBS curriculum, two years faster than most other children. By that time, he had grown passionate about astronomy and biology, subjects which he already intensively pursued as an amateur, and which would remain his most loved fields of study throughout his life. In his astronomical autobiography, Pannekoek described his first acquaintance with astronomy at the age of twelve. Using star charts from the astronomy schoolbook of his elder brother and a German school atlas, he was able to spot an extra star in the constellation of Gemini in the fall of 1885. After much thinking, he realized that this must be a planet, which turned out to be Saturn. After this experience, he learned more about astronomy from his physics teacher Dr J.M. Smit. Smit had socialist sympathies and was fired in 1887, because he had argued in favour of general suffrage at a political meeting – at that time, only property owners and tax payers were allowed to vote.[4]

In Pannekoek's diaries, which he began at the age of fifteen, astronomical observations were alternated by notes on species of plants and flowers that he had found, and hearing the first nightingale on a spring night.[5] As

3 Of the nineteen Dutch-born Nobel laureates in the sciences (physics, chemistry, physiology or medicine, and economical sciences), twelve had attended the HBS, see Willink 1998.

4 For more on J.M. Smit, see Luikens 2001.

5 These diaries are now stored at API.

**Figure 2.1 First page of Anton Pannekoek's diary, 31 August 1888**

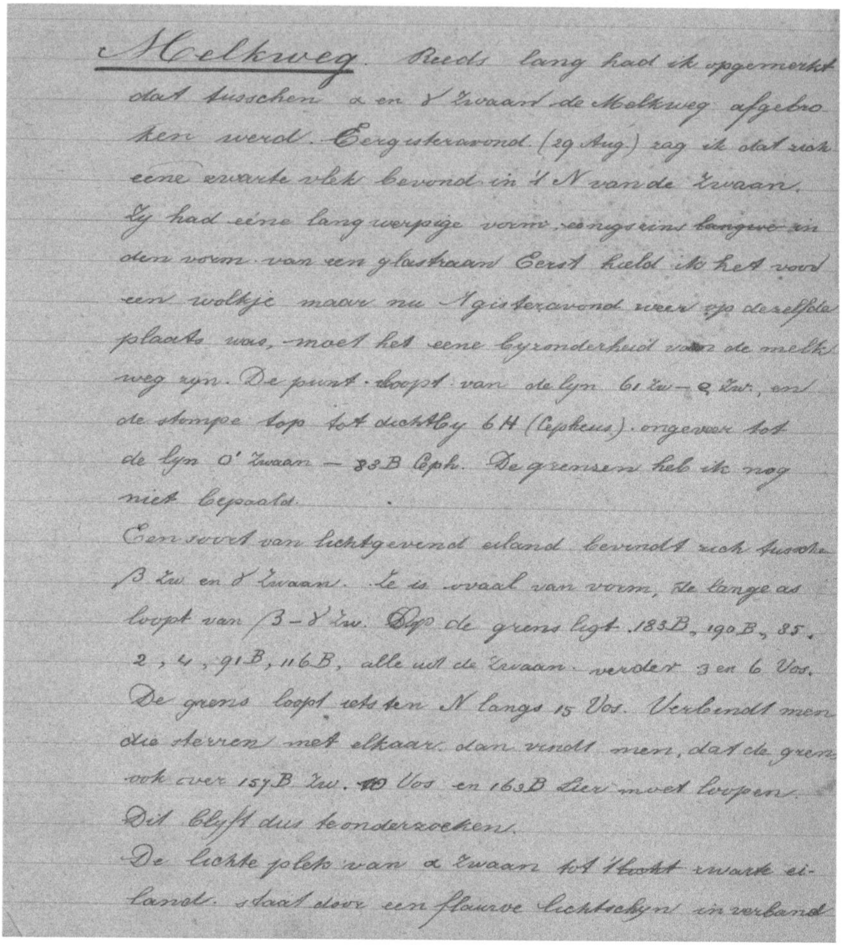
Melkweg. Reeds lang had ik opgemerkt
dat tusschen α en γ Zwaan de Melkweg afgebro-
ken werd. Eergisteravond (29 Aug.) zag ik dat zich
eene zwarte vlek bevond in 't N van de Zwaan.
Zy had eene langwerpige vorm, eenigszins langwe- in
den vorm van een glastraan Eerst hield ik het voor
een wolkje maar nu 't gisteravond weer op dezelfde
plaats was, moet het eene byzonderheid van de melk-
weg zyn. De punt loopt van de lyn 61 Zw – ρ Zw., en
de stompe top tot dicht by 6 H (Cepheus) ongeveer tot
de lyn o' Zwaan – 83 B Ceph. De grenzen heb ik nog
niet bepaald.
Een soort van lichtgevend eiland bevindt zich tussche
β Zw en γ Zwaan. Ze is ovaal van vorm, de lange as
loopt van β – γ Zw. Op de grens ligt 183 B, 190 B, 85,
2, 4, 91 B, 116 B, alle uit de Zwaan verder 3 en 6 Vos.
De grens loopt iets ten N langs 15 Vos. Verbindt men
die sterren met elkaar dan vindt men, dat de gren
ook over 157 B Zw. ~~10~~ Vos en 163 B Lier moet loopen.
Dit blyft dus te onderzoeken.
De lichte plek van α Zwaan tot 't ~~licht~~ zwarte ei-
land staat door een flauwe lichtschyn in verband

The page starts with: 'Already for a long time I had noticed that there is a gap in the Milky Way in the constellation Cygnus, between the stars α and γ Cygni.'

Source: Archive of the Anton Pannekoek Institute for Astronomy, University of Amsterdam

an illustration, Figure 2.1 shows the first page of his first diary, written on 31 August 1888. Pannekoek wrote here: 'For a long time, I had noticed that there is a gap in the Milky Way in the constellation Cygnus, between the stars α and γ Cygni.' A little further, he wrote that in the Milky Way 'there is an oval island of light between the stars β and γ Cygni'.[6] Both these features are clearly visible on current photographic images of this area. The Milky

6 Anton Pannekoek, dagboek 1, API, p. 1.

Way would remain a lifelong love of Pannekoek, and later in his scientific career he would devote much effort in attempting to unravel its structure.

Pannekoek's initial career goal had been to become a teacher, but his secondary school teachers convinced his parents to send him to university. However, with a HBS-certificate one could not immediately enter university. At the time, Dutch universities demanded of their students a fluent knowledge of Latin and Greek, languages that were not taught at the HBS. Furthermore, his parents considered him too young for university. He therefore stayed at home to study for the compulsory Latin and Greek exams, which he obtained only three years later, after failing the exams twice. In his autobiography, he explained this by claiming that he studied these languages 'rather unsystematically and inadequately'.[7] During the same period, he spent most of his energies on self-study, focusing on astronomy and biology. Following the advice of his physics teacher Smit, Pannekoek had begun to buy second-hand books at auctions during his secondary-school years, among them a guide for amateur observations by German astronomer Friedrich Wilhelm Argelander.[8] With this book, he taught himself to accurately estimate, with the naked eye, the brightness of a star by interpolating between the known brightnesses of both brighter and fainter neighbouring stars. He became very skilled in this technique and in his diaries he noted long series of brightness measurements of stars and of the Milky Way. At the age of seventeen, in the very clear and cold nights of December 1890 (the winter of 1890/1891 was, as my grandparents told me, one of the coldest of their lives), he discovered that the brightness of Polaris (α Ursae Minoris, the North Star) varies by a small amount over a period of about four days. His diary contains long series of observations of this star, extending until March 1891, documenting this variation. He continued to observe the variations until 1900, when he was already a professional astronomer. After seeing that Campbell at Lick Observatory in 1898 had found the radial velocity of the star to be variable with a period of 3,968 days, he published his findings in 1906 in a footnote in an article about the luminosities of stars of a different type.[9] Two years later, in 1908 (see below) this variation of the North Star was photographically confirmed by Hertzsprung, who found that it is a pulsating star of the Cepheid type.[10]

7 Pannekoek 1982, 229.
8 Argelander 1855.
9 Pannekoek 1906, 148.
10 Pannekoek 1982, 241; Hertzsprung 1911.

**Figure 2.2 Anton Pannekoek and his wife Anna Pannekoek-Nassau Noordewier in the garden of their Leiden home around 1903-1906**

Source: Private collection, courtesy of Laurence A. Marschall

In the fall of 1891, Pannekoek entered Leiden University and one year later he obtained his bachelor degree in physics and mathematics. In 1895, he graduated in these sciences, with astronomy as major. In 1892, he held the chair of the Leiden physics-student society 'Christiaan Huygens'. That same year, he started publishing his findings in research journals. His first scientific paper discussed the light variations of the eclipsing binary Algol (β Persei) in the German journal *Astronomische Nachrichten*.[11] In total, he authored well over a hundred peer-reviewed scientific publications.[12]

In the first two years after completing his undergraduate studies in 1895, Pannekoek worked as a geodetic engineer for the triangulation of the Netherlands, a national project with the aim of precisely measuring the dimensions of the country. For this, he travelled from the province of

11 Pannekoek 1892.

12 The SAO/NASA Astronomy and Astrophysics Data System (http://adsabs.harvard.edu) lists 110 publication by Pannekoek (accessed 22 August 2017), while a bibliographical list in API, compiled by David Koelbloed, lists 136.

Zeeland in the South to the island of Ameland in the North. As he mentioned in his autobiography, he enjoyed studying the different flora and fauna in the different parts of the country, and meeting people with different habits and dialects. After these two years, he was offered an assistantship at Leiden University Observatory, which allowed him to start research for a PhD. In 1902, he defended his PhD thesis on the light variations of the eclipsing binary Algol, on which he had published his first research paper ten years earlier.[13]

## Pannekoek's Early Career in Socialism

While working at Leiden Observatory, Pannekoek first was a member of the Liberal Party – like most university staff. However, in 1899, he read the book *Equality* by the American Christian-socialist Edward Bellamy, which converted him to socialism.[14] The book made him realize the unfairness of the highly stratified society of that time, with miserable conditions for most of the workers and their families. In the first half of the twentieth century, Bellamy was popular throughout the world. My parents were great fans of him too, and they named me Edward after him. Bellamy, by the way, did not want to be called a 'socialist'; he called himself a 'nationalist', because he believed that giving every person an equal share in the wealth of the country would be in the best interest of the USA as a nation. At the time, all over the USA, more than one hundred 'nationalist' societies were founded aimed at realizing Bellamy's ideas.

Pannekoek joined the Dutch Social Democratic Workers' Party (Sociaal-Democratische Arbeiderspartij, SDAP) in 1899 and became one of the founders of its Leiden chapter. He started reading the works of Karl Marx and became an active teacher of Marxism for different branches of the party all over the country. During visits to Germany, he addressed meetings of the German Social Democratic Party (Sozialdemokratische Partei Deutschlands, SDP), at that time the largest and most respected socialist party in the world, with over 700,000 members. In the years in which Pannekoek became politically active, he also became more and more frustrated with how astronomy was done at the Observatory, and the way he had to do research himself. While the director of the Observatory, H.G. van den Sande Bakhuyzen (1838-1923) and his younger brother, and later successor as director, E.F.

13 Pannekoek 1902. A summary was published in Astronomische Nachrichten, Pannekoek 1903.

14 Pannekoek 1982, 72.

**Figure 2.3 Opening of the SDP Party School in Berlin on 15 November 1906**

From left to right: Anton Pannekoek, August Bebel (chair of the SDP), Arthur Stadthagen, Simon Katzenstein, Hugo Heinemann. Like Pannekoek, the latter three were also teachers at the Party School.

Source: International Institute for Social History, call nr. IISG BG A10/805

van den Sande Bakhuyzen (1848-1918), had been good astronomers in their time, by the turn of the century, they were old and did not want to hear of the new and more accurate observing methods and instruments that Pannekoek proposed. Moreover, the observations he had painfully carried out night after night with the highest possible precision disappeared in drawers and were never published. Pannekoek wrote in his autobiography that he felt more and more depressed and unfit for the job.[15] By 1906, he decided to abandon his position at the Observatory and started to apply for a position of secondary-school teacher in physics and mathematics. Just around that time, he received a letter from Karl Kautsky, the main theorist of the German SDP, who invited him to become teacher of Marxism at the new Party School that the SDP was establishing in Berlin. The party felt a need for such a school to train the leaders of the local party chapters throughout the country to improve their knowledge of the theoretical aspects of socialism and Marxism. Such a background, it was felt, was needed to properly lead the chapters and convince others to join the SDP.

15 Pannekoek 1982, 236-237.

Pannekoek accepted the invitation and quit his Leiden job. In his astronomical memoirs, he wrote that he thought that he had abandoned astronomy forever.[16] The official inauguration of the Party School took place on 15 November 1906, starting Pannekoek's career as a professional theorist of socialism (Figure 2.3). In a letter to his family in Holland, he described the large house he had rented in Berlin, with an interesting detail for a socialist: 'On the loft there also is a nice little room for the maidservant.'[17]

After a year, the German authorities forbade foreigners to continue their work at the Party School. In order to help him continue his socialist work, Kautsky helped Pannekoek to set up a weekly column ('Zeitungskorrespondenz') to which thirty socialist newspapers subscribed. This provided him with sufficient income during the next three years.[18] In addition to the weekly column, Pannekoek travelled the country and even visited Switzerland to give lectures for the local divisions of the party. Because of these activities, Pannekoek became a well-known socialist in both countries, as well as the in the Austrian empire. In 1910, he was invited by the Bremen chapter of the SDP to become lecturer at their new local party school. As a Free City (Freistadt), the German government had no authority in Bremen and therefore could not prohibit Pannekoek to teach there.

While Pannekoek was initially on good terms with his German mentor Kautsky, they later clashed on several issues. Pannekoek was much more radical than Kautsky and in 1912/1913, he wrote that all 'old' government officials and members of the civil service should be fired and replaced by socialists once the latter had come to power. Kautsky was much more pragmatic and did not want the country to turn into chaos, and called Pannekoek an anarchist. (Lenin agreed with Pannekoek's views and in 1913 wrote an article in his support.)[19]

In the years between 1906 and 1914, Pannekoek wrote ten brochures on varying subjects related to Marxism and socialism. Among these is the remarkable *Darwinisme en Marxisme* (1909), which contains a very clear explanation of Darwin's theory of evolution.[20] Pannekoek explained that the 'bourgeois' idea that Darwin's theory of natural selection would support capitalism was a wrong and vulgar interpretation of Darwinism. According to Pannekoek, humans have evolved as social beings. We thus owe our speech

16 Pannekoek 1982, 241.

17 Private communication from Dr David Koelbloed

18 Pannekoek 1982, 126-140.

19 For more detail, see Gerrit Voerman, 'Anton Pannekoek: A "Principled Theorist"', in this volume, 51-74.

20 Pannekoek 1909. An English translation was published in 1912 as *Marxism and Darwinism*, Pannekoek 1912.

and language faculties, as well as our abilities to make tools to our (biological) evolution. Over millions of years, our ancestors cooperated and supported each other, rather than competing and struggling with each other. With the rise of script and advanced tools, the biological evolution of humans ended. The only human evolution nowadays is the evolution of societies, which spans a relatively short timescale compared to the evolution of species. According to Pannekoek, this societal evolution proceeds according to the laws discovered by Marx. These laws were not yet completely known, as social science was still in its initial phase of development, like astronomy in the time of the Babylonians and Greeks, but just like with astronomy, continual development of these would eventually lead to the discovery of the definitive laws. Marxism, then, was a 'work in progress'. Pannekoek stated that Darwinism and Marxism had both destroyed old and rigid worldviews, but that each is valid only in its own domain; Darwinism in biology and Marxism in the social sciences.[21]

Much later, in 1944, he resumed the subject of Darwinism and Marxism, and wrote his book *Anthropogenese. Een studie over het ontstaan van den mensch* (*Anthropogenesis: A Study of the Origin of Man*), which was published in 1945.[22] It remains until this day an amazingly modern and far-sighted view on how humans and their brains and speech have evolved, due to toolmaking and social communication.

In 1909, to his great surprise, Pannekoek was visited by Ejnar Hertzsprung of the Imperial Observatory in Potsdam. Hertzsprung told him that, from a large number of photographic observations, he had fully confirmed Pannekoek's discovery of the variability of the Pole Star. He had found a regular period of about four days for this variability and an amplitude of 0.17 magnitudes, which corresponded to a maximum change of 16% in brightness of the star (this is a very small amount and means that Pannekoek must have had very good eyes to detect it in 1890/91). The conclusion was that the Pole Star is a pulsating star of the Cepheid type. When Hertzsprung invited him to visit the Potsdam Observatory, Pannekoek accepted because he wanted to renew his acquaintance with the director Karl Schwarzschild. As students, Pannekoek and Schwarzschild – who had studied in München – had regularly exchanged scientific articles. Meeting Schwarzschild was a great pleasure, and this visit and the meeting with Hertzsprung revived his interest in astronomy.[23]

21 For more on Pannekoek's writings on Marxism and Darwinism, see Omar W. Nasim, 'The Labour of Handwork in Astronomy', in this volume, 249-283.

22 Pannekoek 1945. English edition: Pannekoek [1944] 1953.

23 Pannekoek 1982, 241-242.

Shortly after the meeting, Pannekoek reduced his old naked-eye observations of the Pole Star of the period 1890-1900 and found that the regular period of four days that Hertzsprung had found was clearly present in his own observations He published these findings a few years later.[24] In his spare time, Pannekoek resumed his observations of the Milky Way and increasingly enjoyed astronomy again. He also renewed his interest in Babylonian astronomy, building on the new work by Franz Xaver Kugler. Pannekoek and his wife started to follow the motions of constellations, the sun, moon, and planets on a daily basis, to get a feeling of how the Babylonian astrologers had made their naked-eye observations. This inspired Pannekoek to resume the writing of a popular astronomy book, in German, which he had already started in 1903 in Leiden. The book was finished in the summer of 1914, a few months before World War I started.[25]

## Return to the Netherlands and Pannekoek's Appointment at the University of Amsterdam

In the summer of 1914, Pannekoek's book had been typeset in lead, with over 140 copper engravings of the figures and star charts, and was ready to be printed in Dresden, when the war broke out. The German government abruptly confiscated all the lead and copper in the country, including the typesetting of Pannekoek's book, to be melted for bullets. As a result, the German edition of the book was never published.[26]

At the outbreak of the war, Pannekoek happened to be on holidays in the Netherlands, staying with his in-laws in Arnhem. He could not return to Germany, because the German government had banned all foreign socialists from the country. Pannekoek decided to translate his book into Dutch and have it published in the Netherlands. He translated the text in two months and it was published under the title *De Wonderbouw der Wereld* ('The Miraculous Construction of the World'), a title which Pannekoek did not like very much, but was conceived by the publisher.[27] It is a beautiful book, providing a lucid history of astronomy, starting from the discoveries of the regularities in the motions of the moon and planets by the Babylonians, which allowed them to predict eclipses of the Sun and Moon and the positions of the planets. The

24 Pannekoek 1913.
25 Pannekoek 1982, 239-240.
26 Pannekoek 1982, 240.
27 Pannekoek 1916.

**Figure 2.4 Pannekoek in the library of Leiden Observatory in 1916**

Source: Private collection, courtesy of Laurence A. Marschall

book impressed the leading Dutch astronomer Jacobus C. Kapteyn (professor in Groningen), as well as Kapteyn's former student Willem de Sitter (professor in Leiden). It convinced De Sitter in 1916 to offer Pannekoek an unpaid position at Leiden University (in Dutch: 'privaatdocent') to lecture on the history of astronomy. Pannekoek, who was earning his living as a secondary school teacher of physics at the time, gladly accepted this position (Figure 2.4).[28]

28 Pannekoek 1982, 240-241.

During Pannekoek's time as a social activist in Germany, Kapteyn had once mentioned to Pannekoek's father-in-law, Dr Hendrik J. Nassau Noordewier (linguist and principal of the Latin secondary school in Delft), how he regretted Pannekoek's departure from astronomy very much. Pannekoek wrote in his memoirs: 'I still know how this astonished me, and I had thought that I had left astronomy forever.'[29]

Because of his communist convictions, secondary schools in the Netherlands were hesitant to give Pannekoek a permanent job. Initially he could only get temporary jobs in various parts of the country as a replacement teacher, before being hired in 1916 for a longer-lasting job as a physics teacher in Bussum, a village located some twenty kilometres southeast of Amsterdam. During World War I, he remained politically active, although he refrained from attending public meetings and wrote most of his essays under pseudonyms. The war was a catastrophe for the international socialist movement. The German SPD supported the government's war effort in parliament, while the French socialists did the same with respect to their government. The situation was the same in Russia and England. In 1915, socialists from various countries who opposed the war held a meeting in Zimmerwald in Switzerland. Socialists from neutral nations, like the Netherlands, also participated in an effort to resurrect international solidarity. The conference was notably attended by three Russian communists living in exile in Switzerland: Vladimir Lenin, Grigory Zinoviev, and Karl Radek. From Holland, the famous poet Henriette Roland Holst attended, and Lenin approached her to start a new international communist journal together with Pannekoek. Both agreed and in 1916 the first issue of *Vorbote* appeared as the journal of the Zimmerwald Left – with Pannekoek and Roland Holst as editors (and financed by the wealthy Roland Holst). The first issue included articles by Lenin, Zinoviev, Radek, and Roland Holst. Even so, the journal did not last long and only two issues appeared. The Zimmerwald initiative ended among others due to Lenin's power grab in 1917 and the creation of the Soviet Union. Pannekoek was unaware of the horrors that accompanied the Russian Revolution and, like many socialists elsewhere, initially thought that the Russian communist takeover was a positive development that could start a socialist revolution all over the world.

Around the same time, a promising new opportunity arose for Pannekoek in astronomy. In 1917, Leiden Observatory director Ernst F. van den Sande Bakhuyzen suddenly passed away and Willem de Sitter was invited to become the institute's new director. After seeking advice from Kapteyn, de

29 Pannekoek 1982, 241.

Sitter put forward his conditions for acceptance: a drastic reorganization of the observatory. This included the appointment of two new professors, who would both act as deputy directors: Ejnar Hertzsprung and Anton Pannekoek. His plan was accepted by the Board of Leiden University and by the Minister of Science in 1918, and Hertzsprung was soon appointed. The minister, however, hesitated with the appointment of Pannekoek. At the end of the war, communist revolutions broke out in several places in Europe: in Munich, Berlin, and Brussels. From Russia, Lenin encouraged the revolutionaries to incite a 'world revolution'. Pannekoek and Henriette Roland Holst were heavily involved in debates about revolutionary politics, although Pannekoek refrained from doing so publicly, since he felt it his responsibility as a teacher, with respect to the parents of his pupils, to remain politically neutral. Even so, when, in 1919, a communist revolution broke out in Budapest, its leader Béla Kun declared Lenin, Pannekoek, and several other prominent socialists 'honorary members of the Hungarian communist republic'. This news received attention from Dutch newspapers, who added that Pannekoek was assigned to become deputy director of Leiden Observatory. De Sitter was angry at Pannekoek, because he felt that Pannekoek was responsible for sabotaging his own appointment, thus embarrassing De Sitter, but nonetheless he immediately went to The Hague to try to save the situation. But the government, led by the Catholic Prime Minister Ruys de Beerenbrouck, declared that Pannekoek under no condition would be appointed at a state university.[30]

In the meantime, however, another opportunity had arisen. At the University of Amsterdam – a municipal institution independent of the national government – mathematics professor Diederik J. Korteweg, whose appointment included lecturing astronomy, would soon retire. At the initiative of mathematician Luitzen Egbertus Jan Brouwer, the University offered the position to Pannekoek, under the condition that half his lectures would be in mathematics, teaching students who needed it as a support science, e.g. chemists, geologists, etc. The teaching load was nevertheless not too large, perhaps six to eight hours per week – far less than the 26 hours a week he had taught at the secondary school in Bussum. This left Pannekoek with substantial time to do research and establish an astronomical research unit. Pannekoek was appointed in 1919.

Shortly after, in January 1920, Pannekoek was visited by Dutch engineer Sebald Rutgers, who was Lenin's right hand in Moscow. Rutgers had earlier

30 Pannekoek 1982, 245-246. For an elaborate discussion of this episode, see Baneke 2004; and his 'Pannekoek's One Revolution', in this volume, 87-108.

worked as Head of Public Works on Sumatra in the Dutch East Indies. After hearing about the Russian revolution in 1917, he quit his job and travelled through Japan and Siberia to Moscow to join the revolution. Rutgers came to Amsterdam to invite Pannekoek, on behalf of Lenin, to come and work as a party theorist for the new Soviet Union in Moscow. Pannekoek refused, arguing that he did not want to be in a similar position as he had been in his German years when his income was dependent on his relation with a party and a government. He preferred to keep his job at the university but remain politically active. This way, he could develop his socialist ideas independently.

Although Pannekoek maintained a strict separation between his political and his astronomical activities, one clearly sees that in reality, there were strong connections between the two. In fact, the University of Amsterdam would not have had an astronomical institute if Pannekoek had not been a communist, because in that case he would have been appointed in Leiden. Secondly, his job as a lecturer – and later professor – of astronomy gave him a position from which he could develop his own ideas about socialism and revolutionary politics, leading ultimately to the development of council communism.

Already in 1920, Pannekoek realized that Lenin's 'dictatorship of the proletariat' and the dissolution of the soviets (revolutionary councils of workers and soldiers) was leading to a dictatorial regime of state capitalism, rather than a communist society. In his brochure, 'The New Blanquism' (1920), he criticized Lenin's politics. Blanqui was one of the leaders of the 1870 Paris commune, who had argued that a small group of skilled leaders should lead the revolution in a strictly centralized way, on behalf of the workers. According to Pannekoek, this was the same model envisioned by Lenin.[31]

Lenin replied with a booklet called *'Left-Wing' Communism: An Infantile Disorder*, in which he attacked Pannekoek and his associates. According to Lenin, they had 'shown most plainly that they consider themselves sound Marxists, but talk incredible nonsense in a most ridiculous manner and reveal their failure to understand the ABC of Marxism'.[32] Pannekoek's friend and comrade, the famous Dutch poet Herman Gorter, made one last attempt to convince Lenin and other Soviet leaders by travelling to Moscow in 1920, but he was unsuccessful, and returned to the Netherlands disappointed.

31 Pannekoek [1920] 1969.

32 Lenin [1920] 2008, 33.

## Astronomical Research in Amsterdam

In the year of his clash with Lenin, Pannekoek published his major work on the Northern Milky Way in the Annals of the Leiden Observatory. It included the beautiful drawing 'The real aspect of the Milky Way, as true as possible', which was derived as a 'mean' of drawings by different observers, with much weight given to the work of Easton and to his own observations of 1897-1899 and 1910-1913.[33] In the previous year, Pannekoek had published an important paper in the *Monthly Notices of the Royal Astronomical Society* on the earth's distance from the centre of the Milky Way. By studying the distribution of star clouds in the Galaxy, Pannekoek calculated a distance to the galactic centre of 60,000 lightyears in the direction of the constellation Sagittarius.[34] This result supported the work of American astronomer Harlow Shapley, who claimed in 1918 that the Galactic Centre is located at a large distance in the direction of Sagittarius.[35] In 1920, Pannekoek received from Dr Herko Groot, his successor as physics teacher at the secondary school in Bussum, a few reprints of papers from Indian physicist Meghnad Saha (1893-1956), which Groot had received in exchange for his own work.[36] In these papers, Saha had derived and articulated the famous 'Saha law' on the ionization of gases as a function of temperature and density. This law would turn out to become a cornerstone of astrophysics. Thanks to this law, together with Boltzmann's laws for the excitation of atomic energy levels, astronomers finally understood why stellar spectral types depend on the temperature of the stellar atmosphere. In 1920, Saha's law was completely new, of course, and Pannekoek saw a beautiful opportunity to start a new field of research: stellar spectroscopy and the study of the physics of stellar atmospheres. In 1922, he published his first paper on the ionization in stellar atmospheres, in the *Bulletin of the Astronomical Institutes of the Netherlands*.[37] It was the beginning of astrophysical research in the Netherlands, and, as described above, the coincidence of the meeting between Pannekoek and his successor at the Bussum secondary school, Groot played a key role in this start.

Pannekoek, then almost 50 years old, quickly acquainted himself with the modern atomic physics, statistical physics and quantum mechanics, needed

33 Pannekoek 1920; for more on Pannekoek's Milky Way drawings, see Chaokang Tai, 'The Milly Way as Phenomenon', in this volume, 219-247; and Omar W. Nasim, 'The Labour of Handwork in Astronomy', in this volume, 249-283.

34 Pannekoek 1919.

35 For a detailed report, see Tai 2017, 235-237.

36 Pannekoek 1982, 251.

37 Pannekoek 1922.

**Figure 2.5 Technician David Koelbloed (1905-1977)**

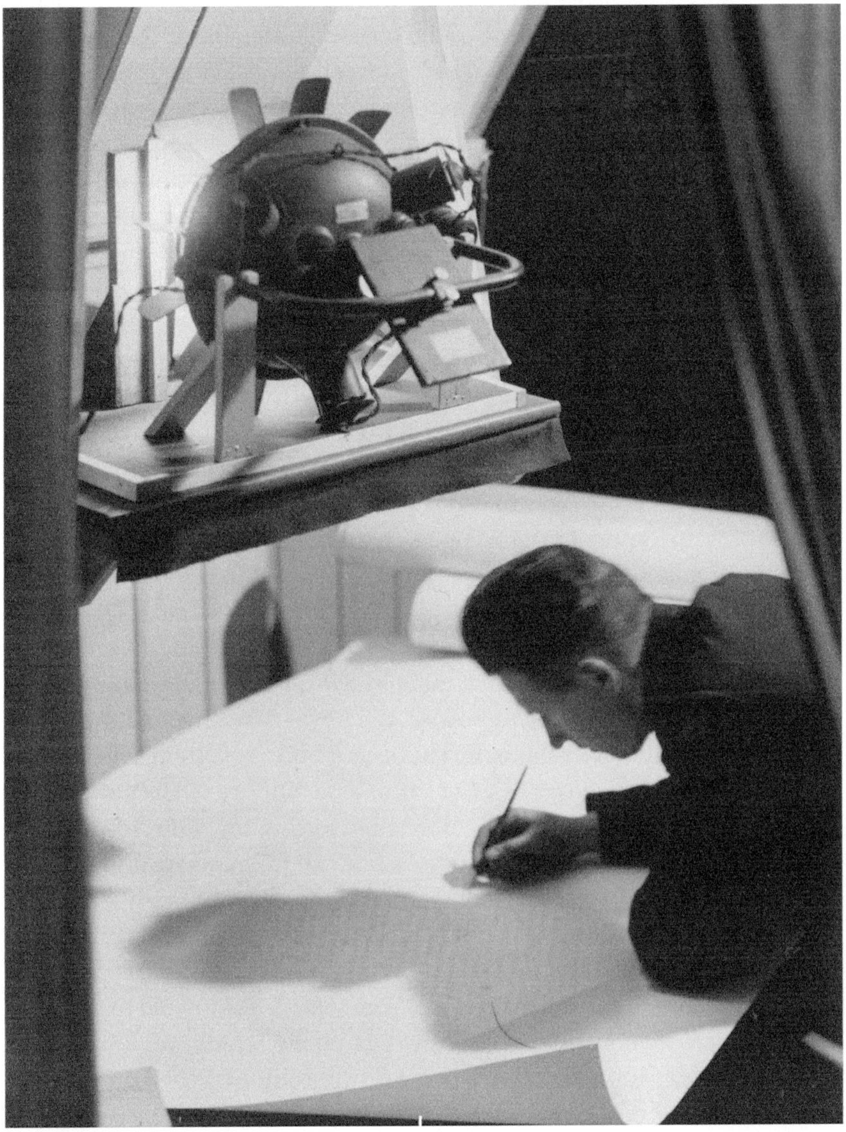

Working on photometry of the Northern Milky Way at the astronomical institute (then located in the loft of the Oudemanhuispoort), in May 1932. Koelbloed had been hired at age fifteen in 1921 as the first employee of the institute, and retired as lecturer in Astronomy in September 1974.

Source: Archive of the Anton Pannekoek Institute for Astronomy, University of Amsterdam

for these astrophysical studies. The late professor Jan de Boer, founder of the Institute for Theoretical Physics of the University of Amsterdam, told me in the 1970s that in the 1920s none of the physics professors at the university taught this 'new physics'; he and his fellow students learned their atomic physics and quantum mechanics from Pannekoek. Other Amsterdam physics and chemistry professors, such as experimental physicist Frank Arnoud Muller, who had followed mathematics courses from Pannekoek, told me in the 1970s that they had very much enjoyed Pannekoek's brilliantly clear mathematics lectures.

In 1921, the university was able to support Pannekoek with a modest research budget, which allowed him to hire his first two technical staff members, the fifteen-year-old 'computers' David Koelbloed (Figure 2.5) and Hendrik Reus, who – with a three-year HBS certificate – were selected from a considerable number of candidates. With this, the Astronomical Institute of the University of Amsterdam was founded. As Pannekoek wrote in his autobiography, his aim was to establish an astronomical laboratory after the example of Kapteyn in Groningen. In such a laboratory, one carries out the reduction and theoretical interpretation of observational data taken elsewhere with telescopes, such as photographs, spectra, etc.[38]

In 1925, Pannekoek was elected to the Royal Netherlands Academy of Sciences – at that time an appointment by the queen – and was appointed associate professor. In earlier decades, the Royal Academy had supported expeditions to observe solar eclipses, because during a total solar eclipse the hot and tenuous layers of the outer atmosphere of the sun, called *chromosphere* and *corona*, become visible for the few minutes that the eclipse lasts. During these few minutes, one can take spectra and pictures of the eclipsed sun in different colours, which can later be analysed to gain important information on the physical state of the outer layers of the sun; like temperature, density, and chemical composition. Because of his new expertise on the ionization of hot gases, Pannekoek was invited to take part in the 'Eclipse Committee' of the Royal Academy, which was responsible for planning expeditions for observing future eclipses. When, in 1925, the Committee's chairman and Utrecht Professor Willem Julius died, Pannekoek was invited to become its new chairman. The first expedition was to the January 1926 total solar eclipse in Sumatra in the Dutch East Indies. Pannekoek was accompanied by solar physicist Marcel Minnaert (who in 1937 was to become the director of Utrecht Observatory), astronomer Jan van der Bilt and student Johanna Cornelia Thoden van Velzen, all from Utrecht. They

38 Pannekoek 1982, 248-249.

**Figure 2.6 The solar eclipse team on Sumatra in January 1926, with soldiers of the Netherlands Indies Colonial Army**

In the picture to the left of Pannekoek, Utrecht solar physicist Marcel Minnaert and to the right of Pannekoek, Utrecht PhD student Johanna Cornelia Thoden van Velzen

Source: Archive of the Anton Pannekoek Institute for Astronomy, University of Amsterdam

are all visible in the picture in Figure 2.6. Unfortunately, at the moment of the eclipse, it was cloudy, and the expedition was a failure. But Pannekoek had arranged to stay longer in Indonesia to study the Southern Milky Way, and to make drawings of it with the same care with which he had earlier mapped the Northern Milky Way. To this end, he had contacted Joan Voûte, director of the new Bosscha Observatory in Lembang (near Bandung) on Java, and had arranged for a three-month stay there, after the eclipse (Figure 2.7). The observatory, financed by tea-planter millionaire Karel Albert Rudolf Bosscha, was still being constructed at the time. A house at the observatory grounds was built especially for Pannekoek and is nowadays still called 'Rumah Pannekoek' (Indonesian for 'Pannekoek House').

During their three-month stay on Java, Pannekoek and his wife worked hard to make careful drawings of the Southern Milky Way, using his decades of experience in drawing the Northern Milky Way. The result was published in the *Annals of the Bosscha Observatory* in 1928 (the reduction

**Figure 2.7 Anton Pannekoek and his wife Anna Pannekoek-Nassau Noordewier**

Anton Pannekoek (standing left) and his wife Anna Pannekoek-Nassau Noordewier (sitting right) with the director Joan Voûte of the Bosscha Observatory and his wife, Frieda J.G.E. Voûte-Adloff on the veranda of the director's house, Lembang, Java, spring 1926

Source: Stichting Indisch Thee- en Familiearchief Van der Hucht c.s., courtesy of Dr Karel A. van der Hucht

of the observations had taken much time).[39] Next to manual drawings, Pannekoek also used photographic exposures to capture the star clouds in the Milky Way – taken by colleagues elsewhere, e.g. in Heidelberg, at Harvard University Observatory, and in Lembang – to unravel its structure.[40] One of the things he noticed during his studies of the Southern Milky Way was that certain parts of it were bluish in colour while others were more whitish or yellowish. He also found the clumping of blue B-type stars in certain parts of the Milky Way when he investigated their distribution using the Henry Draper Catalogue in 1929. These clumps were much less concentrated than star clusters, each clump extending over a considerable number of degrees on the sky.[41] As Groningen astronomer Adriaan Blaauw would remark

39 Pannekoek 1928.

40 See Chaokang Tai, 'The Milky Way as Optical Phenomenon', in this volume, 219-247.

41 Pannekoek 1929.

later, when he visited our Institute and saw Pannekoek's papers, this was already a hint to the existence of OB associations, which after World War II were recognized by Soviet astronomer Viktor Ambartsumian and Blaauw: expanding groups of very young massive O and B stars.

In the 1920s and 1930s, next to his Milky Way research, Pannekoek also continued his study of stellar spectra, to derive the physical conditions of stellar atmospheres. He started in 1923 with photographs of spectra that were taken in observatories elsewhere, such as Lick Observatory near San Jose, California and Dominion Astrophysical Observatory near Victoria, British Colombia. When, in 1928, Pannekoek again asked his Canadian colleague John S. Plaskett to take some more photographs of spectra for him, the answer was that he had no time to do this, but that Pannekoek was welcome to come to Victoria and take the spectra himself. Thus, Pannekoek went to Canada in 1929 for half a year to photograph spectra of a variety of stars with the 180cm reflecting telescope of the Dominion Astrophysical Observatory. As he described in his autobiography, he tremendously enjoyed this observing work with a large telescope: the beauty of the dark night sky when he was observing, the hard work after the night, the careful developing of the photographic plates of the spectra taken, etc.[42]

The spectra were analysed in Amsterdam by Pannekoek and his PhD students, and led to several PhD theses, including those of Sijtze Verweij, Gale Bruno van Albada, Theodore Walraven, and David Koelbloed.[43] The latter, after being hired as a fifteen-year-old computer in 1921, had completed a full secondary-school education in his spare time, then obtained teacher certificates for mathematics and physics, followed by an MSc, and then in the 1950s obtained his doctorate degree. He ended his career in 1974 at the age of 68, as a lecturer in astronomy. His career, with a length of 53 years in the service of the University of Amsterdam, is still an all-time record in the history of the university.

In the 1920s and 1930s, Pannekoek was the pioneer in numerically calculating the structure of stellar atmospheres, and the spectra produced by these atmospheres. This put him on the map as an international expert in the physics of stellar atmospheres and led to the invitation to co-author in 1930 the world's foremost standard astrophysics handbook, the *Handbuch der Astrophysik*, published by Springer.[44] Subsequently, he was invited by Shapley to teach at the 1935 Harvard University Astronomy Summer School.

42 Pannekoek 1982, 257-264.

43 Verweij 1936; van Albada 1945; Walraven 1848; Koelbloed 1953.

44 Pannekoek 1930.

It was a great success, and in 1936, at the tercentenary celebration of Harvard University, he was invited again and was awarded an Honorary Doctorate of this university for his pioneering works in Milky Way research and in astrophysics. In 1952, Pannekoek was awarded the Gold Medal of the Royal Astronomical Society, arguably the highest international award in astronomy.

## Work after Retirement and Concluding Remarks

Pannekoek's retirement should officially have taken place in 1943, at the age of 70. But due to World War II, it was postponed until 1945. In 1946, he was succeeded by Herman Zanstra. After his retirement, he wrote his beautiful *De groei van ons wereldbeeld* ('the growth of our picture of the world'), which came out in Dutch in 1951 and, as *A History of Astronomy* in English in 1961.[45] The book was translated into English by American astronomer Priscilla Bok, the wife of Dutch-born Harvard astronomer Bart Bok, who had obtained his PhD with Professor Pieter J. van Rhijn in Groningen, the successor of Kapteyn. Bok told me in the 1970s that during his PhD work in Groningen in the late 1920s, he had wanted to go speak with Pannekoek in Amsterdam about determining stellar distances by means of 'spectroscopic parallaxes'. The ultra-conservative Van Rhijn, who knew nothing about stellar spectroscopy, thought that 'spectroscopic parallaxes' were nonsense, and forbade Bok to go speak with 'that man Pannekoek', whom he apparently hated because of his political ideas.[46] Bok told me that he nevertheless went to visit Pannekoek, whom he found to be very kind, and that he learned much from him that was useful for his PhD thesis. Bok himself was a wonderful man and a great popularizer of science. After obtaining his PhD in 1929, he obtained a position at Harvard University and worked the rest of his life in the USA and Australia.[47]

Pannekoek's *A History of Astronomy* is a special book. It not only discusses how astronomy developed, but also places this development in the context of the larger development of human societies. In Pannekoek's own words:

> When the astronomer looks back at his predecessors, he finds Babylonian priests and magicians, Greek philosophers, Mohammedan princes,

45 Pannekoek 1951; 1961.

46 In Dutch universities, a professor will normally always refer to his/her colleagues as 'colleague', never as: 'that man'.

47 For a biography of Bok, see Levy 1993.

> medieval monks, Renaissance nobles and clerics – until in the scholars of the seventeenth century he meets with modern citizens of his own kind. To all these men astronomy was not a limited branch of science but a world system interwoven with the whole of their concept of life. Not the traditional tasks of a professional guild, but the deepest problems of humanity inspired their work.[48]

This wonderful book is itself a great source of inspiration for all of those who are interested in how our science developed and for those who lecture astronomy.

Pannekoek was an unusually productive astronomer. His work on astronomy, leading to some 136 scientific publications, covers almost 70 years: from the 1888 diaries of his youth until 1957, when he published his last paper on colour differences in the Milky Way.[49] It is hard to fathom that this is indeed the work of only one man. It is even harder to believe that this man still devoted such a large part of his life to matters outside astronomy. I hope that, with this article, I have been able to convince the reader that, while Pannekoek is widely known for his political activities and contributions, he was an outstanding astronomer of the highest international calibre. And that, while to his political activities he was driven more by sense of responsibility for society and mankind, throughout his life, astronomy was his real love.

## Archives

API Archive of the Anton Pannekoek Institute, University of Amsterdam.

## Bibliography

Albada, Gale Bruno van. 1945. 'Een onderzoek van lijnintensiteiten in eenige sterspectra van het tweede type'. PhD Thesis, Amsterdam: University of Amsterdam.

Argelander, Friedrich. 1855. *Handleiding voor vrienden der sterrekunde, tot het volbrengen van belangrijke waarnemingen, die geene werktuigen vorderen*. Ed. by Frederik Kaiser. Trans. by W.F. Kaiser. Zwolle: De erven J.J. Tijl.

48 Pannekoek 1961, 13.

49 Pannekoek 1957.

Baneke, David. 2004. '"Hij kan toch moeilijk de sterren in de war schoppen". De afwijzing van Anton Pannekoek als adjunct-directeur van de Leidse Sterrewacht in 1919'. *Gewina* 27(1): 1-13.

Hertzsprung, Ejnar. 1911. 'Nachweis der Veränderlichkeit von α Ursae minoris'. *Astronomische Nachrichten* 189(6): 89-104.

Koelbloed, David. 1953. 'Line Spectra of Some Giant and Dwarf K-Type Stars'. PhD Thesis, Amsterdam: University of Amsterdam.

Lenin, Vladimir I. (1920) 2008. *'Left-Wing' Communism, an Infantile Disorder.* Rockville, MD: Wildside Press.

Levy, David H. 1993. *The Man Who Sold the Milky Way: A Biography of Bart Bok.* Tucson, AZ: University of Arizona Press.

Luikens, Elze. 2001. 'Smit, Jan Martinus'. Biografisch Woordenboek van het Socialisme en de Arbeidersbeweging in Nederland. http://hdl.handle.net/10622/7E4C72B1-747D-4900-9057-03D941C12636 (accessed 4 April 2018).

Pannekoek, Anton. 1892. 'Minima von Algol am 2. und 5. Nov. 1891'. *Astronomische Nachrichten* 130(28): 436-436.

—. 1902. 'Untersuchungen über den Lichtwechsel Algols'. PhD Thesis, Leiden: Leiden University.

—. 1903. 'Untersuchungen über den Lichtwechsel Algols'. *Astronomische Nachrichten* 161(12): 213-224.

—. 1906. 'The Luminosity of Stars of Different Types of Spectrum'. *Proceedings of the Section of Sciences, Koninklijke Akademie van Wetenschappen te Amsterdam* 9(1): 134-148.

—. 1909. *Darwinisme en Marxisme.* Amsterdam: Brochurehandel S.D.P.

—. 1912. *Marxism and Darwinism.* Trans. by Nathan Weiser. Chicago, IL: Charles H. Kerr & Co.

—. 1913. 'The Variability of the Pole-Star'. *Proceedings of the Section of Sciences, Koninklijke Akademie van Wetenschappen te Amsterdam* 15(2): 1192-1199.

—. 1916. *De wonderbouw der wereld. De grondslagen van ons sterrekundig wereldbeeld populair uiteengezet.* Amsterdam: S.L. van Looy.

—. 1919. 'The Distance of the Milky Way'. *Monthly Notices of the Royal Astronomical Society* 79(7): 500-507.

—. 1920. *Die Nördliche Milchstrasse.* Annalen van de Sterrewacht te Leiden, 11:3. Haarlem: Joh. Enschedé en Zonen.

—. 1922. 'Ionization in Stellar Atmospheres'. *Bulletin of the Astronomical Institutes of the Netherlands* 1(19): 107-118.

—. 1928. *Die Südliche Milchstrasse.* Annalen van de Bosscha-Sterrenwacht, 2:1. Amsterdam: De Bussy.

—. 1929. *Researches on the Structure of the Universe: 2. The Space Distribution of Stars of Classes A, K and B, Derived from the Draper Catalogue; 3. The Cape*

*Photographic Durchmusterung*. Publications of the Astronomical Institute of the University of Amsterdam 2. Amsterdam: Stadsdrukkerij.

—. 1930. 'Die Ionisation in den Atmosphären der Himmelskörper'. In *Handbuch der Astrophysik*, ed. by G. Eberhard, A. Kohlschütter, and H. Ludendorff, 7 vols. Berlin: Springer, III.1: 256-350.

—. 1945. *Anthropogenese. Een studie over het ontstaan van den mensch*. Verhandelingen der Koninklijke Nederlandsche Akademie van Wetenschappen, Afdeeling Natuurkunde. Tweede sectie, 42:1. Amsterdam: North-Holland.

—. 1951. *De groei van ons wereldbeeld*. Amsterdam: Wereldbibliotheek.

—. (1944) 1953. *Anthropogenesis: A Study of the Origin of Man*. Amsterdam: North-Holland Publishing.

—. 1957. 'Colour Differences in the Milky Way?' *The Observatory* 77(901): 241-242.

—. 1961. *A History of Astronomy*. London: George Allen and Unwin.

—. (1920) 1969. 'Der neue Blanquismus'. In *Organisation und Taktik der proletarischen Revolution*, by Anton Pannekoek and Herman Gorter, ed. by Hans Manfred Bock. Frankfurt am Main: Neue Kritik, 118-122.

—. 1982. *Herinneringen: herinneringen uit de arbeidersbeweging; sterrenkundige herinneringen*. Ed. by Ben A. Sijes, Johanna M. Welcker, and J.R. van der Leeuw. Amsterdam: Van Gennep.

Tai, Chaokang. 2017. 'Left Radicalism and the Milky Way: Connecting the Scientific and Socialist Virtues of Anton Pannekoek'. *Historical Studies in the Natural Sciences* 47(2): 200-254.

Verweij, Sijtze. 1936. 'The Stark Effect of Hydrogen in Stellar Spectra'. PhD Thesis, Amsterdam: University of Amsterdam.

Walraven, Theodore. 1848. 'The Line Spectrum of Delta Cephei'. PhD Thesis, Amsterdam: University of Amsterdam.

Willink, Bastiaan. 1998. *De tweede Gouden Eeuw: Nederland en de Nobelprijzen voor natuurwetenschappen, 1870-1940*. Amsterdam: Bert Bakker.

## About the Author

**Edward P.J. van den Heuvel** is Professor of Astronomy, Emeritus, at the University of Amsterdam and the Free University of Brussels. He was director of the Anton Pannekoek Institute for Astronomy of the University of Amsterdam until 2005. For his studies of the formation and evolution of neutron stars and black holes in binary systems he was awarded an honorary doctorate of the Catholic University of Leuven (1994), the Spinoza Prize of the Netherlands Science Foundation NWO (1995), and the Viktor Ambartsumian International Science Prize (2018). As leader of the European collaboration for the study of gamma-ray bursts he was awarded the Descartes Prize of the European Commission (2002). He is member of the Royal Netherlands Academy of Arts and Sciences, of the European Academy of Sciences, Honorary Fellow of the Indian Academy of Sciences (Bangalore), and of the Royal Astronomical Society.

# 3 Anton Pannekoek: A 'Principled Theorist'

*Gerrit Voerman*

**Abstract**

Anton Pannekoek was not only an astronomer, but also a Marxist theorist. He developed a form of anti-authoritarian socialism in which the workers had to liberate themselves rather than follow their political parties and the trade unions. His anti-authoritarian opinions and his emphasis on spontaneous actions of the masses went too far for many leaders of the labour movement: Pannekoek came into conflict with Troelstra, the leader of the Dutch Social Democratic Workers' Party, with Kautsky, the leading theorist of the Social Democratic Party of Germany and the Second International, and with Lenin, the revolutionary Russian leader. These clashes of Pannekoek with the establishment of the labour movement were not solely the result of his radical theoretical views, but also of his rigourous personality.

**Keywords:** Anton Pannekoek, P.J. Troelstra, Karl Kautsky, Lenin, council communism

When Anton Pannekoek (1873-1960) told his father in the summer of 1899 that he had joined the *Sociaal-Democratische Arbeiderspartij* (Social Democratic Workers' Party, SDAP), his father replied: 'If you had joined the Freemasons I would have been pleased, but this!'[1] Pannekoek senior, manager of a small iron foundry, did not think much of socialism and preferred not to see his son going down that path. This was not only for political reasons, but also because he was afraid that this political choice would damage Anton's career. In those days, it was 'not done' to be a socialist, especially for someone from a middle-class milieu.

1 Pannekoek 1982, 73.

Tai, Chaokang, Bart van der Steen, and Jeroen van Dongen (eds), *Anton Pannekoek: Ways of Viewing Science and Society*. Taylor & Francis Group, 2019
DOI: 10.5117/9789462984349_CH03

Was Pannekoek senior's fear justified? Did his son's career suffer because he chose to follow the path of socialism? The answer to this question is somewhat ambiguous: it depends on what you understand by Pannekoek's career. Surprisingly, his academic career did not suffer in the end, as in 1925 he became full professor of astronomy in Amsterdam and member of the Royal Academy of Arts and Sciences of the Netherlands, and in 1936 Harvard University bestowed on him an honorary doctorate for his astronomical research. Paradoxically, however, Pannekoek became increasingly marginalized within the socialist movement. This contribution tries to explain this eventual marginalization by analysing the development of Pannekoek's role and position within the international working-class movement.

Pannekoek senior's disappointed reaction was partly due to the fact that his son had been doing so well in society. Anton had graduated from Leiden University with a degree in Astronomy. A few years later, he became observer at the Leiden Observatory. After he was appointed, Pannekoek joined the local Liberal electoral association, mainly because he thought this was expected of him, given his social status. He followed in the footsteps of his father, who was a staunch Liberal. After a personal struggle, however, Pannekoek converted to socialism. During a public Liberal meeting he put forward the standpoint of the SDAP. Later he wrote in his memoirs: 'Now all acquaintances, notables, and academic colleagues knew where I stood.'[2] There is little reason to doubt this account, given the principled stance he was to take later in life. Pannekoek did not shy away from the consequences of the sometimes sharp political choices he made, even though they occasionally cost him the sympathy of the people around him or resulted in major conflicts, ending in complete breaks with former kindred spirits – like Pieter Jelles Troelstra, the leader of the SDAP; Karl Kautsky, the most prominent theorist of the *Sozialdemokratische Partei Deutschlands* (Social Democratic Party of Germany, SPD); and Vladimir Ilyich Lenin, the leader of the Russian Bolsheviks.

## Troelstra

After Pannekoek had joined the SDAP, he was actively involved in the foundation of the local party branch in Leiden, which he also chaired. He regarded himself not so much as an organizer or a propagandist, but foremost as a

2 Pannekoek 1982, 73.

theorist. He thus contributed regularly to the social democratic journal *De Nieuwe Tijd* ('The New Era').

Pannekoek's fellow party members soon became familiar with his principled attitude. At the beginning of the twentieth century, the SDAP, as social democratic parties in other countries, was embroiled in an internal struggle between revolutionary Marxists and so-called 'revisionists'. Following the German theorist Eduard Bernstein, the revisionists preferred a practical path, geared to steady but gradual social reforms, over a socialist revolution. They argued that the working class in the Netherlands was weak, since it made up only a minority of the population and was not well organized. Therefore, the SDAP's main goal should be to broaden its electoral base through ideological adaptions. This way, the party's political power could be increased through electoral victories. Indeed, the SDAP soon oriented itself not only on factory workers, but also on small farmers, tenants, and Christian workers. In addition, given the Dutch electoral majority system, the party did not a priori want to exclude making deals with bourgeois parties in order to gain more legislative seats.

The most prominent exponent of this moderate and pragmatic line was Pieter Jelles Troelstra (1860-1930), who had been a charismatic leader of the party since its foundation in 1894. He was much esteemed within the international labour movement and was part of the inner circle of the Second International, the organization of socialist parties and trade unions founded in Paris in 1889. He firmly believed in the parliamentary road to socialism, although he did not give up on the revolution altogether. The orthodox leftist minority strongly opposed Troelstra's position. In agreement with the German orthodox Marxist theorist Karl Kautsky, its view was that socialism would not be achieved gradually and peacefully, but could only be the outcome of a revolution. The inevitable concentration of capital in the hands of a few and the *Verelendung* (pauperization) of the large majority of the masses resulting from this, would lead to a sharpening of the class struggle. Capitalism would thus collapse due to the inevitable intensification of these internal contradictions and make way to socialism.

Pannekoek became one of the main voices of the Marxist opposition in the SDAP, along with the poets Herman Gorter (1864-1927) and Henriette Roland Holst (1869-1952), who both became his close friends. All three were editors of the critical journal *De Nieuwe Tijd*, which considered itself the guardian of revolutionary Marxist politics. They enjoyed a warm and friendly relation with their intellectual guide Kautsky, with whom they frequently corresponded. Partly under his influence, Pannekoek, Gorter, and Roland Holst became committed to a rather principled kind of Marxism. They made

Troelstra their scapegoat for what they perceived as the political degeneration of the SDAP into a moderate non-revolutionary party. Pannekoek wrote in 1944 in his autobiography that when he met the SDAP-leader for the first time, he noticed something 'demagogic' in the party leader's demeanour; he experienced 'the lawyer, the politician, which made me suspicious'.[3] Pannekoek's image of Troelstra may have grown more negative as the years advanced, but it is clear that in the internal party struggle, Pannekoek was at the forefront criticizing Troelstra's opportunist and reformist positions. At the same time, Pannekoek championed theoretical purity on party congresses and in articles in *De Nieuwe Tijd*. Sometimes, his fellow opponents thought Pannekoek was going too far; Gorter once called him a 'hyper-Marxist', and Roland Holst, in a letter to Kautsky, would criticize him for being a 'Hitzkopf' (hothead).[4]

According to the Marxist opposition, too much emphasis on parliamentary and trade union work would only 'weaken the principled revolutionary character of the party'.[5] In their view, political action – directed against the bourgeoisie and the state – was too often confused with parliamentary action, which was only part of it. According to Pannekoek, parliament had to be the arena where social democracy criticized capitalism continuously, defended the interests of the working class, forged proletarian unity and raised the level of insight of the workers into the workings of capitalism.[6] Because the SDAP-politicians did not act in a 'purely proletarian' fashion, the more class-conscious workers had become frustrated and consequently remained at a distance from the party.[7] In Pannekoek's opinion, striving for social reforms as goals in themselves would 'obscure the fundamental difference between our party and all bourgeois parties'. This, then, undermined class-consciousness.[8] The focus should be on class struggle and raising the proletarian awareness of the workers; the evolution of capitalism, as revealed by Marx, would naturally lead to the victory of the working class.

Pannekoek clashed with Troelstra for the first time in the wake of the railway strike of 1903 and the subsequent unsuccessful political strike against the introduction of legislation banning further strike action of railway workers by the ruling conservative cabinet. The SDAP-leader had initially spoken out in favour of a political strike, but then abruptly and obstinately called it off.

3 Pannekoek 1982, 77.
4 Buiting 1989, 249-250.
5 Pannekoek and Gorter 1906, 24.
6 Buiting 2003, 225.
7 Pannekoek 1903.
8 Buiting 2003, 260.

Pannekoek, who now began to believe that the masses were more important than the party in accomplishing social transformation, pointed out the demoralizing effect of Troelstra's sudden change of heart.[9] He accused Troelstra of 'little less than betrayal' – not in a moral sense, but as an 'objective lack of insight'.[10] This nuance understandably got lost in the heat of the battle, which became increasingly ferocious and personal. Pannekoek accused Troelstra of conspiring and stirring up fellow party members against him and the party's left wing, while Troelstra attacked Pannekoek for not only offending him as party leader, but also for consequently diminishing the standing of the party, thus effectively giving ammunition to political enemies of the labour movement.[11] Troelstra criticized Pannekoek's 'hunt for heretics' and publicly ridiculed his opponent.[12] 'It is possible for someone to be an excellent philosopher, a competent observer of celestial things, and yet still behave like a perfect lout in political and parliamentary matters', he stated.[13] Troelstra blamed Pannekoek, whom he saw as dogmatic, for being the main instigator of inner party strife.

The conflict would ultimately end in a schism. In 1909, the majority of the Marxist opposition, including Pannekoek, left the SDAP and founded the *Sociaal-Democratische Partij* (Social Democratic Party, SDP). Initially, Pannekoek regarded the schism as a 'disaster', but soon he spoke of a 'liberation', bringing 'so much new and free thinking and action'.[14] Eventually, however, Pannekoek clashed with the new party leadership – as did Gorter and, a little later, Roland Holst.

After the schism of 1909, Pannekoek, Gorter, and Roland Holst became even more critical of the parliamentary orientation of the SDAP. According to them, it was the growing bureaucracy of the SDAP and the trade unions that made them afraid of conflict and instead geared both organizations towards policies of compromise and practical reforms. As countermeasures, the three started 'to emphasize the political mass strike, radical democratic organization and the encouragement of the revolutionary-creative potential of the masses'.[15] These priorities made Pannekoek, Gorter, and Roland Holst not only stand out as radically left of the SDAP, but also of the SDP. Their emphasis on the spontaneous, creative self-action of the masses was inspired by the epistemology of the German philosopher Joseph Dietzgen,

9 Gerber 1989, 38.
10 Buiting 1989, 168-169; Sijes 1982, 24-25.
11 Troelstra 1906, 93.
12 Quoted in Buiting 1989, 254.
13 Quoted in Pannekoek 1982, 312 n. 173.
14 Buiting 1989, 580.
15 Buiting 1989, 607; see also Gerber 1989, 38.

which they regarded as a necessary addition to Marxism.[16] As a result, they became known as the 'Dutch Marxist School', which emphasized ideological purity, 'self-action' by the masses, and a voluntarist interpretation of classical Marxism, thus challenging the mechanistic and determinist character the latter had taken.[17] In the words of Pannekoek's biographer Gerber, 'they firmly rejected theoretical revisionism and practical reformism, maintained a deep mistrust of parliamentarianism and advocated an active strategy of confrontation with the state and capital'.[18] In the first two decades of the twentieth century, Pannekoek, Gorter, and Roland Holst were influential thinkers in the European labour movement.[19]

## Kautsky

At the same time as the conflicts within the SDAP escalated, Pannekoek had come to dislike his work at the Observatory. In November 1906, after receiving an official invitation by Kautsky and the leadership of the SPD, he left for Berlin to teach at their party school. Karl Kautsky (1854-1938) had been the guiding light of the Dutch revolutionary Marxists in their struggle against Troelstra. He was known as the 'Pope of Marxism' and was respected and admired by many, including Pannekoek, who regarded himself a 'pupil' of him.[20] The two were close political allies as well as close friends. They had met in 1900, when Kautsky gave political lectures in Amsterdam and Delft, after which they started to correspond.[21] Kautsky asked Pannekoek to contribute to *Die Neue Zeit* ('The New Era'), the theoretical journal of the SPD, of which he was the founder and chief editor. Kautsky soon acknowledged how theoretically gifted the Dutch astronomer was. Their relationship was amicable and Pannekoek was very respectful. In due course, however, personal frictions surfaced. Pannekoek did not mince words and complained, for instance, that Kautsky had deleted a critical remark from one of his book reviews. He also expressed his frustration when, in 1909, the prominent German ideologist 'abandoned' the Marxist opposition in the Dutch SDAP

16 Bock 1992; Buiting 2003, 165-175.

17 Bock 1992; van der Steen 2006; Gerber 1989, 22.

18 Gerber 1989, 42.

19 Gerber 1989, 32.

20 Pannekoek to Kautsky, n.d., KK, D XVIII, 368.

21 Gerber 1989, 43.

by not publicly supporting them when Troelstra was trying to have them thrown out of the party.[22]

Back in 1906, however, Pannekoek was excited to be invited by Kautsky and to work for the SPD. When he arrived in Berlin, Kautsky took care of him. He made sure that his Dutch confidant received a decent salary, and initially Pannekoek stayed with the Kautsky family. Later, Pannekoek and his wife frequently paid social visits to the Kautsky's and return visits were made as well. Kautsky also introduced his friend to the German party leadership. At his home, for instance, Pannekoek met Rosa Luxemburg (1871-1919), a revolutionary socialist of Polish-Jewish origin. In 1907, when the Prussian government prohibited Pannekoek from working as an instructor at the SPD's party school, Kautsky helped him to find an alternative source of income. Pannekoek started to write a weekly column (*'Zeitungskorrespondenz'*), which was published in various socialist papers. This development further increased his influence in the German labour movement. In 1910, Pannekoek and his family moved to Bremen, outside the Prussian state, where he worked for the local SPD branch, delivering courses and giving lectures. In Berlin, he had missed the daily life of the party; 'we only saw the big shots, but not the workers themselves'. He later claimed that he felt that he had lived in a 'special world, not in the real world', by being confined only to higher party circles, among only officials.[23] In Bremen, this was different: here he was a part of the ordinary party life.

In the bitter conflict between revisionists and reformists on the one hand and the orthodox Marxists on the other, Kautsky and Pannekoek had been on the same side. Despite their political like-mindedness and mutual affection, however, Pannekoek and his German friend hit upon a fundamental difference of opinion about the importance of mass actions in 1911. Pannekoek followed Luxemburg, who held that the masses should develop their own forms of struggle, beyond the stifling control of the party and its leadership. Both Luxemburg and Pannekoek stated that mass actions were the expression of the will of the proletariat.[24] Kautsky, on the other hand, saw this weapon as a last resort of the proletariat, only to be used under dire circumstances and under the direction of the party. He mistrusted unorganized and spontaneous masses, which he felt were unpredictable and therefore uncontrollable. Instead, he emphasized strong organization, discipline, and parliamentary action, thereby reasserting the primacy of the

22 Pannekoek to Kautsky, 4 February 1909, KK, D XVIII, 408.

23 Pannekoek 1982, 138.

24 Pannekoek 1982, 140; Pannekoek to Kautsky, n.d., KK, D XVIII, 372.

party and the trade unions. For Kautsky, organization was indispensable in order to channel the undirected energy of the masses and to strategically plan which struggles ought to be waged by the masses. Moreover, Kautsky feared the repercussions of a frontal attack on the powerful Prussian state for the social democratic organizations.[25]

As a result of this debate, a divide grew within the orthodox-Marxist camp, with Pannekoek, preceded by Luxemburg, emerging as a figurehead of the 'neo-radical' faction.[26] Pannekoek had been impressed by the Russian revolution of 1905, with its barricade struggles and mass strikes, and the Prussian suffrage demonstrations of a few years later. Like Luxemburg, he saw mass action as the ultimate means to revolutionize the masses and destroy the capitalist state. Unlike Kautsky, Pannekoek came to expect less and less from parliamentary and trade union tactics that aimed at strengthening labour organizations step by step into a strong political power. He had seen with his own eyes in both the Netherlands and Germany how small-time local officials, as well as the big wigs of the labour movement bureaucracy – the 'labour aristocracy' –, opted for caution and moderation out of self-interest. According to Pannekoek, parliamentary tactics had their benefits for as long as the working class was weak, but it was not a way to prepare for revolution.[27] For that, the workers had to take the initiative in their own hands and propel the party and the trade union organizations forwards by means of mass action. Mass protests, rallies, and strikes would advance their knowledge, insight, and political experience, and thus raise their social and political conciousness. The revolutionary potential of the masses would lead to new forms of organization of the working class, based on democratic self-government. This would render the leaders of existing organizations superfluous – even if only partly. At this time, Pannekoek still believed that mass actions would be most productive if the party and trade unions were involved, not in a leading but in a supporting role, encouraged and driven by the revolutionary energy of the masses.[28]

Pannekoek's views, in which not the leaders but the masses had to take the initiative, were of course strongly opposed by the establishment of the SPD and the trade unions. Kautsky, who in this period started to identify his own positions as 'centrist' (being opposed to revisionism as well as the 'new

25 Buiting 2003, 399-400; Salvadori 1979, 154.
26 Salvadori 1979, 143.
27 Salvadori 1979, 156.
28 Sijes 1982, 28-37; Buiting 1989, 607; Gerber 1989, 98-99.

radicalism'), also turned against him.[29] They clashed in lengthy articles in *Die Neue Zeit*.[30] Their polemic addressed several issues, such as the proper role of the state in the process of social transformation. While Kautsky was bent on conquering and taking over the state by achieving a parliamentary majority, Pannekoek claimed that the state had to be destroyed.[31] Yet, the central theme of their discussion was the role of mass action.

When Kautsky started the discussion with his 1911 article 'Die Aktion der Masse', Pannekoek's initial response was quite mild. In a letter to Kautsky he proposed an 'objective discussion', which would be in the interest of the party.[32] Later, Pannekoek announced a 'sharp presentation' of their differences, because he was convinced that in order to come together, it was necessary to know exactly and clearly the differences between each other's point of view.[33] In his reply to Kautsky ('Massenaktion und Revolution'), Pannekoek was indeed sharp and critical. He not only pointed out their differences in all their facets, as he already had announced he would; he also reproached Kautsky for not using the Marxist analytical method properly. In Pannekoek's eyes, Kautsky was more or less approaching revisionism; a very harsh critique.[34] At one point Pannekoek even wrote 'that one could hardly believe that these sentences have flowed from Kautsky's pen'.[35]

To be castigated by his former pupil in his own journal proved to be too much for Kautsky. The internationally distinguished theorist seemed personally hurt – something Pannekoek later also sensed, when he wrote Kautsky that it appeared to him that the latter felt personally wronged.[36] Kautsky's public response was venomous and sarcastic: 'Aye, does comrade Pannekoek really believe, that I have forgotten the ABC of Marxism, ideas to whose gaining acceptance I have spent the best part of my life? [...] Fortunately we have comrade Pannekoek, who exposes my "bourgeois misunderstanding"'. Kautsky found Pannekoek simplistic ('a one-size-fits-all blueprint'), obscure ('all this is abundantly unclear and mysterious, reminds one more of the

29 Salvadori 1979, 149.
30 See Kautsky 1911-1912a; 1911-1912b; 1912-1913; Pannekoek 1911-1912; 1912-1913a; 1912-1913b.
31 Buiting 2003, 399-400.
32 Pannekoek to Kautsky, 6 November 1911, KK, D XVIII, 413.
33 Pannekoek to Kautsky, 8 May 1912, KK, D XVIII, 417.
34 Pannekoek 1911-1912, 589, 592.
35 Pannekoek 1911-1912, 611.
36 Pannekoek to Kautsky, 30 December 1912, KK, D XVIII, 423.

Delphic oracle and Sybilline books than the substantiation of a new tactic'), and engaged in nitpicking ('Talmudic hair-splitting').[37]

Even though Pannekoek must have been upset by Kautsky's article, he seemingly remained calm. 'I certainly have never believed that it would be possible that you would distort my views in such a way', he wrote to Kautsky in a personal letter. Pannekoek further wrote that he found Kautsky's article 'so bourgeois [*bürgerlich*] and un-Marxist, that I more than anything else regret to find this under your name' – which is difficult to interpret in any other way than that Kautsky had fallen from his pedestal in the eyes of Pannekoek. Kautsky's criticisms hardly made an impression on Pannekoek, on the contrary, the validity of his opinions were only confirmed.[38]

Early in 1913, the two once befriended opponents ended their polemic. In Pannekoek's final remarks and Kautsky's very short reply, there were hardly any signs of mutual appreciation or affection.[39] Their political collision also meant the end of their friendship. Later, Pannekoek would write in his autobiography that their relationship had cooled, 'not so much personally but rather from a theoretical perspective'; an obvious understatement.[40] In April of 1912, Pannekoek already had let Kautsky know that it was 'painful' for him to see them drifting away from each other.[41] Pannekoek felt that Kautsky had made no effort to understand him, but instead had simply dismissed him as a 'half-syndicalist or anarchist antiparliamentarian'.[42] After their clash in *Die Neue Zeit*, Pannekoek's letters to Kautsky stopped. Pannekoek's last letter to be found in Kautsky's archive is dated 5 February 1913. After that, Pannekoek no longer published in *Die Neue Zeit*.[43] Shortly after, after the outbreak of World War I, the SPD voted in the German parliament in favour of extending war credits to the government, a move that was incomprehensible for Pannekoek, and he publicly scolded Kautsky for supporting it.[44] In his autobiography, Pannekoek recollected that he had had 'the greatest pleasure' in 'finishing off' his former teacher in an article.[45]

37 Kautsky 1911-1912b. The English quotations are taken from the English translation at https://libcom.org/library/new-tactic-karl-kautsky (accessed 25 May 2016).

38 Pannekoek to Kautsky, 18 August 1912, KK, D XVIII, 419.

39 Pannekoek 1912-1913b, with a footnote by Kautsky.

40 Pannekoek 1982, 165.

41 Pannekoek to Kautsky, 14 April 1912, KK, D XVIII, 416.

42 Pannekoek to Kautsky, 17 October 1912, KK, D XVIII, 421.

43 Welcker 1986, 91.

44 Salvadori 1979, 181.

45 Pannekoek 1982, 183.

Pannekoek declared the Second International dead, after other social democratic parties in Europe had decided to support the war too. He announced the formation of a new one, 'more Socialist than the one that perished'.[46] Russian Bolshevik leader Lenin was delighted: 'The only one who has told the workers the truth – although not loudly enough, and sometimes not quite skilfully – is Pannekoek.'[47]

After the outbreak of World War I, Pannekoek and his family returned to the Netherlands. Pannekoek tried to earn a living, first as a journalist, and then as a secondary school teacher. He did not want to upset his pupils' parents, so he refrained from lecturing at SDP-meetings. He continued writing for publications such as *De Nieuwe Tijd*, because this did not attract much attention, as he saw it. After Pannekoek had become a full-time lecturer in mathematics and astronomy at the University of Amsterdam in 1919, he started to write under the pseudonym of Karl Horner, in order not to harm his academic career.

## Lenin

After Troelstra and Kautsky, the third prominent leftist leader with whom Pannekoek clashed was Vladimir Iliych Lenin (1870-1924), leader of the Russian Bolsheviks and the October Revolution of 1917, and the founder of the Soviet Union. Other than with Troelstra and Kautsky, Pannekoek never met Lenin. He did, however, maintain relatively close contacts with Lenin and other Russian Bolsheviks around the time of World War I, just like some other Dutch orthodox Marxists.[48] The Bolsheviks were rather isolated within the international Socialist movement and were glad to welcome kindred spirits. Lenin himself was greatly interested in the publications of Pannekoek and Gorter. With the help of a German-Dutch dictionary he could 'understand about 30-40%' and initially he had a high opinion of both.[49] He welcomed their publications and sided with Pannekoek in his polemic with Kautsky.[50]

Lenin could use the support of his Dutch comrades. In his struggle against the 'social chauvinists' – his label for social democrats who supported the war efforts of their national governments – he tried to forge closer links

46 Pannekoek 1914, 688.
47 Quoted in Gerber 1989, 109.
48 Bauman 1988; Voerman 2001; 2007.
49 Quoted in de Liagre Böhl 1996, 361.
50 de Liagre Böhl 1996, 362; Lenin [1917] 1964, 488-496.

between the Bolsheviks and other leftist internationalist groups. In the summer of 1915, Lenin proposed to SDP-leader David Wijnkoop to draft a joint declaration aimed against the 'imperialist' war, which was to be presented in September at the Conference of Zimmerwald, where delegates of leftist anti-war groupings would gather. This plan came to nothing, as Wijnkoop and his fellow party leaders had no confidence in the political outcomes of the conference. Despite Lenin's insistence, they decided not to participate. Pannekoek had been strongly in favour of attending the conference, but was not able to go himself. However, he also did not think highly of the common declaration Lenin wanted to draft. 'I do not have high expectations of him', Pannekoek wrote to Wijnkoop. 'The intellectual guidance has to come from the SDP.'[51]

In this period, Pannekoek was involved, along with Roland Holst, in founding a left-wing Marxist theoretical journal called *Der Vorbote* (The Herald), which was meant to combat Kautsky and other moderates. They collaborated with Lenin, who supported Pannekoek as a 'trusted representative' of the Bolsheviks.[52] The journal was supportive of the revolutionary socialist wing of the Zimmerwald conference, and aimed for the foundation of a third, radical International, as the successor of the 'bankrupt' Second International.

Only two issues of *Der Vorbote* were published, partly as a result of frictions between Pannekoek and Lenin. Politically, they were quite close: both could be found at the left wing of the international labour movement, and opposed the 'imperialist' war as well as the revisionist 'deviations' in the labour movement. Furthermore, both were in favour of a new, radical international. However, signs of an impending rift were also soon visible. In his book *State and Revolution*, which was published in 1917, Lenin welcomed Pannekoek's earlier attack against Kautsky, but disapproved of the Dutchman's preoccupation with mass action. Lenin believed that the term obscured the concept of 'revolution'.[53] Pannekoek, on the other hand, admired Lenin's dedication, but did not think of him as a 'high-flyer'. In 1915, in a letter to SDP-leader Wijnkoop, he described the Russian revolutionary leader in theoretical respect as 'a curious chap who, moreover, sees Western Europe too much from a Russian perspective'.[54] One year later, he wrote to Wijnkoop 'that Lenin is still, to a large extent, an old revolutionary conspirator, and he has

51 Pannekoek to Wijnkoop, 22 July 1915, CA, 581/1/35.

52 Pannekoek to Van Ravesteyn, 22 (quote) and 24 October 1915, WvR, 15.

53 Bauman 1988, 166.

54 Pannekoek to Wijnkoop, 12 July 1915, CA, 581/1/35.

no clear understanding of imperialism'.[55] There was a huge theoretical and strategic difference between the two, which would later become all too visible. While Pannekoek had great confidence in the creative potential and the revolutionary energy of the masses, Lenin – just as Kautsky – believed that one could not rely on the spontaneity of the masses. He saw it as the task of the disciplined communist vanguard to raise the proletarian consciousness of the masses. Clearly, such a fundamental difference of opinion could only lead to a clash between Pannekoek and Lenin.

Pannekoek welcomed the Russian October Revolution of 1917 with great enthusiasm. A new period had started, he stated, not only for Russia, but also for the European proletariat. Pannekoek embraced the Bolsheviks, led by Lenin and Trotsky, as 'those who courageously go before us on the road to socialism'.[56] He saw the 'Soviets', the workers' and soldiers' councils, as 'the new institution of power of the proletarian masses', being far more democratic than the parliamentary system.[57] Again, Pannekoek emphasized the importance of the self-mobilization of the masses, who should act independently and liberate themselves, instead of working as the extension of other actors.[58] Nevertheless, Pannekoek was also amongst those who warned early on about the difficulties that the Bolsheviks might encounter due to the agrarian character of Russia. Lenin and Trotsky invited Pannekoek and Gorter to come to Moscow in 1918, because they believed that their theoretical and practical work could contribute to the revolutionaries' cause. Nothing came of the invitations, however, because the situation in Europe was too dangerous to travel.[59]

The engineer Sebald Rutgers (1879-1961), a Dutch communist, who had attended the founding congress of the Communist International (Comintern) in Moscow in March of 1919, was then instructed by Lenin to set up an outpost in the Netherlands to facilitate communication between Moscow and the Western European communist groups. Upon his arrival, Rutgers stated that Lenin expected much of the Dutch orthodox Marxists, 'especially of Gorter and Pannekoek'.[60] The Executive Committee of the Comintern (ECCI) appointed Rutgers, Roland Holst, Pannekoek, Gorter, and the SDP-leaders Wijnkoop and Willem van Ravesteyn to head the Comintern Bureau in Amsterdam, and placed twenty million rubles at its disposal. Rutgers received the bulk of the amount in the form of precious stones. Moscow

55 Pannekoek to Wijnkoop, 4 January 1916, CA, 581/1/35.
56 Pannekoek 1917, 560.
57 Buiting 2003, 585.
58 Sijes 1982, 45-46.
59 Voerman 2001, 50.
60 Rutgers 1935, 397.

had reserved additional funds for Pannekoek and Gorter, for the purpose of appointing them in the service of the Comintern. According to Pannekoek himself, Rutgers had asked him to go to Moscow, 'to assist in theoretical work, as adviser, etc.'. He declined a second time, because he had had poor experiences with being on the payroll of the labour movement during his German years. If he were to be employed by the party or the government and differences of opinion would arise, 'I would either have to resign, penniless, to be able to stand firm, or, out of fear for such a situation, to bend, and work and speak contrary to my belief. I do not want to have to make that choice'.[61]

Rutgers arrived in the Netherlands in November of 1919. Activities organized by the Bureau included an international conference to discuss political strategy and the best approach to revolutionary agitation. The meeting was held in February of 1920, with delegates mainly from the United States, England, Germany, and Belgium. The conference ended in a complete failure, as the police managed to monitor the deliberations and to arrest several participants. The Dutch communists had no conspiratorial experience at all: in his memoirs, Pannekoek recalled that during lunchtime conference attendants would enjoy the hospitality of a nearby beer garden, where discussions were loudly continued in various languages.[62]

The Amsterdam Bureau soon found itself in a complicated political situation. Pannekoek had drafted a resolution about parliamentarism for its Amsterdam conference, while another on trade unions was written by the Dutch communist Henk Sneevliet, who one year later would be involved in the foundation of the Chinese Communist Party.[63] At its founding congress in March 1919, the Comintern had been dismissive of participation in elections, but later that year it softened its stance. Comintern-president Gregory Zinoviev declared that although under the dictatorship of the proletariat, parliament would have to make way for the Soviets, under capitalism it might be desirable to utilize popular representation to further the revolution. Whether one would participate in an election, depended on the circumstances.[64] Pannekoek's resolution on parliamentarianism was largely based on Zinoviev's opinions, although the former emphasized more strongly the possibility of an election boycott at times of revolutionary turmoil and to concentrate in that case all forces on direct mass action. This resolution could not be discussed at the conference, due to the police

61 Pannekoek 1982, 196.

62 Pannekoek 1982, 198.

63 Communist International 1920; Buiting 2003, 615.

64 Degras 1971, 69.

intervention, but the parliamentary issue was also addressed in Sneevliet's resolution on trade unions. The delegates declared unanimously that the aim of the revolutionary proletariat was to seize state power.

> Neither parliaments nor trade unions are suited means to that end, but mass action and workers' councils are; mass action should bring together all workers, organized and non-organized, and unite them in an open and direct struggle for power. The councils should be the organs of the revolutionary workers' state – of the proletarian dictatorship.[65]

The conference went further than Zinoviev and Pannekoek in its categorical rejection of parliamentary politics, but the latter would certainly have agreed to it. The Amsterdam conference also adopted a radical stance on trade unions. Its resolution expressed a deep suspicion of traditional unions. It suggested that the reformist trade union movement had become subservient to capitalism and was no longer in a position to take decisive action against it. Therefore, the existing trade unions had to be either 'revolutionized' from within, or replaced by completely new, powerful, anti-capitalist and anti-bureaucratic factory organizations.

Unfortunately for the Amsterdam Bureau, it was out of tune with Moscow right from its inception. Lenin had hoped for a revolution in Europe, which would have helped the Bolsheviks in consolidating their power in Russia. There were some efforts, as in Germany and Hungary, but ultimately, a revolution did not take place in Europe. As a result, Lenin decided on a new strategy: the communists should 'go where the masses are' and try to connect to them, by taking part in parliamentary elections and working within the 'reactionary' trade unions in order to take them over. The resolutions of the Amsterdam conference were clearly contrary to Moscow's new strategy. In February of 1920, the ECCI roundly criticized the policies of the Amsterdam Bureau.[66] A few months later, it revoked its mandate due to its presumed 'sectarian politics': its views on parliamentarianism and trade unions were now in direct opposition to Moscow.[67] The announcement of the Amsterdam Bureau's closure came like a bolt out of the blue. Rutgers was stunned; Roland Holst resigned, but was also a little relieved. Pannekoek responded rather laconically, as he had already seen the writing on the wall. He realized that

65 *De Tribune* 1920.

66 Minutes ECCI session, 2 February 1920, CA, 495/1/2.

67 ECCI decision dated 25 April 1920, sent to Trotter [S.J. Rutgers] by Fritz [F. Ström], 12 May 1920, CA, 497/1/ 9.

the decision to dissolve the Bureau reflected a strategic change of heart in Moscow, which implied that the ideas of the Amsterdam Bureau – which were also his ideas – were now behind the times.[68]

In addition to being condemned by the ECCI, the Amsterdam Bureau was also ridiculed by Lenin, whose pamplet *'Left-Wing' Communism: An Infantile Disorder* appeared shortly before the second Comintern congress in July of 1920. Lenin publicly condemned the 'left-wing communists', whom he labelled sectarians because they were supposedly turning their backs on the masses. According to Lenin, the Dutch 'leftists' argued 'like doctrinaires of the revolution, who have never taken part in a real revolution'.[69] Lenin also sharply criticized their leftist aversion to a disciplined vanguard party led from above.

> Certain members of the Communist Party of Holland, who were unlucky enough to be born in a small country with traditions and conditions of highly privileged and highly stable legality, and who had never seen a transition from legality and illegality, probably fell into confusion, lost their heads, and helped create these absurd inventions.[70]

Lenin appears to have had a personal hand in reversing the decision – in part his own – to set up the Amsterdam Bureau. He insisted that the ECCI and the forthcoming world congress of the Comintern would roundly condemn leftist deviations and 'in particular, the line of conduct of some members of the Communist Party of Holland, who – whether directly or indirectly, overtly or covertly, wholly or partly, it does not matter – have supported this erroneous policy'.[71] His attack on 'Karl Horner' was particularly vicious. According to Lenin, he produced 'incredible nonsense in a most ridiculous manner', and failed 'to understand the ABC of Marxism'.[72] As pointed out before, Horner really was Pannekoek, as Lenin most probably did not know. Even though a few years earlier Pannekoek had been one of Lenin's favourites, now he had fallen from grace.

Pannekoek was 'stunned' by Lenin's sudden change of tactics. 'The acknowledged leader of the world revolution here chose the side of opportunism', he later wrote in his memoirs.[73] Pannekoek was not impressed

68 Pannekoek to Rutgers, 31 January 1920, CA, 495/172/5.
69 Lenin [1920] 1970, 59.
70 Lenin [1920] 1970, 34.
71 Lenin [1920] 1970, 51.
72 Lenin [1920] 1970, 36.
73 Pannekoek 1982, 200.

by Lenin's pamphlet, which he found 'very weak'.[74] He outlined his own views in the run-up to the second Comintern congress, with the intention of weighing in on the discussions in Moscow. Pannekoek's resulting text *Weltrevolution und kommunistische Taktik* was indeed circulated among the congress delegates, but according to Zinoviev only as an example of how not to approach the revolution. Pannekoek sharply criticized Lenin. He argued that Lenin was trying to win over the hesitating and half-hearted Western European masses with his new 'opportunistic' tactic, due to the absence of a European revolution. He pointed out that the situation in Western Europe was very different from the Russian circumstances. The Comintern should thus be fully independent from Moscow and tactics in Western Europe had to be entirely different: not a Leninist vanguard party, but the masses themselves should carry out the revolution.[75]

The outcomes of the second Comintern congress made Pannekoek pessimistic. Initially, Pannekoek had supported Lenin's 'New Economic Policy', which to a certain extent allowed free trade in the new communist state. But soon, he got even more disappointed, especially when Moscow tried to move closer to the West in order to economically reconstruct Soviet Russia.[76] By doing so, Moscow became a stakeholder in the economic development of capitalist countries and an interested party to avoid revolutionary stirrings there. After the third congress of the Comintern in the summer of 1921, Pannekoek had enough: Bolshevik dominance, growing centralization and disciplining, suppression of dissent, the strategic 'shift to the right', and the concessions to the capitalist West were too much for him.[77] Furthermore, after the congress the Comintern expelled the leftist *Kommunistische Arbeiterpartei Deutschlands* (Communist Workers' Party of Germany, KAPD), which in April 1920 had broken away from the *Kommunistische Partei Deutschlands* (Communist Party of Germany, KPD), which was founded in December 1918. Pannekoek strongly identified with the KAPD: both were in favour of workers' councils and opposed to parliamentary politics, and both also wanted to replace trade unions by revolutionary company organizations (*Arbeiter Unionen*).

In September of 1921, Pannekoek broke with Lenin and the *Communistische Partij in Nederland* (Communist Party in the Netherlands, CPN), as the SDP was now called. He no longer saw Moscow as the leading light and forerunner

74 Pannekoek to Rutgers, 25 August 1920, CA, 495/172/5; see also 'Anhang' in Pannekoek 1920.
75 Pannekoek 1920.
76 Buiting 2003, 585.
77 Pannekoek 1921b; Buiting 1989, 648.

of world revolution.[78] With Gorter, he was the first communist theorist who denied Lenin's claim that the Bolshevik Revolution had universal validity, an argument that legitimized Moscow's hegemony of the international communist movement. They also rejected the principal identification of the interests of the world revolution with the general interests of Soviet Russia.[79] By challenging these fundamental premises of the communist regime, both dissidents constituted an ideological threat to the new Russian authorities. At the third congress of the Comintern, Pannekoek and Gorter were excommunicated, when Karl Radek, a confidant of Lenin, ridiculed them: 'One of them is astronomer, gazing only at the stars, and never at a living worker, while the other is a philosopher, and what is more, a poet. (*Laughter*)'.[80]

Pannekoek subsequently became the most important theorist of council communism, but would no longer align himself with any particular party. The proletarian 'party' as a militant organization disappeared completely from his thinking about the self-mobilization of the masses: 'work groups', emerging from within the working class, had to clarify and educate the masses through propaganda and debate. Through such means, the workers themselves should find the proper road to liberation.[81] In 1938, Pannekoek published *Lenin as Philosopher*, in which he argued that 'Lenin never knew real Marxism'.[82] Leninism as an ideology served only to legitimize the Soviet Union's economic system of 'state capitalism', a system in which the workers were again exploited.

## Conclusion

Pannekoek never held a prominent organizational position in the labour movement, but he also never aspired to hold one. He likely did not want a professional career in politics. He did not clash with leaders of the Dutch, German, and Russian labour movements out of a desire for personal advancement of any kind. Instead, he described himself as a 'principled theorist'.[83] Pannekoek believed that he could only function properly if he was completely autonomous, both materially and mentally. 'I want to have

78 Pannekoek 1921a.

79 Voerman 2001, 440-441; Gerber 1989, 146.

80 Riddell 2015, 268; emphasis in the original.

81 Sijes 1982, 65-66.

82 Quoted in Gerber 1989, 190.

83 Pannekoek to Rutgers, 14 January and 27 April 1920, CA, 495/172/5.

a completely open mind when forming and developing my views', he wrote in his memoirs.[84] After his experiences in Germany, Pannekoek no longer wanted a salaried position in the labour movement, because this would make him dependent on others. He thus declined Lenin's invitation to go to Moscow as an adviser.

Pannekoek was a man of principles, and in his view it was hardly acceptable to depart from these in practice. Such ideological purity by itself was enough to bring him into conflict with party leaders like Troelstra, Kautsky, and Lenin, who needed to compromise or change course in day-to-day politics. That Pannekoek increasingly called into question the traditional labour movement organizations with their extensive apparatuses, made his potential for conflict even greater. In his thinking about the 'self-mobilization of the masses', the role of traditional political parties and trade unions became increasingly smaller, to completely disappear by 1920.[85] Pannekoek's growing criticism of the bureaucratization and oligarchization of parties and unions, and of their self-serving leadership did not win him any friends among the labour movement's establishment – neither social democratic, nor Bolshevik.

But how is it possible that Pannekoek eventually always clashed, even with his close friend Kautsky, while others described him as 'serene', a 'modest and mild-mannered' astronomer, and 'an extraordinary modest man without the slightest trace of self-conceit'?[86] Indeed, Pannekoek wrote to both Troelstra and Kautsky in more or less the same words that he was not 'a great fighter by nature', only to subsequently seriously clash with them.[87] Perhaps, Pannekoek was not such a mild-mannered gentleman after all, when he acted as a custodian of socialist purity and principle. Even more so, in this role, he could be unpleasant, nasty, and even rude. Troelstra called him bold and arrogant, but he was of course a political opponent of Pannekoek. Close friends, however, would sometimes express criticism. Roland Holst once called him a hothead, as already mentioned, and at another occasion wrote that the letter she had received from him suggested 'a berating, hostile, and ill-mannered spirit'.[88] In his publications Pannekoek was usually not unhinged or personal (except perhaps with

84 Pannekoek 1982, 196.

85 Gerber 1989, 182-183.

86 Respectively, Gerber 1989, 30; van Berkel 2001, 387; Henk Canne Meijer, quoted in Gerber 1989, 169.

87 Pannekoek to Troelstra, 10 December 1901, PJT, 65/1 ('doordat ik zo weinig vechtnatuur heb'); Pannekoek to Kautsky, n.d. KK, D XVIII, 374 ('bin ich nicht allererst Kampfnatur').

88 Buiting 1989, 580.

Troelstra), but he certainly was not always the mild-mannered astronomer in his correspondence with comrades: 'he does not mince words, writing point-blank… There he lets himself go…, there one reads how he conspires, insinuations do not lack – although they remain scarce –, there one finds contempt of his political adversaries'.[89] In his autobiography too, Pannekoek frequently, sharply, and at times harshly condemned opponents such as Friedrich Ebert, who was prominent in the SPD, or at that time allies such as Radek and Van Ravesteyn.[90] He was convinced that 'opportunism in politics always leads to inferiority in personal actions'. Such a bold maxim is not easy to observe without ending up a in a lot of conflict.[91]

That Pannekoek did not avoid conflicts might have to do with his approach to socialism. Instead of a politician in search of majorities and expansion of his power and sphere of influence, he was a theorist, a 'principled theorist', who was not primarily interested in what was feasible, but in what was true. Furthermore, he considered himself more a schoolmaster than a diplomat, as he confided to Kautsky.[92] According to his one-time party colleague Van Ravesteyn, he had a 'rigorous and mathematical mind […] reasoning mainly in a logical way'.[93] Pannekoek studied the development of society and its consequences for the revolutionary process from an orthodox Marxist and, in his view, scientific perspective.[94] There are similarities between his 'scientific methodology and his socialist straightforwardness'.[95] Pannekoek had no sympathy for someone who diverted from the proper ideology. When the, in his view, 'objective outcomes' of his analyses were not accepted and led to a conflict, this could prove frustrating or disappointing – but it was ultimately inevitable, and Pannekoek was willing to accept the personal consequences. Such a stance, for instance, clearly manifested itself in his assessment of his friend Roland Holst. Pannekoek sided with the opposition in the SDAP and thus joined the new SDP in 1909, directly after its founding. In the same manner, Pannekoek broke with Lenin and the CPN in 1921. Roland Holst, on the other hand, was only

89 Sijes 1982, 14.
90 Pannekoek 1982, 162-163, 166, 189.
91 Pannekoek 1982, 200.
92 Pannekoek to Kautsky, n.d., KK, D XVIII, 374.
93 van Ravesteyn 1948, 27, 136.
94 Sijes 1982, 16-17. See also Tai and van Dongen 2016; and Klaas van Berkel, 'Utopianism in Anton Pannekoek's Socialism and Astronomy', in this volume, 75-86; Annemarie Rullens, 'From Science to Science', in this volume, 151-172; and Bart van der Steen, '"A New Scientific Conception of the Human World"', in this volume, 137-155.
95 van Berkel 1984, 466.

able to take these steps after years of hesitation and doubt. 'Her sense of duty and loyalty was much stronger than rational insight would demand', Pannekoek wrote in his memoirs.[96]

The above reveals a great deal about Pannekoek: the rational imperative should come before the heart. He expressed a similar attitude in his assessment of Luxemburg: 'She does not dare to stand alone, does not want to be an individual, a loner, always wants to be surrounded by a crowd, a group, the party, so criticism is always within the party framework.'[97] Pannekoek himself was not burdened by such sentiments; in contrast to Luxemburg, he accepted isolation. In a letter to Rutgers from January 1920, he characterized his own role in the labour movement: 'I am no good as a representative person: I always side with the minority, who has been marginalized, and I only attach importance to people's proper understanding, which can be taught by theoretical clarification'.[98]

## Archives

| | |
|---|---|
| CA | Comintern Archives. Russian State Archive of Social Political History (RGASPI), Moscow. |
| KK | Karl Kautsky Papers. International Institute of Social History, Amsterdam. |
| PJT | Archief Pieter Jelles Troelstra, International Institute of Social History, Amsterdam. |
| WvR | Archief Willem van Ravesteyn, International Institute of Social History, Amsterdam. |

## Bibliography

Bauman, German. 1988. *De tribunisten. De revolutionaire marxisten van Nederland*. Moscow: Progress.

Berkel, Klaas van. 1984. Review of *Herinneringen. Herinneringen uit de arbeidersbeweging. Sterrenkundige herinneringen*, by Anton Pannekoek. *BMGN – Low Countries Historical Review* 99(3): 465-467.

—. 2001. 'Pannekoek, Antonie'. In *Biografisch Woordenboek van Nederland*, ed. by Jac Bosmans, Mineke Bosch, Klaas van Berkel, J.B.J. Teeuwisse, F. van Vree, H.

96 Pannekoek 1982, 206; see also 1982, 189.
97 Pannekoek 1982, 182.
98 Pannekoek to Rutgers, 14 January 1920, CA, 495/172/5.

van Gelder, Herman de Liagre Böhl, and A.L. Mooij. The Hague: Instituut voor Nederlandse Geschiedenis, 5: 384-387.

Bock, Hans Manfred. 1992. 'Die Marx-Dietzgen-Synthese Anton Pannekoeks und seines Kreises'. In *Die Rezeption marxschen Theorien in den Niederlanden*, ed. by Marcel van der Linden. Trier: Karl-Marx Haus, 106-123.

Buiting, Henny. 1989. *Richtingen- en partijstrijd in de SDAP. Het onstaan van de Sociaal-Democratische Partij in Nederland (SDP)*. Amsterdam: Stichting Beheer IISG.

—. 2003. *De Nieuwe Tijd: sociaaldemokratisch maandschrift ,1896-1921. Spiegel van socialisme en vroeg communisme in Nederland*. Amsterdam: Aksant.

Communist International. 1920. 'Propositions and Theses for an International Conference'. *Bulletin of the Provisional Bureau in Amsterdam of the Communist International* 1(1): 1-4.

*De Tribune*. 1920. 'De internationale communistische conferentie te Amsterdam', 20 March 1920.

Degras, Jane, ed. 1971. *The Communist International, 1919-1943: Documents*. London: Frank Cass & Co.

Gerber, John P. 1989. *Anton Pannekoek and the Socialism of Workers' Self-Emancipation, 1873-1960*. Dordrecht: Kluwer.

Kautsky, Karl. 1911-1912a. 'Die Aktion der Masse'. *Die Neue Zeit* 30(2-4): 43-49, 77-84, 106-117.

—. 1911-1912b. 'Die neue Taktik'. *Die Neue Zeit* 30(44-46): 654-664, 688-698, 723-733.

—. 1912-1913. 'Der jüngste Radikalismus'. *Die Neue Zeit* 31(12): 436-446.

Lenin, Vladimir I. (1917) 1964. 'The State and Revolution: The Marxist Theory of the State and the Tasks of the Proletariat in the Revolution'. In *V.I. Lenin: Collected Works Volume 25, June-September 1917*, ed. by S. Aspresyan and J. Riordan. Moscow: Progress, 385-498.

—. (1920) 1970. *'Left-Wing' Communism: An Infantile Disorder*. Moscow: Novosti Press Agency Publishing.

Liagre Böhl, Herman de. 1996. *Met al mijn bloed heb ik voor U geleefd. Herman Gorter 1864-1927*. Amsterdam: Balans.

Pannekoek, Anton. 1903. 'Een belangrijk oogenblik'. *De Nieuwe Tijd* 8: 162-171.

—. 1911-1912. 'Massenaktion und Revolution'. *Die Neue Zeit* 30(41-43): 541-550, 585-593, 609-616.

—. 1912-1913a. 'Marxistische Theorie und revolutionäre Taktik'. *Die Neue Zeit* 31(8, 10): 272-281, 365-373.

—. 1912-1913b. 'Zum Schluß'. *Die Neue Zeit* 31(17): 611-612.

—. 1914. 'De ineenstorting van de Internationale'. *De Nieuwe Tijd* 19: 677-688.

—. 1917. 'De Russische revolutie'. *De Nieuwe Tijd* 22: 438-452, 548-569.

—. 1920. *Weltrevolution und kommunistische Taktik*. Vienna: Arbeiter-Buchhandlung.

—. 1921a. 'Overgang'. *De Nieuwe Tijd* 26: 714-725.

—. 1921b. 'Rusland and het kommunisme'. *De Nieuwe Tijd* 26: 640-660.

—. 1982. *Herinneringen: herinneringen uit de arbeidersbeweging; sterrenkundige herinneringen*. Ed. by Ben A. Sijes, Johanna M. Welcker, and J.R. van der Leeuw. Amsterdam: Van Gennep.

Pannekoek, Anton, and Herman Gorter. 1906. *Marxisme en revisionisme*. Amsterdam: J.A. Fortuyn.

Ravesteyn, Willem van. 1948. *De wording van het communisme in Nederland, 1907-1925*. Amsterdam: Van Kampen.

Riddell, John, ed. 2015. *To the Masses: Proceedings of the Third Congress of the Communist International, 1921*. Leiden: Brill.

Rutgers, S.J. 1935. 'Een ontmoeting met Lenin'. *Kommunisme* 1(11): 390-397.

Salvadori, Massimo L. 1979. *Karl Kautsky and the Socialist Revolution, 1880-1938*. London: NLB.

Sijes, Ben A. 1982. 'Anton Pannekoek 1873-1960'. In *Herinneringen: herinneringen uit de arbeidersbeweging; sterrenkundige herinneringen*, by Anton Pannekoek, ed. by Ben A. Sijes, Johanna M. Welcker, and J.R. van der Leeuw. Amsterdam: Van Gennep, 11-69.

Steen, Bart van der. 2006. 'Anton Pannekoek en het orthodoxe marxisme'. *Vlaams Marxistisch Tijdschrift* 40(2): 73-82.

Tai, Chaokang, and Jeroen van Dongen. 2016. 'Anton Pannekoek's Epistemic Virtues in Astronomy and Socialism: Personae and the Practice of Science'. *BMGN – Low Countries Historical Review* 131(4): 55-70.

Troelstra, Pieter Jelles. 1906. *Inzake partijleiding: toelichtingen en gegevens*. Rotterdam: Wakker.

Voerman, Gerrit. 2001. *De meridiaan van Moskou: de CPN en de Communistische Internationale, 1919-1930*. Amsterdam: Veen.

—. 2007. 'Proletarian Competition: The Amsterdam Bureau and its German Counterpart 1919-1920'. In *Jahrbuch Für Historische Kommunismusforschung*, ed. by Ulrich Mählert. Berlin: Aufbau-Verlag, 201-220.

Welcker, Johanna M. 1986. 'Pannekoek, Antonie'. In *Biografisch Woordenboek van het Socialisme en de Arbeidersvereniging in Nederland*, ed. by P.J Meertens et al., 8 vols. Amsterdam: Stichting beheer IISG, I: 90-93.

## About the Author

**Gerrit Voerman** (1957) is a historian and director of the Documentation Centre Dutch Political Parties of the University of Groningen and, since 2011, Professor of Development and Functioning of the Dutch and European

Party System at this university. He has published extensively on Dutch political parties, Dutch politics, and European integration. He wrote his dissertation on the relationship between the Dutch Communist Party and the Communist International in the 1920s.

# 4 Utopianism in Anton Pannekoek's Socialism and Astronomy

*Klaas van Berkel*

**Abstract**

During World War II, Anton Pannekoek wrote two separate memoirs, one about his life in astronomy, another about his life in the socialist movement. Historians have often wondered what the deeper unity behind these two different accounts of Pannekoek's life may have been. Here it is claimed that the answer is utopianism. Pannekoek encountered utopianism in the work of the American novelist Edward Bellamy and this utopianism strongly resonated with the ideas of the socialist thinker Joseph Dietzgen, who was much admired by Pannekoek. Careful reading of the memoirs reveals that the ideal of purity, so common around the turn of the nineteenth century, was common to both Bellamy's picture of future society in Boston and Dietzgen's and Pannekoek's ideas about how to arrive at the socialist state.

**Keywords:** Anton Pannekoek, Utopianism, two cultures, Edward Bellamy, Joseph Dietzgen, purity

During the last, dark winter of World War II, Anton Pannekoek wrote two separate autobiographies, one on his involvement in the socialist movement, the other dealing with his career in astronomy.[1] By completely separating his scientific and his socialist memoirs, Pannekoek suggested that his work in astronomy was totally unrelated to his work in politics. On reading the

This contribution is a revised version of a presentation at the annual General Meeting of the Royal Netherlands Academy of Arts and Sciences in 1999. Present at that meeting was Anton Pannekoek's son, the geologist Antonie Johannes (Ton) Pannekoek (1905-2000), who in general agreed with my interpretation of the life and work of his father.

1 Pannekoek 1982.

Tai, Chaokang, Bart van der Steen, and Jeroen van Dongen (eds), *Anton Pannekoek: Ways of Viewing Science and Society.* Taylor & Francis Group, 2019
DOI: 10.5117/9789462984349_CH04

memoirs, one gets the impression that Pannekoek actually lived two different lives. On the one hand, he was a representative of the radical left, a critic of the communist Lenin and the Dutch social democrat Pieter Jelles Troelstra, on the other he was a noted astronomer, one of the founders of astrophysics in the Netherlands. Confronted with this curious phenomenon, historians of science will be tempted to think of Pannekoek as an almost pathological example of C.P. Snow's concept of 'the two cultures'.[2] Snow introduced this concept only in 1959, in the context of the ongoing debate about science policy in Great Britain after World War II, but the idea behind it dates back to the nineteenth century at least and therefore might indeed apply to Pannekoek. Of course, Pannekoek is an exceptional case. Although he strictly separated the two sides of his life, he was a representative of both the literary and the scientific culture, whereas Snow complained that in modern culture people belonged to either the one or the other.[3] Still, Pannekoek's writing of two distinct memoirs raises some interesting questions: Did he indeed exemplify the rift between the two cultures, or was there, hidden somewhere beneath the surface of his memoirs, a common ground for his dealings with politics and science?

There are several ways to answer this question. First of all, we could try to demonstrate that Pannekoek the astronomer and Pannekoek the socialist had something in common. We could try to argue that both shared some fundamental ideas or that there was a common method in both Pannekoek's socialism and his astronomy, or that in the end Pannekoek the socialist and Pannekoek the astronomer were actually pursuing a common goal. We could, however, also try to analyse why Pannekoek wrote a double autobiography, what his motives were for doing so. In asking these questions, we do not concentrate on the contents of what he wrote, but on the literary means Pannekoek used to present his memoirs and the historical context in which he wrote them. Both of these ways of reading the memoirs, the internal and the external so to say, have their merits and the one is not inherently better than the other. In this essay, I will start with the biographical method, but I will also pay some attention to the 'common-ground' approach, as I believe that both approaches should be integrated.

2 Snow 1993.

3 Whether or not Snow's analysis was correct, is an altogether different matter. Snow introduced a catchy phrase for a widespread feeling (at least among scientists) that there was a fundamental lack of mutual understanding between the representatives of the humanities and those of science and technology. Scientists felt misunderstood and underrepresented in the political elite that decided about the future of England. Yet the actual situation was much more complicated and scientists were much more part of Britain's elite than Snow was prepared to acknowledge. See Edgerton 2006.

The basic facts of Pannekoek's life are well known and can be found elsewhere in this volume. Still, a quick recap, with special attention to those episodes that might illuminate how to interpret Pannekoek's two memoirs, may be helpful.[4] Pannekoek was born in 1873 in the village of Vaassen, went to secondary school in Apeldoorn until 1888 and then, in 1891, took the state examination in Greek and Latin that was required to be admitted to university. A three-year period before passing this examination was rather long for a bright boy like Pannekoek. Most students needed only one or two years. The reason for this may have been Pannekoek's growing passion for a new kind of natural history that was becoming a favourite pastime for young schoolboys and students. In the Netherlands, this movement is usually associated with the names of Eli Heimans and Jac. P. Thijsse. These two young schoolteachers from Amsterdam wrote short accounts of their walking tours in the pastoral surroundings of Amsterdam for several newspapers, and in 1896 started their own journal called *De Levende Natuur* ('Living Nature'). Yet this new wave of emotionally experiencing nature, whether aesthetically or somewhat mystically inclined, had already been well on its way before the appearance of Heimans and Thijsse, and had taken hold of the young Pannekoek around 1890.[5]

In 1891, Pannekoek went to Leiden and studied mathematics, physics, and astronomy. He graduated in 1895, became an engineer at the national geodetic committee, switched to a position as observer at the Leiden Observatory, and defended his dissertation on the star Algol in 1902. In Leiden, Pannekoek at first moved in liberal circles, but his reading of the utopian books of the American author Edward Bellamy, especially his second book, *Equality*, published in 1897, convinced Pannekoek of the superiority of socialism over liberalism as a political philosophy. He became the central figure in the small and struggling Leiden chapter of the Sociaal-Democratische Arbeiderspartij (Social Democratic Workers' Party, SDAP), studied classic socialist texts, especially Joseph Dietzgen's books, and started to write articles for the socialist journal *De Nieuwe Tijd* ('The New Era'). He also published articles in newspapers and in German journals like *Die Neue Zeit* and the *Leipziger Volkszeitung*. The general strike of April 1903 was something of a watershed for Pannekoek. He disagreed strongly with the cautious tactics of the leader of the social democrats, Troelstra, and favoured a more radical course of action. After the strike, which ended in a heavy defeat for the labour movement, Pannekoek was for the first time in his

4 Sijes 1982; van Berkel 2013; van den Heuvel 1992.

5 On Heimans and Thijsse, see Theunissen 1993; van Berkel 1998.

life confronted with the political repercussions of his radicalism. He had urged workers to resist the government 'with all possible means',[6] which was interpreted by government officials and the director of the Observatory as a call for violence and revolution, and thus inadmissible for a civil servant. All over the country, people were fired because of their support of the strike and Pannekoek, too, was faced with the prospect of dismissal. Pannekoek defended himself by saying that he had meant all possible *legal* means, but still Pannekoek had to go the Minister of the Internal Affairs (and Prime Minister), Abraham Kuyper, to explain his actions. Pannekoek was not fired, but urgently requested not to do anything that was against the law.[7] Although in his memoirs Pannekoek does not make a great deal of this episode – in fact, he seemed rather proud of the way he defended his position towards Kuyper – it must have been an unsettling affair, for in the following years he distanced himself from party politics and restricted himself to theoretical work. A few months later, he married Johanna Maria Nassau Noordewier (1871-1957), with whom he had a son and a daughter.

Pannekoek loved astronomy, but working at the Leiden Observatory under the very conservative director H.G. van de Sande Bakhuyzen was depressing. Because of this, Pannekoek decided to accept an offer from the social democrats in Germany to become a lecturer at the party school in Berlin. This did not last long: already after a year, the Prussian government threatened to extradite him should he, as a foreigner, remain at the school. Pannekoek changed to journalism and became a correspondent for several socialist newspapers. In 1910, he moved to the free city of Bremen, where there were no objections against his theoretical instructions of members and leaders of the Sozialdemokratische Partei Deutschlands (Social Democratic Party of Germany, SPD). The trade unions, however, put a stop to this. They found him too radical and did not like his promotion of mass strikes as a political tool. As a result, Pannekoek had to resign again, once more becoming a newspaper correspondent.

The outbreak of World War I effectively ended any ideas Pannekoek might have had about making a career in Germany. In the summer of 1914, he was staying in the Netherlands with his family. He immediately returned to Bremen to arrange his affairs, while his wife and two children remained in Arnhem, with his father-in-law. He came back to the Netherlands in 1915 and became a secondary-school teacher in Bussum, not far from Amsterdam. By all appearances, he was finally settling down to a proper bourgeois existence.

6 Pannekoek 1982, 92

7 Pannekoek 1982, 92-93.

He contributed in discussions on the disciplines he was teaching (especially cosmography) and wrote several popular accounts of astronomy.[8] In this period, too, he wrote a popular introduction to astronomy, *De wonderbouw der wereld* [*The Wonderful Structure of the World*].[9] This contribution would prove instrumental in Pannekoek's return to academia. On the recommendation of Willem de Sitter, professor of astronomy at Leiden, Pannekoek was admitted as *privaatdocent* in the history of astronomy there (that is, as lecturer without remuneration). Two years later, a bigger opportunity offered itself, when De Sitter initiated a complete reorganization of the work at the Observatory and asked Pannekoek to become assistant-director. As science historian David Baneke has explained, in May 1919 the Prime Minister and the Minister of Education (Charles Ruijs de Beerenbrouck and J.Th. de Visser) unexpectedly blocked Pannekoek's appointment, because of some newspaper clipping which had informed them of Pannekoek's supposedly close ties to the socialist revolutionaries in Hungary, headed by Bela Kun.[10] The source of this news has never been revealed, but I learned that the Pannekoek family later suspected that it had been a disgruntled brother-in-law of Pannekoek, Hendrik Jan Nassau Noordewier, who had passed the news to the government. Nassau Noordewier was a journalist in Berlin with an intimate knowledge of international politics (any news about Eastern Europe reached the Netherlands through news agencies in Berlin).[11] Still, whatever the source of the news, the government's refusal to appoint Pannekoek at Leiden's Observatory must have been a great disappointment to Pannekoek; it resembled what the Germans would later call a 'Berufsverbot'.

In his memoirs, Pannekoek is again remarkably mild about this episode in his life. In retrospect, he stated that he was glad that he did not get the job in Leiden because sooner or later he would have fallen out with De Sitter, who had struck him as a rather authoritarian person. Furthermore, instead

8 The *Weekblad voor Gymnasiaal en Middelbaar Onderwijs* (a weekly bulletin for secondary-school teachers) in the years 1916-1917 features a number of articles by Pannekoek discussing the method of teaching cosmology and the issue of salary increases for teachers. Apparently he took his teaching job very seriously.

9 Pannekoek 1916.

10 Baneke 2004. See also David Baneke, 'Pannekoek's One Revolution', in this volume, 87-108.

11 Stoop 1988, 14-16. Noordewier (whose family called itself Nassau Noordewier because one of their ancestors supposedly had been an illegitimate child of the Dutch king William II) had been a teacher at a secondary school in Leiden, but after a love affair with one of his students he was fired. After this affair, relations with his family became strained. Noordewier turned to journalism and in 1911 was hired as a news editor of the *Nieuwe Rotterdamsche Courant* (*NRC*). He was responsible for reporting on World War I. From 1920 onwards, he was the NRC correspondent in Berlin.

of becoming deputy-director of the Leiden Observatory, he soon became first lecturer and then full professor at the University of Amsterdam, where he was completely free to decide on his own line of research. Yet, in April and May 1919, not having the advantage of foresight, the government's refusal must have been a great blow to him, if not a traumatic experience. Pannekoek must have felt hurt by politics intervening, once again, in his work as an astronomer. He had always dealt with astronomy and politics as two separate occupations, but now he had to conclude that others were not able to make this distinction. As a result, he decided to separate the two sides of his career even more strictly than before. Before 1919, Pannekoek signed both his astronomical and his socialist articles in his own name, but after that year, he increasingly used pseudonyms when publishing socialist articles, pamphlets or books. Shortly beforeWorld War II, Pannekoek wrote a book on Lenin under the pseudonym of J. Harper and even after the war, so after his retirement as professor of astronomy, he published his well-known book on the workers' councils under the pseudonym of P. Aartsz. Some years later, in 1951, Pannekoek published his widely read *De groei van ons wereldbeeld* ('The Growth of our Worldview'), which contains only superficial hints of his convictions. He does indeed pay some attention to the social context of astronomy, but only at the very end he gave the reader a hint of his socialist point of view: It is about time that man, by establishing a free, self-governing world community of productive labour and by assuring itself of material prosperity in abundance, liberates all spiritual forces for the perfection of its knowledge of nature and especially the science of the universe.[12]

When plans were developed to translate and publish this book in English (which eventually happened in 1961), it was rumoured that some Americans protested that it would be out of the question to translate a work written by a communist, although these anonymous critics would have been hard-pressed to point out any passage that was clearly communist by inspiration.[13] Because of the events in 1919, it had become second nature to Pannekoek to act as if the astronomer and the socialist were two completely different persons. This, then, resulted in the two separate identities presented in the *Memoirs*.

12 Pannekoek 1951, 432. For a different reading of the book see Bart Karstens, 'Anton Pannekoek as a Pioneer in the Sociology of Knowledge', in this volume, 197-217; Jennifer Tucker, 'Popularizing the Cosmos', in this volume,173-195; and Omar W. Nasim, 'The Labour of Handwork in Astronomy', in this volume, 249-283.

13 Personal communication of D.J. Struik. In reviews it was pointed out that the translation was sometimes rather clumsy and that Pannekoek himself may have been the translator (the name of the translator is not mentioned).

As is evident from the above, the strict separation of Pannekoek the astronomer and Pannekoek the socialist is a historical construction. Several historians have pointed out that in fact there were important similarities between Pannekoek's socialist theorizing and his astronomical research.[14] The most fundamental similarity, however, is the *emotional* drive behind both his astronomy and his radical socialism. At the heart of both, I would like to argue, is Pannekoek's utopian longing for purity and community in both nature and society, of which his almost mystical approach to nature and his speculations about a utopia of worker's councils are just two manifestations.

Utopianism was a crucial element in Pannekoek's social ideas from the very beginning.[15] I already mentioned his reading of *Equality*, the second utopian novel by Bellamy, published in 1897 and immediately translated into Dutch. *Equality* is a sequel to Bellamy's first novel, *Looking Backward 2000-1887* (1888), which was translated into Dutch in 1890 as *In het jaar 2000* and became an instant success. In his two novels, Bellamy imagined late twentieth-century Boston as a harmonious community, where people were free from capitalist exploitation, worked only until they reached middle age, were placed in the occupation they were best equipped for, and money was equally distributed among all; a community in which every material need was met, where women had the same rights as men, where everyone spoke a universal language in addition to their native language, and where everybody was a vegetarian. As a result of this system, all the problems of nineteenth-century America would be solved by the year 2000: all production would be regulated by the 'Great Trust', thus ridding society of wasteful competition. The world would be governed by technical experts, which would make political strife superfluous. Under these circumstances, the noble character of man would fully shine through, while the remaining criminals or offenders would be treated in a hospital. *Looking backward* was a short publication, in which the author was unable to address all the issues related to his grand scheme. Because of this, Bellamy published a more extensive second novel, in which he more fully developed the ideas of the original book.

It is not clear whether Pannekoek read *In het jaar 2000*, but he did read the sequel, which in Dutch appeared with a title that seemed to resonate

14 On this point see also Chaokang Tai, 'The Milky Way as Optical Phenomenon', in this volume, 219-247, and Omar W. Nasim, 'The Labour of Handwork in Astronomy', in this volume, 249-283.
15 The literature on utopianism is immense. For a recent survey, with a great deal of attention given to Bellamy, see Beaumont 2012. Very helpful is also: Kemperink and Vermeer 2010, which actually has the cover of the 1919 printing of the Dutch translation of Bellamy's *Looking Backward* on its cover.

with socialist ideas, *Gelijkheid voor allen* ('Equality for all'). In his memoirs, Pannekoek reveals that upon reading Bellamy's book in 1899, he suddenly understood the deceptive nature of capitalist logic and that – by implication – he now fully grasped the true meaning of socialism. Bellamy did not explicitly discuss socialism in his books – it would have weakened their appeal to the American public – but many readers drew the conclusion that the utopia pictured by Bellamy was actually a socialist society. Bellamy had not said much about how exactly the Bostonians had abolished capitalism and created their ideal society; he only suggested that instead of protracted negotiations and street violence a sudden mass conversion, triggered by a general strike, had taken place, resulting in a peaceful 'Revolution' or 'Awakening'.[16] This did not bother Pannekoek: on the contrary, it was in line with his own temperament. In any case, the in reality rather dull book by Bellamy was a revelation to Pannekoek. 'All of a sudden I was no longer blindfolded', Pannekoek noted in his memoirs.[17] In less than a month, after reading some additional literature, he had become fully committed to socialism. And he was not the only one in the Netherlands who upon reading Bellamy converted to socialism. The businessman Floor Wibaut (1859-1936), who would later become a prominent figure in the socialist movement in Amsterdam, was another example.[18]

Utopianism is also a key element in the work of Joseph Dietzgen, the German philosopher and socialist whom Pannekoek valued even more than Marx and Engels.[19] What attracted Pannekoek to Dietzgen was the latter's conviction that the human mind had a certain autonomous role to play in bringing about the future socialist state.[20] Human thinking was not entirely determined by social circumstances, people had some degree of freedom. Socialist utopia would not be the product of impersonal social and economic developments alone; it would come about through people taking action against exploitation and injustice. In fact, Pannekoek was convinced that

16 Bellamy 1897, chapters 25, 28, 33, and 34.

17 Pannekoek 1982, 72.

18 On the reception of Bellamy's books in the Netherlands: Krips-van der Laan 2001. The author focuses completely on Bellamy's first book, *Looking Backward*, and mentions *Equality* only in passing. *Equality* was translated as *Gelijkheid voor allen* (The Hague: A. Abrahams, 1897) by A.C.B. Vincent. Crone 2007 refers to another edition, translated by A. Nolles-Heuff and also published in The Hague in 1897. I suspect that these two editions are identical. There was yet another edition published when Bellamy once again became popular in the 1930s: Edw. Bellamy, *Gelijkheid voor allen*. Nieuwe bewerking naar de Amerikaansche uitgave van 'Equality' door H.N. (Amsterdam: E. & M. Cohen, 1934).

19 Pannekoek 1903.

20 On this point, see also Tai and van Dongen 2016.

the workers of the world had to take action *themselves* in order to become truly free. They did not need the assistance of parties, bureaucrats, or trade unions. It remained unclear how exactly this utopia would come about, but this, of course, is typical of all utopias, including Bellamy's.

Pannekoek's memoirs display an increasing dissatisfaction with organizational structures. At the same time, they exhibit his deep love for unspoiled nature. During his years in Berlin, when he lived in the then still rural town of Zehlendorf between Berlin and Potsdam, he was fond of hiking through the forests and farmlands of Brandenburg. This landscape, he wrote, was 'larger, more carefree, and primitive with its large farms, than the well-kept Dutch garden-nature. It had huge fields surrounded by pine trees, birch wood, and country roads lined with old trees'.[21] Also during his extensive travelling through Germany, from one series of lectures to yet another meeting, he took his time to roam around the countryside while the discussions of these lectures and party meetings were on his mind. 'This combination of brainwork and new impressions of living nature has always been a source of the greatest joy', he wrote.[22]

This combination of thinking about society and directly experiencing nature was not coincidental, but points to a deeply felt need for Pannekoek. When he first discovered Dietzgen, back in 1900, he would seek out similar settings: 'I still remember how on Sunday mornings I went for a cycling tour, how somewhere in the dunes of Wassenaar, under a tree, I read [some of the smaller works of] Dietzgen and how intense this combination of a wonderful experience of nature and a growing sense of understanding was to me.'[23] In nature, Pannekoek experienced a sense of purity that was lacking in society. The impurity of the wheeling and dealing that is inherent in politics, with its revisionism and opportunism, was continuously criticized by Pannekoek. In doing so, he did not attribute revisionism and opportunism to political expediency, but to moral impurity. When he spoke of the communist Karl Radek, Pannekoek declared that opportunism in politics always led to inferiority on a personal level.[24] Yet, he of course also meant that opportunism was a result of inferiority on the personal level. Pannekoek's ideal of a society organized through workers' councils was to be attained by one giant leap, from the present situation of exploitation, to a future utopia of perfect equality and this would bypass all the intermediate imperfections

21 Pannekoek 1982, 121.
22 Pannekoek 1982, 130.
23 Pannekoek 1982, 76.
24 Pannekoek 1982, 200.

of representation, negotiations and tactics. It was an ambitious, perhaps helpless, attempt to hold on to purity under all circumstances.

After the defeat of the ideals of the worker's movement in 1914, after missing the appointment as assistant-director at the Leiden Observatory in 1919, and after his move to Amsterdam, Pannekoek did not dwell any further on the ideal of a combined experience of nature, scientific work, and social progress. Apparently, he had learned to suspend them in separate mental compartments. In 1929, however, when he visited the Observatory of Victoria in Western Canada to take photographs of a particular segment of the sky, his old unified ideal reawakened. In his *Memoirs*, Pannekoek mused about the advantages of taking photographs over observing with the eye through an optical telescope. When one took photographs, the only thing one had to do was to take care that the instrument was constantly directed at the same star. The mind, then, had the opportunity to freely contemplate the star of interest:

> This star is not the meaningless dot of light that has to be kept on the thin black line, it is an object of meaning, with a long history, with a purpose, important for us too; an object of discussion perhaps for some, an object of study for others, a link in a long chain. This is how one feels, while this dot of light dances to and fro along this line, as a part, as a little organ, of the great community of astronomical researchers, who all together push forward in the unknown forest, a community in which each has his assigned role to play. And when, during a short break, one glances at the starry sky as a whole, with all the glittering, which reminds us of earlier hope, work and strivings, then, under the beautiful silent sky, with the sleeping earth all around us, one feels like a guard in the night, like a soldier in the enormous army of progressing mankind.[25]

The sublime experience of observing the night sky once again went hand in hand with a sense of solidarity with all of mankind, as it had done in the times when Pannekoek had been a simple observer at the Leiden Observatory. Apparently, doing research in Canada rekindled thoughts and emotions he had stored away for a long time.

Our conclusion, then, is simple and straightforward. The notion of Pannekoek's split personality – Pannekoek the astronomer and Pannekoek the socialist – is misleading. Both his socialism and his science originated from

25 Pannekoek 1982, 262-263.

the same source, the utopian longing for wholeness and purity that was so popular at the end of the nineteenth century. In Pannekoek's particular case, it was triggered by reading Bellamy's utopia. The subsequent separation between the two domains of his thought is a historical construct, provoked by the dynamics of political repression and scientific specialization. Yet, in the pages of Pannekoek's double memoirs, the original unity of his thinking still shines through.

## Bibliography


Baneke, David. 2004. '"Hij kan toch moeilijk de sterren in de war schoppen". De afwijzing van Anton Pannekoek als adjunct-directeur van de Leidse Sterrewacht in 1919'. *Gewina* 27(1): 1-13.

Beaumont, Matthew. 2012. *The Spectre of Utopia: Utopian and Science Fictions at the Fin de Siècle*. Oxford: Peter Lang.

Bellamy, Edward. 1897. *Equality*. New York: D. Appleton.

Berkel, Klaas van. 1998. *Citaten uit het boek der natuur. Opstellen over Nederlandse wetenschapsgeschiedenis*. Amsterdam: Bert Bakker.

—. 2013. 'Pannekoek, Antonie (1873-1960)'. Biografisch Woordenboek van Nederland. http://resources.huygens.knaw.nl/bwn/BWN/lemmata/bwn5/pannekoe (accessed 12 November 2013).

Crone, Vincent C.A. 2007. *De kwetsbare kijker: een culturele geschiedenis van televisie in Nederland*. Amsterdam: Amsterdam University Press.

Edgerton, David. 2006. *Warfare State: Britain, 1920-1970*. Cambridge: Cambridge University Press.

Heuvel, Edward P.J. van den. 1992. 'Antonie Pannekoek (1873-1960). Socialist en sterrekundige'. In *Een Brandpunt van geleerdheid in de hoofdstad: de Universiteit van Amsterdam rond 1900 in vijftien portretten*, ed. by J.C.H. Blom. Hilversum: Verloren, 229-245.

Kemperink, Mary G., and Leonieke Vermeer, eds. 2010. *Utopianism and the Sciences, 1880-1930*. Leuven: Peeters.

Krips-van der Laan, Hilde. 2001. 'De verschijning van de toekomstroman van Edward Bellamy in Nederland'. *De Negentiende Eeuw* 25(4): 215-232.

Pannekoek, Anton. 1903. 'Die Stellung und Bedeutung von Jozef Dietzgens philosophischen Arbeiten'. In *Das Wesen der menschlichen Kopfarbeit. Ein abermalige Kritik der reinen und praktischen Vernunft*. Stuttgart: J.H.W. Dietz.

—. 1916. *De wonderbouw der wereld. De grondslagen van ons sterrekundig wereldbeeld populair uiteengezet*. Amsterdam: S.L. van Looy.

—. 1951. *De groei van ons wereldbeeld*. Amsterdam: Wereldbibliotheek.

—. 1982. *Herinneringen: herinneringen uit de arbeidersbeweging; sterrenkundige herinneringen*. Ed. by Ben A. Sijes, Johanna M. Welcker, and J.R. van der Leeuw. Amsterdam: Van Gennep.

Sijes, Ben A. 1982. 'Anton Pannekoek 1873-1960'. In *Herinneringen: herinneringen uit de arbeidersbeweging; sterrenkundige herinneringen*, by Anton Pannekoek, ed. by Ben A. Sijes, Johanna M. Welcker, and J.R. van der Leeuw. Amsterdam: Van Gennep, 11-69.

Snow, C.P. 1993. *The Two Cultures*. Cambridge: Cambridge University Press.

Stoop, Paul. 1988. *De geheime rapporten van H.J. Noordewier: Berlijn 1933-1935*. Amsterdam: Sijthoff.

Tai, Chaokang, and Jeroen van Dongen. 2016. 'Anton Pannekoek's Epistemic Virtues in Astronomy and Socialism: Personae and the Practice of Science'. *BMGN – Low Countries Historical Review* 131(4): 55-70.

Theunissen, Bert. 1993. 'Natuursport en levensgeluk. Hugo de Vries, Eli Heimans en Jac. P. Thijsse'. *Gewina* 16(4): 287-307.

## About the Author

**Klaas van Berkel** is Rudolf Agricola Professor of History at the University of Groningen and member of the Royal Netherlands Academy of Arts and Sciences. His research focuses on the history of science from the sixteenth century to the present and on the history of universities. His most recent books include *Isaac Beeckman on Matter in Motion. Mechanical Philosophy in the Making* (2013) and *Universiteit van het Noorden. Vier eeuwen academisch leven in Groningen, vol. I (1614-187) and II (1876-1945)* (2014, 2017).

# 5 Pannekoek's One Revolution

## Anton Pannekoek and the Modernization of the Dutch Astronomical Community

*David Baneke*

**Abstract**

When Anton Pannekoek left Dutch astronomy in 1905, he left a stagnating, uninspiring research community. When he returned a decade later, things started to change in the Dutch astronomical community. By the mid-1920s, De Sitter, Hertzsprung, Oort, Minnaert, and Pannekoek had built a flourishing discipline. Through their work and students, they shaped Dutch astronomy for the rest of the twentieth century.

This paper focuses on Pannekoek's return to astronomy and his role in Dutch astronomy in the Interwar period. First, I will provide a detailed reconstruction of his failed appointment at Leiden Observatory in 1918-1919. After that, I will analyse how he could play an influential role, even though he had little staff, students, or facilities at the University of Amsterdam.

**Keywords:** Anton Pannekoek, history of astronomy, science and politics, discipline formation, astronomy education

When Pannekoek left Leiden in 1906 to teach socialist theory in Germany – which he later referred to as his 'literary work'[1] – he did not only do so out of ideological zeal. He was also deeply disappointed in the way astronomy was done at the Leiden Observatory, where he worked. It was characterized by routine work and an obsession with precision that stifled all interesting research. In his memoirs, Pannekoek described the atmosphere at the institute as 'tomb-like, full of stagnation and boredom'.[2]

1 Pannekoek to De Sitter, 15 April 1919, WdS inv. 45.1: 'de tijd, dat ik in Duitschland literair werkzaam was'.

2 Pannekoek 1982, 237: 'katakombenlucht van doodse verstarring en verveling'.

Tai, Chaokang, Bart van der Steen, and Jeroen van Dongen (eds), *Anton Pannekoek: Ways of Viewing Science and Society.* Taylor & Francis Group, 2019

DOI: 10.5117/9789462984349_CH05

Two decades later, in the 1920s, the situation had changed radically. The new director of the Leiden Observatory, Willem de Sitter, had assembled a new team of first-rate scientists, including the famous Ejnar Hertzsprung and a young Jan Oort. New research programmes were implemented following the insights of J.C. Kapteyn, the most prominent Dutch astronomer since Christiaan Huygens. Pannekoek also planned to work there after returning from Germany in 1914, but his appointment was blocked for political reasons. Instead, he went to Amsterdam. Together with Marcel Minnaert in Utrecht, he now established a new research school in astrophysics. All these men – Kapteyn, De Sitter, Hertzsprung, Oort, Minnaert, and Pannekoek – would receive the Gold Medal of the Royal Astronomical Society, one of the most prestigious recognitions in the field. They truly formed a remarkable generation.[3]

In this paper, I will describe how this generation changed Dutch astronomy, and what Pannekoek's role in these changes was. How was it possible that Pannekoek, would-be German revolutionary, could return to the career of an, eventually, highly respected Dutch astronomer? Did his political convictions influence his professional life in astronomy? What was his position within the astronomical community once he had become professor in Amsterdam? How did he manage to exert his influence with so few facilities and without any graduate students?

To answer these questions, we will analyse changes in the Dutch astronomical community in the decade after World War I. In this period, Pannekoek and his contemporaries reorganized the discipline on all levels: research, teaching, institutions, journals, and the discipline's international relations were all reconstituted by a new generation of astronomers. In the span of a few years, they laid the groundwork for a modern disciplinary infrastructure that essentially would last until well into the twenty-first century.[4] They also reoriented themselves internationally towards American astronomy, which was emerging as the new superpower in the field. Pannekoek's story thus illustrates the consequences of the rise of American astronomy for a European community – and vice versa, as Dutch astronomers contributed in various ways to American astronomy as well, as we will see.

I will start by discussing Pannekoek's attempts to return to astronomy in the context of the reorganization of Leiden Observatory. I will recount in some detail the circumstances of Pannekoek's failed appointment there. I will subsequently discuss how the loosely knit astronomical community

3 Baneke 2010; 2015, chapters 3 and 4.

4 Baneke 2015.

changed into a well-organized scientific discipline with a new international outlook. Finally, I will address the legacy of Pannekoek and his contemporaries in Dutch astronomy.

## Reviving Leiden Observatory

The revival of Dutch astronomy started in Leiden. Leiden Observatory had been founded as a state-of-the-art observatory by Frederik Kaiser in 1861.[5] His research programme of fundamental astrometry was continued by the brothers Hendrik and Ernst van de Sande Bakhuyzen, who explicitly regarded the 'new astronomy' of the late nineteenth century, which was based on spectrographic methods, as oversold. Surely the steady stream of easy discoveries would dry up soon, they thought, requiring serious astronomers to go back to the hard labour of precision measurements.[6] High-precision astrometry was their ultimate goal; so much so, that the number of publications dwindled, since there were always more corrections to be made and potential errors to be checked. When the younger Bakhuyzen died in 1918, his successor Willem de Sitter reported that checking and calibrating the new photographic refractor from 1898 had taken so long that it had not been used for astronomical observations yet.[7] This was the observatory that Pannekoek had fled in 1906. He was not the only one to leave: around the same time, Joan Voûte also left for a more adventurous life, seeking to establish his own observatory in the Southern Hemisphere.[8]

The most prominent Dutch astronomer at this time was J.C. Kapteyn. Kapteyn was professor in Groningen, famously working without an observatory of his own. The story of how he established an 'astronomical laboratory', collecting observations from observatories elsewhere, has been told many times.[9] First, he established his reputation internationally with the *Cape Photographic Durchmusterung*, a star catalogue listing 454,875 stars, in collaboration with David Gill in South Africa. Later, Kapteyn turned to statistical research into the structure of the stellar system. His announcement in 1904 that he had found two 'star streams' in the Milky Way made him one of the most prominent astronomers of his day. It earned him the admiration and support of George

5 Zuidervaart 2011; van Herk, Kleibrink, and Bijleveld 1983.

6 van de Sande Bakhuyzen 1872.

7 De Sitter, report on the state of the Observatory, April 1918, WdS inv. 224.2.

8 He finally succeeded in Lembang, Java, Dutch East Indies. Zuidervaart 2008; Pyenson 1989.

9 See especially van der Kruit 2015; van der Kruit and van Berkel 2000.

Ellery Hale, who founded the Mount Wilson Observatory and many other American astronomical institutions. Yet however much success he attained, Kapteyn remained desiring an observatory of his own. He was frustrated when a new photographic refractor was installed in Leiden, as he felt he would be able to use it more productively than the Bakhuyzen brothers.

Kapteyn's student Willem de Sitter had been appointed professor of theoretical astronomy in Leiden in 1908, but in the next few years he had little to do with the day-to-day running of the observatory, which was supervised by Ernst van de Sande Bakhuyzen. Instead, he worked on theoretical issues, especially on his cosmological ideas in general relativity.[10] When he was appointed acting director in 1918, it turned out that he had already given the future of the observatory a lot of thought, in close consultation with his mentor Kapteyn. Together, they planned to turn Leiden Observatory into the observatory Kapteyn had wished for but never had.

De Sitter obtained support from the Board of Trustees of the university to expand and reorganize the observatory.[11] One of the key elements of this reorganization was to assemble a new, first-rate staff. De Sitter quickly engaged a new lecturer, Jan Woltjer, to assist him in what he called the 'theoretical department'. Woltjer would become one of the main lecturers to Leiden's astronomy students, who taught a notoriously difficult course on celestial mechanics. He also introduced the new quantum physics to the curriculum.

De Sitter further wanted to appoint two associate directors – senior scholars who could lead research departments. One of his candidates was Pannekoek, who was to take responsibility for the main observing programmes, especially those focused on the meridian circle. De Sitter and Pannekoek had been in touch since Pannekoek had returned to the Netherlands in 1914 and had been trying to return to professional astronomy. De Sitter was impressed by Pannekoek's popular astronomy book *De Wonderbouw der Wereld* (1916). He arranged an appointment as *privaatdocent* (unpaid lectureship) in the history of astronomy, while remaining on the lookout for a better position. That opportunity arrived in 1918 with the permission to reorganize the Observatory and Pannekoek was eager to take it. His only condition was that he wanted to have a house with a garden, which was easy to arrange since directors would live in the observatory, which was situated in Leiden's botanical gardens.[12]

The candidate for a second associate director post was Ejnar Hertzsprung, who was already renowned for his work in astrophysics, and someone with a

10 Kerszberg 1989; Guichelaar 2009.

11 Baneke 2010; 2015, chapter 3.

12 Pannekoek to De Sitter, 29 March 1918, WdS inv. 45.1.

deep knowledge of astronomical instruments and photographic technology (he had started his career in photochemistry).[13] At the time, Hertzsprung worked at Potsdam Observatory, but he was happy to leave Germany. His mentor Karl Schwarzschild had died, and the situation in Germany at the end of World War I was far from easy. Moreover, he was married to a Dutchwoman: Henriette Kapteyn, daughter of the astronomer Kapteyn. They had one daughter, Rigel.

Leiden also had some professional advantages for Hertzsprung, most notably the promise of a vibrant scientific community. De Sitter would be there, and Kapteyn would be close. But by leaving Potsdam, Hertzsprung also made a significant professional sacrifice: he gave up access to the large telescope there. In Leiden, he would mainly use the 10-inch photographic refractor, which was nowhere near comparable. Furthermore, it did not seem likely that a new large telescope would be installed in Leiden. Instead, Hertzsprung requested access to a telescope in the Southern Hemisphere. This was very important to him.[14] De Sitter promised to do his best.

Hertzsprung would bring international attention to Leiden – he was better known than De Sitter at this time.[15] He also became the most important supervisor for students, who learned to do observations and measurements according to his exacting standards. Together with the theoretical lectures by Woltjer and, later, Oort, his teaching would lay the foundation for the famous Leiden school of astronomy.[16]

## The Leiden affair[17]

Hertzsprung looked forward to working with Pannekoek.[18] They had corresponded and met while Pannekoek was in Germany.[19] Hertzsprung admired Pannekoek's work, but he had little patience for his political interests: 'I received several papers from Pannekoek (Nova Aquilae and 11 magn-stars). Pity that all this productivity still leaves time for political action', he wrote to

13 Herrmann 1994; DeVorkin 1984.

14 Hertzsprung to De Sitter, 10 June 1919, WdS inv. 23.2: 'Als er op mijn conto besnoeid en bespaard moet worden, omdat Pannekoek bolschew- of soms ergens-istische artikelen schrijft, laat het dan op een ander punt dan het zuiden zijn.'

15 Adriaan Blaauw, interview with the author, 16 March 2009.

16 Baneke 2010.

17 This section is mostly based on Baneke 2004.

18 Hertzsprung to De Sitter, 24 October 1919, WdS inv. 23.2.

19 Pannekoek 1982, 241-242.

de Sitter.[20] His dislike for Pannekoek's political activities was not (or at least not just) motivated by political considerations. Hertzsprung was mostly worried that they would get in the way of his prime concern: science. Hertzsprung was a notorious workaholic, who valued a good measurement above anything else. In this case, his fear was justified. Obtaining formal approval for the appointment of the new directors turned out to be very difficult.

A procedure for appointing new professors involved several steps. First, the candidate had to be nominated by the *faculteit*, the assembled professors of the department (in this case Science). Then the Board of Trustees of the university had to approve. It usually asked the advice of experts from other universities. Finally, any professorial candidate at one of the state universities (Leiden, Groningen, and Utrecht) had to be approved by the national government.

The problems started already with De Sitter's formal appointment as director. Jan van der Bilt, a former Naval officer who worked at Utrecht Observatory, unexpectedly started a lobby to become director himself. He had influential supporters in Leiden's Board of Trustees and the government.[21] At the time, De Sitter was away from Leiden, spending more than a year in a Swiss sanatorium for a lung condition.[22] This did not help in advancing his own appointment. Hertzsprung's appointment was also delayed, because of his salary demands and because some objections had been raised against appointing a foreigner. The latter were easily countered, however: there clearly was no Dutch candidate of Hertzsprung's stature. Pannekoek's case was more problematic, however. The Utrecht astronomer A.A. Nijland (Van der Bilt's boss) wrote the Board of Trustees that appointing Pannekoek 'would be a somewhat dangerous experiment, given his extraordinary views' – a thinly veiled reference to his political positions.[23] Nevertheless, the Trustees ultimately did approve the candidacy of De Sitter, Hertzsprung, and Pannekoek on 16 May 1918, in no small measure due to Kapteyn's influence.

Still, final confirmation had to come from the government in The Hague. But since new elections were due on 3 July, the responsible minister decided to wait. In the Dutch system of coalition governments, it can take a long time before a new government is formed after elections. Time and again, encouraging signals from The Hague were accompanied by the message that

20 Hertzsprung to De Sitter, 10 June 1919, WdS inv. 23.2.

21 A detailed analysis of the reorganization can be found in Baneke 2005.

22 Guichelaar 2009.

23 Nijland to Board of Trustees of Leiden University, 3 May 1918, AC inv. no. 1840: 'met zijn zeer bijzondere kijk op de zaken een iets [sic] of wat gevaarlijk experiment'.

*formal* approval remained withheld until a new minister was installed.[24] As it turned out, the new cabinet would be the first in Dutch history to have a dedicated Minister for Education and Science. The good news was that this was expected to be J.Th. de Visser, who was a member of the same political party (*Christelijk-Historische Unie*, CHU) as jhr. N.C. de Gijselaar, President of the Board of Trustees of Leiden University, Mayor of Leiden, and member of the Senate. But when De Visser was finally installed, he immediately announced that he wanted to seek parliamentary approval for the appointment of the astronomers. By now it was late October, and again the matter was delayed.[25]

The timing was extremely unfortunate. World War I was ending and Germany seemed on the brink of a communist revolution, after one had already taken place in Russia. Even in the Netherlands, which had remained neutral during the war, political tensions were mounting. On 12 November, Socialist leader Pieter Jelles Troelstra declared revolution in Parliament. His attempt failed within hours – the episode has gone into history as 'Troelstra's Mistake' – but Pannekoek was furious, since Troelstra had violated all his ideas about how a genuine revolution should commence. The most important consequence of Troelstra's action was that the new and rather conservative government, made up of Christian political parties under the leadership of a Catholic nobleman, jhr. Ch.J.M. Ruijs de Beerenbrouck, became convinced of the threat to the political order that communism posed.

Ever since he had decided to return to academia in the mid-1910s, Pannekoek had tried to keep a low profile, publishing his political work mainly under pen names. By the end of the decade, it seemed more important than ever to avoid public association with revolutionary communism. But in February and March 1919, his name turned up in several Dutch newspapers, linking him to rumours about illegal transfers of money from Bolshevist Russia. Someone sent a newspaper clipping to President of the Board of Trustees De Gijselaar, who forwarded it to De Sitter, adding a worried note.[26] Pannekoek denied any involvement and suggested that the rumours may have been spread on purpose to harm him. He even suggested that some of Jan van der Bilt's allies could be behind it.[27] In April, his name turned

24 De Sitter, *dagboek reorganisatie*, WdS inv. 223.

25 Minister De Visser to Board of Trustees of Leiden University, 26 October 1918, AC inv. no. 1840.

26 De Gijselaar to De Sitter, 16 March 1918, WdS inv. 224.1. The clipping came from the *Haagse Post*, 22 februari 1919.

27 Pannekoek to De Sitter, 18 March 1919, WdS inv. 45.1; Pannekoek to De Sitter, 15 April 1919, WdS inv. 45.1. Pannekoek's suspicions were shared by Trustee J. Oppenheim: Oppenheim to De Sitter, WdS inv. 224.1.

up again. This time, a newspaper claimed that a certain 'Panekoch' had been appointed as honorary president of a prestigious committee of the short-lived socialist regime of Béla Kun in Budapest. Surely that meant Anton Pannekoek, the paper added.[28]

At this point, Pannekoek started to expect that the government would object to his appointment after all.[29] According to his memoirs, De Sitter was losing his patience too: 'What are you,' he snapped, 'astronomer or communist?'[30] Pannekoek answered in a letter of 18 April: since his return from Germany, he had completely returned to science. His commitment to the Leiden Observatory was complete – he had even turned down a position as lecturer in mathematics at Amsterdam University. It was not his fault, Pannekoek added, if the government was willing to sacrifice science to politics. In any case, he concluded, he did not think his socialist writings merited all the fuss, since they were only of a purely theoretical nature. Pannekoek attributed all the public attention to the circumstance that he represented a socialist faction that had always been a minority, but was now gaining the upper hand. As a result, new revolutionary leaders regarded him as one of their forebears. He added: 'If I were in touch with these people, I would tell them to omit all these niceties that benefit no one.'[31] De Sitter forwarded Pannekoek's letter to the Board of Trustees, adding that Troelstra's socialist allies had actually accused Pannekoek of not being an active revolutionary.[32]

A few weeks later, two things happened in quick succession. On 29 April 1919, the government approved De Sitter's appointment as director of the observatory, and ministry officials approved the annual salary of Pannekoek (4500 Dutch guilders), starting from the moment he would become deputy director.[33] A few days later, on 3 May, however, Minister De Visser announced that he would 'under no circumstances' appoint Pannekoek, for reasons of state interest.[34] Apparently, something had happened between those two dates that changed the minister's mind.

Pannekoek himself pointed to an article in a local Groningen newspaper on 2 May. It mentioned the expected appointment in Leiden of Hertzsprung,

28 Newspaper clipping without source information, sent by J.E. Boddaert, Secretary to the Trustees, to De Sitter, 16 April 1919, WdS inv. 224.1.

29 Pannekoek to De Sitter, 15 April 1919, WdS inv. 45.1.

30 Pannekoek 1982, 245.

31 Pannekoek to De Sitter, 18 April 1919, WdS inv. 45.1.

32 De Sitter to J.E. Boddaert, 19 April 1919, WdS inv. 45.1.

33 Staff file 'dr. A. Pannekoek', OKW file 52.

34 Minister to Board of Trustees of Leiden University, 3 May 1919, AC inv. no. 1840.

who was married to a Groningen local (Kapteyn's daughter Henriette). The piece also mentioned the other candidate, Pannekoek, 'whose communist political convictions have been written about these days'.[35] The article thus explicitly stated that a known communist was about to be appointed at a government institute. Pannekoek suspected that this had been the last straw for the government, more specifically for Prime Minister Ruijs de Beerenbrouck. The President of Leiden University's Board of Trustees, De Gijselaar, who was usually well informed about intrigues in The Hague, confirmed this.[36] In this scenario, it would have been the second time that a prime minister concerned himself with Pannekoek. In 1903, Prime Minister Abraham Kuyper had already personally reprimanded him for supporting the great railway strike while working as a civil servant at a state university.[37]

Leiden's Board now concluded that the minister's negative decision was final. They withdrew their own support for Pannekoek, referring to Troelstra's actions. De Gijselaar now expressed the widely shared view that 'these red gentlemen never keep their word'. Personally, he could well believe that Pannekoek would not do anything stupid, but he did not see how he might convince the minister of that.[38]

A few months later, the liberal newspaper *De Nieuwe Courant* looked back on the affair. Its editors approved of the government's decision, arguing that universities were teaching institutions, and while a bolshevist astronomer might be harmless, a bolshevist lecturer was not. The popular newspaper *Nieuws van de Dag* disagreed: if scientific appointments were subject to political considerations, this required more explanation from the government.[39] An explanation was also demanded by a communist member of parliament, W. van Ravesteijn, who spoke of a 'political inquisition' in higher education.[40] He was supported by several MPs from other political parties. J.H.A. Schaper, a social democrat, ridiculed the risks posed by a communist astronomer: 'he can hardly throw the stars into disarray'.[41] Minister De Visser answered that political considerations should not play a role in academic appointments, but he argued that Pannekoek

35 *Provinciale Groninger Courant*, 2 May 1919. A newspaper clipping can be found in WdS inv. 224.1.

36 Pannekoek to De Sitter, 8 May 1919 WdS inv. 45.1; De Sitter, *dagboek reorganisatie*, WdS inv. 223.

37 Pannekoek 1982, 92-93.

38 De Gijselaar to De Sitter, 16 May 1919, WdS inv. 224.1.

39 *Nieuwe Courant*, 6 September 1919; *Nieuws van den Dag*, 12 September 1919.

40 *Handelingen van de Tweede Kamer*, 25 November 1919.

41 *Handelingen van de Tweede Kamer*, 25 November 1919: 'Hij kan toch moeilijk de sterren in de war schoppen'.

was a special case, since he publicly promoted the overthrow of the state. Interestingly, De Visser subsequently stated that the formal reason for his rejection was the fact that Pannekoek had been expelled from Germany for political reasons. A small majority of parliament accepted the minister's answer.[42] In a later debate in the *Eerste Kamer* (senate), De Visser added that enemies of the state should not be involved in teaching students in these turbulent times.[43]

Despite everything, De Sitter was taken by surprise by the rejection of Pannekoek. He was furious, both at the government and at Pannekoek, and he even considered stepping down.[44] Hertzsprung was angry too, particularly because he initially believed that Pannekoek himself had triggered the rejection with a political publication. 'It is possible that the world can still be improved, but let us not try to do it in a way that harms the working conditions of our science', he wrote.[45] He also threatened to withdraw himself for consideration if his appointment was not approved soon. Formal approval in his case was finally issued on 21 July.

In hindsight, Pannekoek's appointment in Leiden fell through because of bad timing. Troelstra's rash actions, combined with publications in the media about Pannekoek, led the government to take the rare step of rejecting an academic nomination. Soon after, things calmed down again and Pannekoek's appointment as professor in astronomy in Amsterdam in 1919 raised few eyebrows. Some years later, he was even admitted to the Royal Netherlands Academy of Sciences. Also, in 1924, the communist Dirk Coster was appointed as professor of physics in Groningen, a state university. Nevertheless, in 1920 Albert Einstein's position as visiting professor in Leiden was held up in government circles for close to a year because he had been confused with communist art critic Carl Einstein; Pannekoek's failed appointment a year before was cited by Paul Ehrenfest, Einstein's close friend and professor of physics in Leiden, as a reason why the government took an exceptionally close look at Einstein's credentials before he was finally approved.[46] The appointment of Marcel Minnaert as professor in Utrecht in

42 The vote was 37 against 35. *NRC*, 26 November 1919.

43 *Handelingen van de Eerste Kamer*, 12 February 1920.

44 Pannekoek to De Sitter, 8 May 1919, WdS inv. 45.1; Kapteyn to De Sitter, 8 May 1919; De Gijselaar to De Sitter, 9 May 1919; De Sitter to the Board of Trustees, 2 June 1919; and personal notes, WdS inv. 224.1; see also De Sitter, *dagboek reorganisatie*, WdS inv. 223.

45 Hertzsprung to Pannekoek [draft, undated], EH C46/10: 'Best mogelijk dat de wereld nog te verbeteren is, maar laten ons [sic] het niet op een manier trachten te doen die de werkcondities van onzen wetenschap schaadt.'

46 van Dongen 2012.

1937 was equally controversial, probably because Minnaert, like Pannekoek, was not just a party member but a known activist.[47]

According to his memoirs, Pannekoek's political life interfered with his scientific work on only one other occasion. In 1926, he wanted to join an expedition to the Dutch East Indies, to observe a solar eclipse and to draw the Southern Milky Way. The Governor General demanded that he promised to refrain from political agitation, as the authorities feared communist activism by the local population. Pannekoek was offended by the suggestion that he would use a scientific expedition for political aims, but in the end, he did make the promise.[48]

## Astronomy as a Discipline

Pannekoek could not work at the largest and best-equipped Dutch observatory, but this did not stop him from becoming a prominent researcher. Looking back in 1944, Pannekoek himself thought that the failed appointment in Leiden may even have been a blessing in disguise. No matter how much he respected De Sitter, he expected that conflict would have ensued sooner or later: 'He was like a Pope, who liked to make his authority felt'.[49] A conflict between Hertzsprung and De Sitter did in fact occur for exactly this reason.[50] Pannekoek was fortunate enough to have an alternative, although this did not look like a very appealing option at first. When it was clear that he could not go to Leiden, he quickly accepted a position as lecturer at the University of Amsterdam. Unlike Leiden, it was not a state university: it was supervised by the city of Amsterdam. The municipal council had no qualms about appointing 'red professors'; in 1917 it had already welcomed the known communist Gerrit Mannoury as professor of mathematics.[51]

Pannekoek was the first scientist at the University of Amsterdam who was appointed to exclusively study astronomy: until then, astronomy had been one of the responsibilities of mathematician Diederik Korteweg. This gave Pannekoek the opportunity to start his own research programme – provided that it did not require any resources, since there weren't any, and that he

47 Molenaar 2003, 262-264.

48 Pannekoek 1982, 212.

49 Pannekoek 1982, 246-247 'Hij [De Sitter] was als een paus, die graag zijn meester-zijn naar voren bracht en deed voelen'.

50 Baneke 2010.

51 Knegtmans 1998, 29.

spent a significant portion of his time teaching mathematics and physics to undergraduates.

Just like Kapteyn, Pannekoek was thus dependent on observations from elsewhere. To collect data, Pannekoek spent time at the Dominion Observatory in Canada and the Bosscha Observatory in Lembang (Java) – founded by Joan Voûte, his one-time colleague who had also fled Leiden Observatory in the early 1900s. Pannekoek also joined several eclipse expeditions, including one to Lapland (where he was joined by the students Bart Bok and Gerard Kuiper, who travelled to Lapland by bike).

Aside from collecting observations from elsewhere, Pannekoek also did theoretical work, with great success. Others in this volume have written about Pannekoek's remarkable switch to theoretical astrophysics, in which he built on Meghnad Saha's work.[52] The main point to observe here is that he began research in a subject that was completely new to the Netherlands. But did he also start a research tradition? As has been stated before, he had few students and the University of Amsterdam did not have the facilities to offer a graduate programme in astronomy; students who had ambitions in this direction had to go to Leiden or Utrecht. Nevertheless, Pannekoek still influenced the organization and research of the Dutch astronomical community, because precisely around this time, a national disciplinary infrastructure was being established.

With Pannekoek as a new lecturer in astronomy in Amsterdam, and with an almost entirely reconstituted staff in Leiden, a new generation of astronomers had stepped forward in Dutch astronomy. Around the same time, in 1921, Kapteyn retired from Groningen. He was succeeded by his student Pieter J. van Rhijn. This generational transition had far-reaching consequences for Dutch astronomy. The new generation introduced new research topics and methods, new teaching programmes, and new institutions, such as a professional society and a journal. They also collaborated and coordinated much more closely than their predecessors. This meant that Pannekoek could interact with students and staff members of other research institutes on a regular basis.

Leiden started a graduate programme for astronomy students in the 1920s, which involved both theory (taught by De Sitter, Woltjer, and Oort) and a thorough training in observational practice (Hertzsprung's specialty).

52 See Edward P.J. van den Heuvel, 'Anton Pannekoek's Astronomy in Relation to his Political Activities, and the Founding of the Astronomical Institute of the University of Amsterdam', in this volume, 25-50; and Robert W. Smith, 'Astronomy in the Time of Pannekoek and Pannekoek as Astronomer of his Times', in this volume, 109-136.

This was a new development, since in earlier days future astronomers typically finished a degree in physics or mathematics before starting hands-on astronomical training at an observatory. From 1923 onwards, students could also benefit from an agreement between Leiden and the Union Observatory in Johannesburg, South Africa – in this way, De Sitter had sought to give Hertzsprung access to the Southern sky, building on Kapteyn's South African contacts.[53] The Leiden school of the 1920s and 1930s produced many well-known astronomers, including for example Gerard Kuiper and Adriaan Blaauw.

Pannekoek was not directly involved in the Leiden programme. He did, however, play a role in the introduction of another new member to the astronomical community who would become extraordinarily influential: Marcel Minnaert. If Pannekoek's career was unusual, Minnaert's was even stranger. After a turbulent life, he ended up as the other great pioneer of astrophysics in the Netherlands, founding a renowned research school in Utrecht and inspiring many generations of students.[54] Minnaert was radical in everything he did. He was strictly vegetarian, a teetotaller, and he had strong political convictions. During World War I, he joined the Flemish nationalist movement that founded a Dutch-language university in Ghent, supported by the German occupation. Although trained as a biologist, Minnaert would teach physics. To prepare himself, he spent a year with Paul Ehrenfest in Leiden. After the war, he was convicted in Belgium for collaborating with the Germans. He fled to the Netherlands and ended up in Utrecht, where W.H. Julius was founding a solar physics laboratory.[55] Minnaert had the right combination of theoretical and technical skills to help in building its instruments. Even his background in biology turned out to be relevant: for his original PhD thesis, he had tried to measure the intensity of sunlight to study its influence on plant development. Now, measuring the solar spectrum became his life's work. He initially followed Julius's rather unorthodox theory of solar physics, worshiping him with characteristic intensity. Pannekoek, together with the physicist L.S. Ornstein, advised him to establish his own research projects. In 1926, Minnaert obtained a second PhD, in physics.

In the course of the 1920s, Pannekoek introduced Minnaert to the Dutch and international astronomical community. After being a biologist and physicist, Minnaert became an astronomer. He joined several eclipse

53 Feast 2000.

54 Molenaar 2003.

55 Heijmans 1994; Verbunt and Bleeker 2010.

expeditions, including the one of 1926 to Sumatra, and in 1933 he made a lecture tour of astronomical institutes in the US and Canada. Finally, in 1937, he succeeded A.A. Nijland as professor of astronomy in Utrecht – not without controversy, because the former Flemish nationalist had in the meantime become a radical communist with strong internationalist convictions.

Minnaert turned Utrecht Observatory into a centre for solar physics. He also turned it into a major research school for new generations of astronomers. He was famous as an inspiring teacher and populariser of astronomy. One of his main innovations was the creation of an 'astronomical practicum', in which all first-year students of physics, mathematics, and astronomy learned how to do astronomical observations. It inspired more than one prospective physicist to change course and become an astronomer instead.[56] The expansion of teaching and new research in Utrecht and Amsterdam led to the establishment by Minnaert and Pannekoek of a tradition of astrophysics in the Netherlands.

The generation of De Sitter, Pannekoek, and Minnaert did things differently than their predecessors, not only in their research and teaching, but also in the way in which they cooperated and organized their discipline. Since the turn of the century, Dutch astronomy professors regularly met in the 'Eclipse Commission' of the Royal Academy of Sciences, an initiative of A.A. Nijland of Utrecht.[57] There were no other national astronomical institutions, apart from the amateur society for astronomy and meteorology that had been founded in 1901.[58] That changed in 1918 with the founding of the *Nederlandse Astronomenclub* (Dutch Club of Astronomers, NAC), a professional organization. It is not entirely clear whose initiative this was, but in 1918 Pannekoek had suggested to establish a new astronomical society, comparable to the Royal Astronomical Society in Britain. One of his arguments for increased communication between Dutch astronomers was the difficulty of communicating with colleagues abroad because of World War I. It was to be a *professional* society that explicitly excluded amateurs. Similar plans were also proposed by others around that time.[59]

Before World War I, many Dutch astronomers had been members of the German *Astronomische Gesellschaft*, which had now become politically problematic. It is also possible that talks about joining the new International

56 de Jager, van Bueren, and Kuperus 1993; Baneke 2015.

57 van Berkel 2004.

58 de Boer and van der Brugge 2001.

59 Pannekoek to De Sitter, 29 September 1918, WdS inv. 45.1. I thank Chaokang Tai for this reference. See also Stein 1928; Baneke 2015, 121-125.

Astronomical Union played a role.[60] This required a national organization to represent the Netherlands. Finally, the initiative was likely closely related to the reorganization of Leiden observatory and the momentum this created.

It turned out to be difficult to define the target group of the new society. The combined tenured astronomical research staff at the four universities in 1920 consisted of eleven men (six in Leiden, two in Utrecht, two in Groningen and one in Amsterdam), but there were also research assistants, advanced students, and active astronomers who worked at other institutions – like Pannekoek when he was teaching at a 'HBS' high school between 1915 and 1919. Suggested descriptions included 'professional astronomers or people who can be regarded as their equivalent', or astronomers 'who are actively pursuing research'.[61] Interestingly, NAC president J. Stein SJ (himself a teacher at a gymnasium, although he would later become director of the Vatican Observatory) later, after the NAC's creation, spoke of 'doctoral candidates and their equivalents', making academic training the decisive criterion. By speaking of 'doctoral candidates' (*doctorandi*) instead of PhDs (*doctors*), he explicitly included graduate students. In practice, new members had to be approved by the club, which actually rarely caused controversy. The NAC started with about twenty members in 1918, growing to more than 40 in 1940.

The *Astronomenclub* was a relatively informal society, which gathered two or three times per year to discuss ongoing research and organizational issues, for example the status of Voûte's new observatory in the Dutch East Indies or its relation with amateur astronomers. Pannekoek frequently contributed as a lecturer. Many students were introduced into the professional research community through these meetings – arguably their most important function.

At least as important as the *Astronomenclub* was the creation of the *Bulletin of the Astronomical Institutes of the Netherlands* (*BAN*) in 1921. Until then, the astronomical institutes of Leiden, Utrecht, and Groningen each had their own publication series (in German, French, and English, respectively).[62] Dutch astronomers also published in the proceedings of the Royal Academy of Sciences, and occasionally in the *Monthly Notices* of the (British) Royal Astronomical Society or the (German) *Astronomische Nachrichten*.

60 Blaauw 1994.

61 The notebooks with the proceedings of the Astronomenclub are preserved in Leiden Observatory; see also Stein 1928.

62 *Annalen der Sternwarte in Leiden*; *Recherches astronomiques de l'Observatoire d'Utrecht*; *Publications of the Astronomical Laboratory at Groningen*.

At the end of World War I, however, the *Astronomische Nachrichten* had to fight paper shortages and an international scientific boycott against Germany and its allies. Its dire situation led Hertzsprung to propose to establish a Dutch journal. Hertzsprung added that publishing in the proceedings of the Royal Netherlands Academy of Sciences was not a suitable alternative: he described them as a 'graceful tomb', in which astronomical publications disappeared, never to be read.[63] De Sitter agreed, adding that a new journal should be published in English and have British and American support – an indication of their reoriented international outlook. Pannekoek suggested that it should have the format of a 'bulletin': submissions should have a quick turnover time and issues should appear as soon as there was enough material. His ideas followed the model of the *Lick Observatory Bulletin* and were in fact realized. The Bulletin's fast and cheap production made it an efficient medium to quickly disseminate Dutch astronomical research. The *BAN* became the main journal for Dutch astronomers, including, for example, many of Jan Oort's groundbreaking publications. In 1927, Hertzsprung reported from Harvard that it was well read there.[64]

The Dutch astronomers agreed that Visibility in the English-speaking world had become of great importance. Around this time, the first decades of the twentieth century, America was emerging as the leading country in astronomy, while World War I caused great material and political difficulties for German astronomy.[65] At the same time, George Ellery Hale and others founded an impressive series of large new observatories in the United States. Hale also founded several new journals and astronomical organizations, and pioneered the integration of modern astrophysics into mainstream astronomy.[66] Thanks especially to Kapteyn's friendship with Hale, Dutch astronomers developed close connections to the American astronomical community. Kapteyn introduced many Dutch astronomers at Mount Wilson, for example.[67]

De Sitter's international network was also important. As President of the International Astronomical Union, he hosted its 1928 General Assembly in Leiden, which was attended by Harlow Shapley, Frank Schlesinger, and Henry

63 Hertzsprung to De Sitter, 16 February 1921, WdS inv. 23.2: 'een sierlijk graf'. Correspondence about the founding of the BAN can be found in the WdS inv. 23.2, 23.9, and 45.1.

64 Hertzsprung to De Sitter, [?] January 1927, WdS inv. 23.5.

65 Lankford has also pointed at the *Carte du Ciel* project as a cause of stagnation in European astronomy. See: Lankford 1997.

66 Wright 1994; Sandage 2004.

67 DeVorkin 2000b.

Norris Russell, three of the leading American astronomers (also known as 'the generals').[68] It was the first astronomical conference since World War I to which German astronomers were invited.[69]

American observatories and universities were eager to import European scholarship to help establish professional American research programmes. Kapteyn, Hertzsprung, De Sitter, and Pannekoek all embarked on extensive lecture tours in the 1910s and 1920s. Pannekoek remarked that Americans tried to squeeze all usable knowledge out of European astronomers.[70] His own itinerary in 1929 was typical: Victoria (Canada), Berkeley, Lick, Mount Wilson, Yerkes, Washington, New York, Cambridge, MA. He was also invited to teach at a summer school in Harvard in 1935 and 1936 (Bart Bok reported in a letter that 'Papa Pannekoek nicely did his best'[71]). Harvard even awarded him an honorary doctorate in 1936, communist or not.

Dutch astronomers not only promoted their own research while in the US. They also used their connections to recommend their students for fellowships and research positions. The number of Dutch astronomy graduates increased sharply in the 1920s and 1930s, but there were few career opportunities in the Netherlands. All permanent staff positions had been filled around 1920, so opportunities to succeed retiring professors were not to be expected for some time. As Bart Bok recalled, De Sitter's advice to astronomy students was: 'Boys, when you get your PhD, you can either become a secondary school teacher, or you go to the United States'.[72] Especially Hertzsprung was actively promoting his students. For instance, he met with representatives of the Rockefeller Foundation's International Education Board to suggest their names for fellowships.

Dutch graduates were particularly welcome in the US due to the new teaching programmes in Leiden and Utrecht that combined theoretical studies with observing skills.[73] The list of young Dutch astronomers who departed for the US in the 1920s and 1930s includes Jan Oort, Jan Schilt, Dirk Brouwer, Bart Bok, Gerrit (Gerard) Kuiper, Pieter Oosterhoff, and Adriaan Blaauw. Hertzsprung also introduced Kaj Strand there, a Danish student who had worked with him in Leiden for several years. Most remained in the US for the rest of their career; only Jan Oort and Adriaan Blaauw returned to the

68 Peter van de Kamp, interview with David DeVorkin, Session II, 18 March 1977, AIP; see also DeVorkin 2000a.

69 Blaauw 1994.

70 Pannekoek 1982, 270. He also described his lecture tour in Pannekoek 1930.

71 Bok to Van de Kamp 12 October 193[?], PvdK: Papa Pannekoek heeft fijn zijn best gedaan.

72 Bart Bok, interview with David DeVorkin, Session I, 15 May 1978, AIP.

73 Herrmann 1994, 50; van der Kruit and van Berkel 2000.

Netherlands. According to historian John Lankford, the Dutch were second on the list of foreign-born astronomers in the US in 1940, after Canadians.[74]

## Legacy

In the decade after 1918, the Dutch astronomical community changed dramatically, and Pannekoek played an active role in initiating these changes. Not, as first envisioned, as a member of De Sitter's all-star cast in Leiden, but as the sole astronomer of Amsterdam University, with few resources and students. Returning to astronomy after his political adventures initially turned out to be more difficult than he had expected, mostly because of unfortunate timing. It probably was the prime minister himself who, in 1919, had vetoed his appointment in Leiden. Soon after, however, Pannekoek was appointed professor in Amsterdam and started working with others towards building a new national disciplinary structure that made it possible for him to contribute significantly to the renaissance of Dutch astronomy. His most important contribution was the introduction of modern theoretical astrophysics to the Netherlands. Indirectly, he also inspired colleagues and students, most notably helping Marcel Minnaert to become an astronomer.

Pannekoek's generation of astronomers was remarkable. In the first place because of their research, but also because of the students that they produced and because of the way they cooperated with each other. A good example of the latter was the 1923 statement on the state of the field that the institute directors submitted to the government. They wanted to ensure that astronomy was represented at all Dutch universities. Interestingly, they added that Leiden Observatory, by far the largest institute, should not become too dominant. Apparently, De Sitter's ambitions were not entirely uncontroversial. The directors agreed that all institutes, including smaller ones like Pannekoek's, should have at least one observer, several (human) computers and a decent budget for instruments.[75]

The statement also listed the research specializations of each institute, demonstrating that there was no overlap between them. Pannekoek's research

74 Lankford 1997, 361; overviews of Dutch astronomers in the US are provided by Oort 1941; van Berkel 2000. The quip that 'Leiden is the place where they grow tulips and astronomers for export' is usually ascribed to Harlow Shapley, for example in van Herk, Kleibrink, and Bijleveld 1983, 85. The oldest version of this remark that I could find was by the South African Minister of Science J.H. Viljoen, in his opening address for the new telescope in Hartbeespoort on 9 September 1957, see Viljoen 1957, 214. He added 'jenever' (gin) to the list of notable Dutch exports.

75 WdS inv. 229.

topic was listed as 'stellar astronomy' and the measuring of photographic plates that had been made elsewhere – much like Kapteyn's programme, even though theoretical astrophysics would soon become his main focus. The principle of dividing research subjects over the institutes through mutual agreement would remain an important feature of the Dutch astronomical community.[76] Other elements of the new disciplinary structure also proved durable. The *Bulletin of the Astronomical Institutes of the Netherlands* remained the main publication medium for Dutch astronomers until it merged with other European journals to create *Astronomy and Astrophysics*, still one of the leading journals in the field. The *Astronomenclub* had its most famous meeting on 15 April 1944, when Henk van de Hulst presented his prediction of a 21 cm hydrogen spectral line. Later, its function was largely taken over by the Nationale *Astronomenconferenties*, informal annual conferences initiated by Minnaert. Close relations with America also remained, as the US emerged as the uncontested leading astronomical nation after World War II. Ambitious Dutch students still found their way to American institutions.

Pannekoek's legacy has remained tangible in the two major research traditions of Dutch astronomy in the twentieth century. One was the collaboration between Groningen and Leiden, which was dominated by Kapteyn's research programme into stellar astronomy and galactic structure. It was the result of careful planning by Kapteyn and De Sitter, and continued by Hertzsprung and Oort. The other was the Amsterdam-Utrecht collaboration in astrophysics, which was not planned at all: it resulted from the unpredictable careers of Pannekoek and Minnaert, in which politics had played such an important role. However dependent on the unpredictable turns of history, its legacy remains visible today. In the 1970s, the Amsterdam astronomical institute was greatly expanded by Edward van den Heuvel, Minnaert's last PhD student, and therefore a direct descendant of this research school. Van den Heuvel was instrumental in continuing the astrophysical research tradition, extending its life for many decades, even after the Utrecht astronomical institute was closed in 2011.

## Archives

AC Archief van Curatoren, 1878-1953 (Archives of the Board of Trustees). Leiden University Library.

76 Baneke 2015.

AIP Oral History Interviews. Niels Bohr Library & Archives. American Institute of Physics, College Park, MD. http://www.aip.org/history/ohilist/transcripts.html
EH Ejnar Hertzprung papers. Center for Videnskabsstudier. Aarhus University.
OKW Archief van het Ministerie van Onderwijs, Kunsten en Wetenschappen: Persoondossiers Rijkspersooneel, c. 1870-c. 1974 (Archives of the Ministry of Education, Culture and Science: Personal files government personnel), catalog no. 2.14.17. National Archives of the Netherlands, The Hague.
PvdK Peter van de Kamp papers. Anton Pannekoek Institute for Astronomy. University of Amsterdam.
WdS Willem de Sitter papers, Leiden Observatory Archives. Leiden University Library.

## Bibliography

Baneke, David. 2004. '"Hij kan toch moeilijk de sterren in de war schoppen". De afwijzing van Anton Pannekoek als adjunct-directeur van de Leidse Sterrewacht in 1919'. *Gewina* 27(1): 1-13.

—. 2005. '"Als bij toverslag". De renovatie en nieuwe bloei van de Leidse Sterrewacht 1918-1924'. *BMGN – Low Countries Historical Review* 120(2): 207-225.

—. 2010. 'Teach and Travel: Leiden Observatory and the Renaissance of Dutch Astronomy in the Interwar Years'. *Journal for the History of Astronomy* 41(2): 167-198.

—. 2015. *De ontdekkers van de hemel. De Nederlandse sterrenkunde in de twintigste eeuw*. Amsterdam: Prometheus Bert Bakker.

Berkel, Klaas van. 2000. 'Growing Astronomers for Export: Dutch Astronomers in the United States before World War II'. In *The Legacy of J.C. Kapteyn: Studies on Kapteyn and the Development of Modern Astronomy*, ed. by Pieter C. van der Kruit and Klaas van Berkel. Dordrecht: Kluwer, 151-174.

—. 2004. 'De Akademie, Indië en de bloei van de sterrenkunde in Nederland. De Eclipscommissie van 1899'. In *De Akademie en de Tweede Gouden Eeuw*, ed. by Klaas van Berkel. Amsterdam: Koninklijke Nederlandse Akademie van Wetenschapen, 107-138.

Blaauw, Adriaan. 1994. *History of the IAU: The Birth and First Half-Century of the International Astronomical Union*. Dordrecht: Kluwer.

Boer, Jan A. de, and Aad H. van der Brugge. 2001. 'Honderd jaar NVWS'. In *Evolutie in weer- en sterrenkunde: 100 jaar Nederlands onderzoek*, ed. by Mat Drummen, Cornelis de Jager, and Hugo van Woerden. Utrecht: Stichting De Koepel.

DeVorkin, David H. 1984. 'Stellar Evolution and the Origin of the Hertzsprung-Russell Diagram'. In *Astrophysics and Twentieth-Century Astronomy to 1950*, ed. by

Owen Gingerich. The General History of Astronomy 4. Cambridge: Cambridge University Press, 90-108.

—. 2000a. *Henry Norris Russell: Dean of American Astronomers*. Princeton: Princeton University Press.

—. 2000b. 'Internationalism, Kapteyn and the Dutch Pipeline'. In *The Legacy of J.C. Kapteyn: Studies on Kapteyn and the Development of Modern Astronomy*, ed. by Pieter C. van der Kruit and Klaas van Berkel. Dordrecht: Kluwer, 129-150.

Dongen, Jeroen van. 2012. 'Mistaken Identity and Mirror Images: Albert and Carl Einstein, Leiden and Berlin, Relativity and Revolution'. *Physics in Perspective* 14(2): 126-177.

Feast, Michael. 2000. 'Kapteyn and South Africa'. In *The Legacy of J.C. Kapteyn: Studies on Kapteyn and the Development of Modern Astronomy*, ed. by Pieter C. van der Kruit and Klaas van Berkel. Dordrecht: Kluwer.

Guichelaar, Jan. 2009. *De Sitter. Een alternatief voor Einsteins heelalmodel.* Amsterdam: Veen.

Heijmans, Henri G. 1994. *Wetenschap tussen universiteit en industrie: de experimentele natuurkunde in Utrecht onder W.H. Julius en L.S. Ornstein 1896-1940*. Rotterdam: Erasmus.

Herk, Gijsbert van, Herman Kleibrink, and Willem Bijleveld. 1983. *De Leidse Sterrewacht: vier eeuwen wacht bij dag en bij nacht.* Zwolle: Waanders/DeKler.

Herrmann, Dieter B. 1994. *Ejnar Hertzsprung: Pionier der Sternforschung*. Berlin: Springer.

Jager, Cornelis de, Henk G. van Bueren, and Max Kuperus. 1993. *Bolwerk van de Sterren*. Amersfoort: Bekking.

Kerszberg, Pierre. 1989. *The Invented Universe: The Einstein-De Sitter Controversy (1916-17) and the Rise of Relativistic Cosmology*. Oxford: Oxford University Press.

Knegtmans, Peter Jan. 1998. *Een kwetsbaar centrum van de geest. de Universiteit van Amsterdam tussen 1935 en 1950*. Amsterdam: Amsterdam University Press.

Kruit, Pieter C. van der. 2015. *Jacobus Cornelius Kapteyn: Born Investigator of the Heavens*. Cham: Springer.

Kruit, Pieter C. van der, and Klaas van Berkel, eds. 2000. *The Legacy of J.C. Kapteyn: Studies on Kapteyn and the Development of Modern Astronomy*. Dordrecht: Kluwer.

Lankford, John. 1997. *American Astronomy: Community, Careers, and Power, 1859-1940*. Chicago: University of Chicago Press.

Molenaar, Leo. 2003. *Marcel Minnaert, astrofysicus 1893-1970. De rok van het universum*. Amsterdam: Balans.

Oort, Jan H. 1941. 'Jongere Nederlandsche sterrekundigen op belangrijke posten in het buitenland'. *Hemel en Dampkring* 39: 355-363.

Pannekoek, Anton. 1930. 'Een vluchtig bezoek aan eenige Amerikaansche sterrewachten'. *Hemel en Dampkring* 28 (1-3): 1-7, 33-45, 81-91.

—. 1982. *Herinneringen: herinneringen uit de arbeidersbeweging; sterrenkundige herinneringen*. Ed. by Ben A. Sijes, Johanna M. Welcker, and J.R. van der Leeuw. Amsterdam: Van Gennep.

Pyenson, Lewis. 1989. *Empire of Reason: Exact Sciences in Indonesia, 1840-1940*. Leiden: Brill.

Sandage, Allan. 2004. *The Mount Wilson Observatory: Breaking the Code of Cosmic Evolution*. Cambridge: Cambridge University Press.

Sande Bakhuyzen, H.G. van de. 1872. 'De methode welke thans bij de beoefening der sterrenkunde moet gevolgd worden'. Inaugural lecture, Leiden: Leiden University.

Stein, Johan W.J.A. 1928. 'De Nederlandsche Astronomenclub 5 oct. 1918 – 5 oct. 1928". *Hemel en Dampkring* 26(11): 335-347.

Verbunt, Frank, and Johan Bleeker. 2010. 'Spectroscopy in Utrecht: A Brief History'. *Space Science Reviews* 157(1-4): 3-13.

Viljoen, J.H. 1957. 'Rede deur sy Edele die Minister van Onderwys, Kuns en Wetenskap, mnr. J.H. Viljoen by geleentheid van die opening van die Leidse Sterrewag te Hartbeespoort op 9 September 1957 om 15 uur.' *Hemel en Dampkring* 55: 213-215.

Wright, Helen. 1994. *Explorer of the Universe: A Biography of George Ellery Hale*. Woodbury, NY: American Institute of Physics.

Zuidervaart, Huib J. 2008. 'Joan Voûte 1879-1963. Een reuzentelescoop op de Bosscha-sterrenwacht te Lembang'. In *De telescoop: erfenis van een Nederlandse uitvinding*, ed. by Dirk van Delft, Albert van Helden, Hans Hooijmaijers, and Huib J. Zuidervaart. Amsterdam: Bert Bakker, 45-51.

—, ed. 2011. *Frederik Kaiser (1808-1872), schepper van de 'nieuwe' Leidse sterrewacht*. Amsterdam: Gewina.

## About the Author

**David Baneke** is Assistant Professor at Utrecht University, where he coordinates the master's programme History and Philosophy of Science. He studied history in Groningen, and obtained his PhD in 2008 in Utrecht. His research addresses the role of science in twentieth-century culture and politics, and the history of modern astronomy. In 2015, he published *De ontdekkers van de hemel*, about the history of modern astronomy in the Netherlands; an English translation is forthcoming. In 2019, he co-authored *The International Astronomical Union: Uniting the Community for 100 Years* about the history of the International Astronomical Union.

# 6 Astronomy in the Time of Pannekoek and Pannekoek as an Astronomer of his Times

*Robert W. Smith*

**Abstract**

The astronomical enterprise underwent enormous changes during Pannekoek's lifetime, including, most importantly, in terms of the technical content and practices of the science, the rise of astrophysics. I suggest that the history of astrophysics between the 1860s and early 1950s can be divided roughly into three stages and that in his later career Pannekoek is best seen as a 'third-stage' astrophysicist. The institutional landscape of astronomy was also transformed during Pannekoek's lifetime, most tellingly with the emergence of the United States as the leading nation for observational astrophysics. However, in the Netherlands, J.C. Kapteyn had shown that it was possible to be an active astronomer without a telescope and Pannekoek would do the same, and fashioned a successful career as an interpretive and theoretical astrophysicist.

**Keywords:** Anton Pannekoek, positional astronomy, astrophysics, Meghnad Saha, Jacobus C. Kapteyn, galaxy

## Introduction

In the period between 1873 and 1960 – the span of Anton Pannekoek's lifetime – the accepted body of astronomical knowledge grew enormously, which coincided with an expansion in the sort of knowledge astronomers regarded as appropriate to pursue. Astronomy in the Netherlands and elsewhere underwent a series of institutional, social, and economic changes too, with the most striking developments in the United States.

Tai, Chaokang, Bart van der Steen, and Jeroen van Dongen (eds), *Anton Pannekoek: Ways of Viewing Science and Society*. Taylor & Francis Group, 2019
DOI: 10.5117/9789462984349_CH06

In this chapter, I will examine essential elements of these shifts, and in so doing I will consider two central problems: What counted as both proper and legitimate astronomy? How did astronomers attempt to reveal the size and structure of the Milky Way? Addressing these two questions will, I argue, help bring aspects of astronomy in the time of Pannekoek into focus that were significant for his career. Other scholars, including in this volume, have examined Pannekoek's astronomical career from various perspectives, including perhaps most fruitfully the investigation by Tai and van Dongen and Tai of the epistemic values exhibited in that life's work.[1] Here the emphasis will be on the changing nature of the astronomical enterprise so we can better understand the institutional possibilities and limits that Pannekoek both confronted and shaped, as well as the effect of new conceptual tools and resources.

Pannekoek, we will see, was doubtful about the broader ambitions of the statistical astronomers who sought to model our galactic system. He showed his willingness to adopt new ideas on the structure of the Galaxy when in 1918 he sided with Shapley when Shapley advanced the controversial 'Big Galaxy' thesis. An advocate of the tight linking between astronomical observation and theory, Pannekoek also enthusiastically embraced and applied atomic theory and quantum mechanics to help develop a new sort of astrophysics in the 1920s. He would be one of the earliest and most successful practitioners of what we will describe as 'Third Stage Astrophysics'. In his application of the latest physical theories to the interpretation of stellar spectra, Pannekoek, then, would fashion himself very much as a model of a modern astronomer.

## What is Legitimate Astronomy? Astrophysics

The notions of what counted as legitimate astronomy for many astronomers, though certainly by no means all, were greatly expanded in the middle of the nineteenth century by the rise of what would come to be known as 'astrophysics' (a term usually credited to Johann Carl Friedrich Zöllner writing in 1865). To see what changed, it is helpful to refer first to John Narrien's *An Historical Account of the Origin and Progress of Astronomy* (1833). Narrien, for many years a lecturer in mathematics at the British Royal Military College at Sandhurst, painted a clear picture of the advances

I am very grateful to Chaokang Tai and David DeVorkin for helpful comments and discussion on earlier versions of this paper.

1 Tai 2017; Tai and van Dongen 2016.

still to be expected from astronomy. According to him, the application of human ingenuity to astronomy would 'be able to accomplish little more than an improvement in the means of making observations, or in the analysis by which the rules of computation are investigated'.[2] At a time when the first reliable parallax determinations to a distant star were still several years away, and so the only star whose distance was known to any degree of accuracy was the Sun, Narrien reckoned the future of the discipline was decidedly limited. That was not because it had failed. To the contrary, it was because astronomy had reached a very high degree of refinement. As William Whewell, one of the great polymaths of the nineteenth century and the person who coined the term 'scientist', put it in the same year as Narrien's book was published:

> Astronomy is not only the queen of sciences, but, in a stricter sense of the term, the only perfect science; – the only branch of human knowledge in which particulars are completely subjugated to generals, effects to causes [...] and we have in this case an example of a science in that elevated state of flourishing maturity, in which all that remains is to determine with the extreme of accuracy the consequences of its rules by the profoundest combinations of mathematics, the magnitude of its data by the minutest scrupulousness of observation.[3]

For Whewell, Narrien, and others with a professional stake in astronomy, the aim of the science was to track the movements of objects in the solar system and then reduce these motions to order by use of Newton's law of universal gravitation. In this vision of the science, the stars were of importance because they provided a background grid against which the motions of planets, minor planets and comets could be plotted. The physical nature of astronomical bodies was hardly the concern of professional astronomers. As Friedrich Wilhelm Bessel, probably the leading astronomer in the world at the time, put it in 1832:

> What astronomy must do has always been clear – it must lay down the rules for determining the motions of the heavenly bodies as they appear to us from the earth. Everything else that can be learned about the heavenly bodies, e.g. their appearance and the composition of their surfaces, is certainly not unworthy of attention; but it is not properly of astronomical interest.[4]

2 Narrien 1833, 520.
3 Whewell 1833, xiii.
4 Quoted in Hufbauer 1993, 43.

Agnes Clerke, the well-known historian and astronomical popularizer, succinctly summarized matters in 1885 in her account of early nineteenth-century astronomy:

> The astronomy so signally promoted by Bessel – the astronomy placed by Comte at the head of the hierarchy of the sciences – was the science of the *movements* of the heavenly bodies. And there were those who began to regard it as a science which, from its very perfection, had ceased to be interesting – whose tale of discoveries was told, and whose farther advance must be in the line of minute technical improvements, not of novel and stirring discoveries.[5]

Clerke's claim does not mean that there were no changes in the methods and procedures of positional astronomy in the nineteenth century. Observatory practice became increasingly routinized as a consequence of both novel forms of mechanical technology, the chronograph most notably, and new versions of what can be described as organizational technologies. These shifts also meant new sorts of observers, as Kevin Donnelly has emphasized.

> At the forefront of both of these developments were observatory directors like George Biddell Airy, Adolphe Quetelet and Friedrich Bessel, who simultaneously revolutionized the practice of astronomy and created entirely new kinds of scientific labour that demanded patience, discipline and attentiveness in place of open-ended observation, reflection and creativity.[6]

At the end of the nineteenth century, when Pannekoek began his training in astronomy, the view of positional astronomy as the only legitimate form of the discipline seemed exceptionally narrow to many, but far from all, astronomers. The critics of astrophysics included, as we shall see, the director of the Leiden Observatory during Pannekoek's time there. The rise of astrophysics from around 1860 and the efforts of the first generation of astrophysicists to chart the spectra of celestial objects, served to produce a new body of knowledge of the heavens. In time, the pursuit of this knowledge would drive the reconstitution of astronomy. Initially, however, the old-style positional astronomy and the newer astrophysics were separate, so that it is misleading to talk about the pre-existing astronomy being 'transformed

5 Clerke 1885, 185; emphasis in the original.

6 Donnelly 2014, 3. See also Schaffer 1988; and Smith 2003.

by the emergence of astrophysics'.[7] Astrophysics instead owed its birth to developments in experimental spectroscopy that sprang from studies in physics and chemistry.

The researches of the chemist Robert Bunsen and the physicist Gustaf Kirchhoff at the University of Heidelberg provided the initial impetus for this occurrence. In the late 1850s, they demonstrated to the satisfaction of other practitioners that particular sets of spectral lines are associated with particular chemical elements and compounds and also explained how such lines are produced.[8] Remarkable and what, by earlier standards, seemed like almost miraculous powers were now handed to students of celestial objects. As one enthusiast for this novel sort of research wrote:

> The physicist and the chemist have brought before us a means of analysis that [...] if we were to go to the sun, and to bring away some portions of it and analyse them in our laboratories, we could not examine them more accurately than we can by this new mode of spectrum analysis.[9]

Mainstream astronomers, however, sometimes dismissed, ignored or were very slow to warm to the new astrophysics.

The use of photography, especially before the introduction of the dry plate in the years around 1880, was also widely seen by many professional astronomers as problematic. As the leading American positional astronomer (and later discoverer of the two moons of Mars) Asaph Hall explained in 1866:

> For one, I shall be glad to see improvements in methods of observing, but for a very large part of the accurate work of astronomy, I don't yet see how photography is to help much [...] It seems doubtful whether it is well to insert such a method between the observer and the result, since new sources of error are brought in.[10]

The suspicious if not hostile attitude towards the newer developments that were held by many positional astronomers meant that the first generation of astrophysicists and astronomical photographers were usually not 'mainstream' positional astronomers who had expanded their interests. One result of this situation was that until the 1890s, talented and driven

7 Lankford 1997, 36.

8 See, for example, Meadows 1984b; 1984a; Becker 2011; and Hentschel 2002.

9 de la Rue 1861, 130.

10 A. Hall to C. Peters, 19 April 1866, quoted in Rothenberg 1974, 6.

practitioners could pursue astrophysics and astronomical photography at the cutting-edge of research even without an extensive mathematical training or very costly equipment or a professional position. Further, astrophysical investigations in the first two or three decades of the new discipline were usually not driven by specific theoretical problems.[11] Instead, they were commonly surveys of stellar spectra or the detailed investigation of the spectrum of a single object, albeit often with the vague hope that such studies might lead in time to an understanding of the evolution of nebulae and stars. Thus both positional astronomers and astrophysicists tended to be, by later standards, very conservative in their scientific goals.

By the early 1890s, astrophysics had nevertheless become much more of a professional activity than it had been even a decade earlier. New sorts of astronomical observatories were also becoming more common. The focus of traditional observatories was positional astronomy. Their telescopes and ancillary instruments were chosen accordingly to centre on accuracy rather than light grasp, and so observatory directors emphasized transit instruments.[12] In the closing decades of the century, however, new astrophysical observatories were established, and some traditional observatories also added astrophysical researches. In 1874, Kaiser Wilhelm I founded the Potsdam Observatory, and it became the first state-sponsored astrophysical observatory. Others soon followed at Meudon in France as well as the Solar Physics Observatory at South Kensington in London.[13]

These developments had relatively little impact on the activities of the Leiden Observatory. At the turn of the century, it was the largest and best supported of the astronomical institutions in the Netherlands and Pannekoek completed his PhD there in 1902 on the light curve of the variable star Algol.[14] After 1899, he also served as the third observer at the Observatory, the lowest of the institution's professional positions, but it was a permanent post. Pannekoek's principal work, however, entailed making and reducing meridian observations to determine stellar locations in very much the old style. The work at Leiden under the leadership of the two brothers

11 There were exceptions of course, and perhaps the leading counterexample is provided by the investigations of Norman Lockyer.

12 Dewhirst noted that earlier in the nineteenth century, in 1843, readers of the *Penny Cyclopedia* who searched for 'Observatory' were directed to 'Transit instrument': Dewhirst 1985, 150.

13 Herrmann 1975; Laurie et al. 1984. On the establishment of the Solar Physics Observatory at South Kensington, see Meadows 1973.

14 On the history of the Dutch astronomical community in the twentieth century, see, among others, van der Kruit and van Berkel 2000; Baneke 2010; and 2015.

E.F. and H.G. van de Sande Bakhuyzen, neither of whom thought much of astrophysics, was also regimented (Pannekoek wrote his dissertation under H.G. van de Sande Bakhuyzen). Such labour was not to Pannekoek's taste, particularly as he struggled to see the social worth of the positional astronomy done at Leiden. In what is now a well-known quotation, he recalled this time:

> In this environment, where everything happened in the traditions of twenty or thirty years earlier, where there was only endless computation and without anything ever being finished, where the new ways of astronomy were hardly appreciated, all enthusiasm must eventually disappear. Later, [Jacobus C.] Kapteyn once remarked to me: I never understood how you kept up with it so long. [...] I dreaded every Monday morning, when I had to attend the weekly conference in the director's office, where there would be some chatter, and every one mentioned what they had done that week – or invented something – and I realized that every week was in large measure the same, just trickled along a bit. I then always felt a smell around me like in catacombs, of deadly rigidity and boredom.[15]

For some of the lower-level practitioners, positional astronomy, even when there was a clear direction and programmes were accomplished, *was* boring in and of itself,[16] and this picture of Leiden, with the complaint about unfinished work, reveals no sense of accomplishment lifted the tedium for Pannekoek.[17] Pannekoek's decision in 1906 to move from Leiden to Berlin to teach Marxism at the new party school of the Social Democratic Party of Germany (Sozialdemokratische Partei Deutschlands, SPD), therefore, has to be seen in the light of both his rejection of what struck him as the stultifying sort of astronomy that constituted his working life at Leiden as well as his political commitments.

15 Quoted in van der Kruit 2015, 582.

16 See Donnelly 2014.

17 The dreariness of meridian observations was also the key point of a story recounted in 1943 by Otto Struve, then a prominent astronomer and the director of the Yerkes Observatory as well as the Macdonald Observatory in Texas. According to Struve, in 1913, he was aboard a German train returning from a meeting of the *Astronomische Gesellschaft* along with an astronomer he called 'Dr X… an assistant in a German observatory' who had worked on a routine programme with a meridian circle. 'His appointment', Struve recalled, 'had expired on December 31, and he was telling with considerable delight how at the exact second of midnight he had interrupted the transit observations of a star and written *finis* in the official record-book.' See Struve 1943, 475. Lankford also told this story in Lankford 1997, 400.

## A Transition

In the early 1910s, then, Pannekoek was in Germany and engaged principally in very different activities than astronomy. But at the start of World War I in the summer of 1914, he was on holiday in the Netherlands. With the outbreak of hostilities, Pannekoek was not allowed to remain in Germany, and so he turned to teaching physics at secondary schools to make a living. When Willem de Sitter assumed the directorship of the Observatory at Leiden in 1918 (he had been a professor of astronomy at Leiden since 1908) after the death of E.F. van de Sande Bakhuyzen, he, as is now well known, tried hard to appoint Pannekoek as one of two assistant directors at the Observatory. Pannekoek, de Sitter thought, should be placed in charge of positional astronomy. If Pannekoek had secured a position at Leiden, he would have had access to working telescopes and significant resources. But despite de Sitter's strong support, Pannekoek, as described elsewhere in this volume, was not appointed because his political views were not acceptable to the incoming government of conservative Christian Democrats led by Charles Ruys de Beerenbrouck that took office in September 1918.[18]

Politics mattered again when in 1920 the University of Amsterdam appointed Pannekoek to a post. The university, as Pyenson has explained, was 'a municipal institution, and so all appointments came at the pleasure of the B & W (*burgemeester en wethouders*), or mayor and town council. In the 1920s, the B & W were stoutly socialist, to which fact Pannekoek, as a long-time left-wing politician, owed his own appointment.'[19] Pannekoek had an astronomical position, but at an institution with no functioning observatory and few resources. He, therefore, confronted a very similar situation to the one J.C. Kapteyn had found himself in when he assumed the professorship of astronomy and theoretical mechanics at Groningen in 1878.

Kapteyn, too, had inherited no staff or facilities. He tried to found an observatory, but, in the face of opposition from Leiden and Utrecht, which already had observatories, and a government unwilling to fund another, he did not succeed.[20] Kapteyn instead made himself relevant by engaging in collaborative efforts with other astronomers. While he was never able to establish an observatory, he did found an 'Astronomical Laboratory'. Aided

18 W.R. de Sitter 2000, esp. 85-93; Baneke 2005; 2010, 170-173; and his 'Pannekoek's One Revolution', in this volume, 87-108.

19 Pyenson 1989, 147.

20 For a recent biography of Kapteyn, see van der Kruit 2015; But see also the essays in van der Kruit and van Berkel 2000.

by a grant from the British Royal Society, David Gill, director of the Royal Observatory at the Cape of Good Hope, aimed to produce a photographic star map of the southern hemisphere. The result was a major collaborative effort between Gill and Kapteyn. Gill and two assistants secured the photographic plates. The plates were then shipped to Groningen for Kapteyn and his assistants to measure them. The resulting catalogue contained the positions and photographic magnitudes of over 450,000 stars. As Lankford has pointed out:

> Kapteyn and Gill were among the first since Bond [at Harvard] to engage in sustained research on such fundamental problems as the determination of photographic magnitudes and the measurement of stellar coordinates on photographic plates, and the [Cape Photographic Durchmusterung] paved the way for the international *Carte du Ciel* project.[21]

But even as late as 1887, such projects were resisted by the practitioners of the older sort of positional astronomy who were afraid that photographic methods would replace meridian instruments and so the Royal Society stopped funding Gill in that year.[22]

Kapteyn later also became the driving force and central figure in the international plan to secure an enormous body of data on stars in a series of 'Selected Areas'. This information would then feed into his pioneering researches on the structure of the galactic system (of which more later). Through these collaborative means, Kapteyn had made himself relevant, indeed a world leader, at Groningen, despite the institutional difficulties. Kapteyn's concern to address a specific problem using a huge collection of data also marks him out as an exemplary practitioner of what we will later term 'Second Stage Astrophysics'.

But how, in 1920, was Pannekoek to make himself relevant at Amsterdam and perform 'cutting-edge' research? Pannekoek also faced a challenge that Kapteyn had not. Whereas Kapteyn in the 1880s and early 1890s did not have to take into account much in the way of competition from the US, American astronomers and telescopes loomed very large by 1910, if not well before.[23] Pannekoek made himself relevant by establishing an Astronomical Laboratory in the manner of Kapteyn, instead of operating an observatory.

21 Lankford 1984, 27.

22 Lankford 1984, 26.

23 For an important discussion of how Swedish astronomers dealt with the issue of making themselves relevant in the face of US telescopes and resources, see Holmberg 1999.

Pannekoek was sufficiently successful that he would win prestigious awards, such as an Honorary Degree from Harvard University in 1936 and the Gold Medal of the Royal Astronomical Society in 1951, widely regarded by astronomers at the time as one of the very top honours in astronomy.

## Collaboration and Competition: The Rise of American Astronomy

Over the course of the nineteenth century, big cooperative projects became an increasingly significant feature of astronomy. Many European observatories were engaged in the single largest of them at the end of the nineteenth century, the *Carte du Ciel*, the aim of which was to photograph the entire sky and to produce both a catalogue and chart of the observed stars. The leaders of the project reckoned that over 88,000 photographic plates would have to be taken, measured, and the results compiled. The participating observatories expected to be engaged for twenty years. The Paris Observatory was the headquarters of the project, and twelve other European observatories were involved over its lifetime. The *Carte*, however, had a negative impact on European astronomy as many observatories and astronomers were locked into a costly project of what proved to be very limited scientific worth for many years.[24]

In contrast, no American observatories participated in the *Carte*, and so the expansion and remaking of astronomy in the United States in the late nineteenth and early twentieth centuries were not impeded by the project's hefty demands. And in these years numerous affluent Americans patronized astronomy. Funding arrived in the form of donations or support from philanthropic foundations. The most notable products were the Lick Observatory on Mount Hamilton in California, the Yerkes Observatory of the University of Chicago, and the Carnegie Institution of Washington that from 1904 onwards funded the building and running of what was initially called the Mount Wilson Solar Observatory.[25]

By the time construction started on Mount Wilson, astrophysics had become well established in many countries. It had secured strong institutional support, and its practitioners pursued a range of programmes of research that followed well-defined methods and standards. Following the founding of the Pulkova Observatory in Russia in 1839, positional astronomers, too, had taken it as the model of how things should be done in terms of the scale and efficiency of its

24 Lankford 1997, 394-400; and Lankford 1984. But see also DeVorkin 1998.

25 Miller 1970. On the early history of Lick, see Wright 2003; and on Yerkes, see Osterbrock 1999.

operations.[26] But by the turn of the century, the United States was jockeying with Europe for leadership in observational astrophysics – which was composed principally of the observational study of the spectra of astronomical objects – and Mount Wilson would soon become widely recognized as the premier astrophysical observatory in the world.[27] American observational astronomy was in fact in the process of supplanting Central European astronomy as the world leader.[28] Thus, by the early 1910s, what Lankford has called the 'political economy' of observational astronomy was radically different from that in the early 1880s, and it would be markedly different again by the end of World War I with the financial problems that engulfed Europe.

## Third Stage Astrophysics and the Influence of Saha

After he arrived at the University of Amsterdam in 1920, Pannekoek pursued some of his established interests, such as the nature and structure of the Milky Way (as we will see in the next section and in other papers in this volume). But, most importantly, he also struck out in a radically new direction by engaging with and developing the investigations of the Indian theoretical physicist Meghnad Saha. By the early 1920s, astronomers working at numerous observatories had collected a vast body of empirical information on stellar spectra. How, though, were these data to be interpreted and processed? Astronomers like the leading American astrophysicist Henry Norris Russell soon acknowledged that Saha had created the 'master key' to understanding stellar spectra by coupling the Bohr theory of the atom (in which negatively charged electrons orbit in shells around a positively charged nucleus) to thermodynamics.[29] Saha argued that the primary determinant of a star's spectrum is its temperature, with pressure as a secondary factor, and some astronomers, like Russell, now followed Saha's lead and applied atomic physics and quantum mechanics to astrophysics.[30]

Another astronomer to do so was Pannekoek. He had been active in the older sort of astrophysics even in his Leiden years. In 1906, for example, he examined '[t]he relation between the spectra and the colours of the stars'.[31]

26 See, for example, Werrett 2010.
27 van Helden 1984, 138.
28 Among other works that make this point, see Lankford 1997, 371-404; and Baneke 2010, 168.
29 DeVorkin 2000, 178-179.
30 DeVorkin and Kenat 1983a; 1983b. See also Naik 2017.
31 Pannekoek 1906.

But in 1920, he was given copies of a few of Saha's papers and he soon after set about mastering, applying, and extending Saha's findings. Pannekoek published his first paper in this new area in 1922, a study in which he tackled 'ionization in stellar atmospheres'.[32] He contended that:

> Spectral analysis has disclosed the chemical constitution of stellar atmospheres by the lines visible in their spectra. As to their physical state we may infer the temperature from these spectra also, as the series of spectral types, at least from [spectral types B to M] corresponds to a series of decreasing temperatures. But this temperature is not deduced directly from the spectral lines. [...] The deduction of the physical conditions in stellar atmospheres from the lines of their spectra has now become possible by the application of the theory of chemical equilibrium on partly ionized gases by Dr MEG NAD SAHA.[33]

For Pannekoek, there was now the exciting prospect that 'a more minute and quantitative investigation of stellar spectra will reveal other characteristics, which in some other way than the state of ionization are connected with diameter, density, temperature, mass and luminosity.'[34] He thereby joined a group of astronomers who had a rigorous training in observational astronomy early in their careers, but who later devoted themselves mainly to interpretation and theoretical researches, with the most prominent other such practitioners being perhaps the British astrophysicist A.S. Eddington and the American Henry Norris Russell. Pannekoek became one of the first practitioners of what I have termed as 'Third Stage Astrophysics' in the tentative periodization of astrophysics given here. Just as Kapteyn, as a practitioner of 'Second Stage Astrophysics', had created and seized opportunities to fashion an Astronomical Laboratory at Groningen, Pannekoek, a practitioner of 'Third Stage Astrophysics', would do the same at Amsterdam, and the output of their respective Astronomical Laboratories exemplified these different stages of astrophysics.

Pannekoek now spent much of his time measuring the relative line intensities in stellar spectra of various spectral types to address issues of spectral classification. For these researches, with no observatory of his own, he measured photographic plates from other observatories,

32 By stellar atmosphere is meant the outer region of a star.

33 Pannekoek 1922, 107; emphasis in the original.

34 Pannekoek 1922, 118.

*The Three-Stage Development of Astrophysics*

The stages described below are designed to be suggestive rather than definitive. The temporal breaks should not be read as firm as there were very significant periods of overlap. Indeed, there are strong echoes of the 'Great Correlation Era' in evidence today, as DeVorkin has argued. There were very major shifts in each of these stages regarding conceptual tools and technologies. For the first two stages, there were also crucial institutional changes.

1) First Stage Astrophysics; c. 1860-1890. Often pursued by non-professional astronomers with limited formal training, who put the emphasis on the identification and charting of spectral lines. Essentially opportunistic, astrophysicists observed what could be observed, although often with the (usually) distant hope of being able to understand the course of stellar evolution. In this period, some, perhaps many, professional positional astronomers were sceptical about, if not hostile towards, astrophysics. Founding of the first observatories devoted to astrophysics. An exemplar of a practitioner of First Stage Astrophysics: William Huggins.
2) Second Stage Astrophysics; c. 1890-1920. Characterized by a growing number of practitioners and increased professionalization. The researchers' emphasis was on large surveys, collecting spectra and radial velocities of stars, with more emphasis on tackling specific problems rather than merely collecting data. Various attempts were made to correlate different bodies of evidence, with the most significant example being the development of what would be called the Hertzsprung-Russell Diagram. This period also witnessed the formation of the International Union for Cooperation in Solar Research in 1905 (its charge was expanded to include stellar research in 1910), and the establishment of *The Astrophysical Journal* in 1895. In 1921, W. Carl Rufus offered a detailed periodization of American astronomy, and he identified a 'Correlation Period' that began in 1890. DeVorkin has instead termed this era 'The Great Correlation Era'.[1] Exemplars of practitioners of Second Stage Astrophysics: Ejnar Hertszprung and J.C. Kapteyn.
3) Third Stage Astrophysics; c. 1920-1950. In this phase, the field was fully professionalized. The great majority of astronomical observatories were now devoted mostly, if not entirely, to astrophysics. This era saw the introduction into astrophysical practice of state-of-the-art physical theory in terms of the new atomic physics and quantum mechanics, as well as the close combination of theory and observation with a new emphasis on the interpretation of spectral lines. Exemplars of practitioners of Third Stage Astrophysics: Pannekoek and H.N. Russell.

1 Rufus 1921; DeVorkin 2010, 140.

including, for example, the Harvard College Observatory. However, during extended stays at the Bosscha Observatory in Java and the Dominion Astrophysical Observatory in Victoria, Canada as well as on eclipse expeditions, he also secured plates for his own use.[35] Indeed, it is telling of the shifting leadership in observational astronomy, that Kapteyn obtained his plates from Gill at what was effectively a colonial observatory in South Africa, while Pannekoek got the majority of his from institutions in North America.[36] His theoretical researches included, for example, reworking Saha's ionization formula to correct it for departures from thermodynamic equilibrium.[37]

Baneke has argued persuasively that Pannekoek's generation changed Dutch astronomy and 'reorganized the discipline on all levels: research, teaching, institutions, journals, and international contacts. In these few years, they built a modern disciplinary infrastructure that would last until the end of the [twentieth] century.'[38] Here, however, our focus is somewhat different. We have seen that through Pannekoek's move to both interpretive and theoretical astrophysics, and his evident talents in these areas, Pannekoek had, like Kapteyn, effectively solved the problem of how to pursue state-of-the-art research as an astronomer without a telescope. He had done so, moreover, in the face of significant changes in the astronomical enterprise, including its shifting political economy and the rise to pre-eminence of American observational astronomy.

## The Milky Way

Pannekoek, as discussed elsewhere in this volume, was fascinated by the Milky Way from an early age and he was an enthusiastic naked-eye observer as well as a very keen student of photographs of the Milky Way. He was deeply impressed by its complexity throughout his career. His researches on the Milky Way have been very well treated by Tai in this volume and elsewhere, as well as by Tai and van Dongen,[39] so here my focus will be on how he became an early advocate of Harlow Shapley's radical new picture of the stellar system, first advanced publicly in 1918.

35 See, for example, Pannekoek 1927.

36 I am grateful to Chaokang Tai for pointing this out to me.

37 Pannekoek 1926.

38 David Baneke, 'Pannekoek's One Revolution', in this volume, 87-108.

39 Tai and van Dongen 2016; Tai 2017; and his 'The Milky Way as Optical Phenomenon', in this volume, 219-247.

Shapley, an astronomer at the Mount Wilson Observatory in California, argued for what became known as the Big Galaxy. Other astronomers placed the Sun close to the centre of our stellar system. For Shapley, it was tens of thousands of light years distant. Shapley, moreover, reckoned the Big Galaxy to be about 300,000 light years in diameter, and so roughly ten times larger than the usually accepted size. In supporting Shapley's radical, and initially often criticized scheme, Pannekoek again put himself into what turned out to be the vanguard of astronomy.

Towards the end of the nineteenth century and in the early years of the twentieth, more professional astronomers turned their attention to galactic structure and the Milky Way.[40] Three changes were crucial for this shift: advances in photography, the growing number of professional astronomers who tackled problems other than those related to positional astronomy and the forging of powerful new mathematical techniques by a small number of astronomers in pursuit of a plausible model of the Galaxy.

Cornelis Easton, a Dutch popularizer of astronomy and a prominent amateur astronomer in his own right, was among those who at the end of the nineteenth century stressed the observational complexities of the stellar system. He sought to connect the overall structure of the Milky Way with the observed fluctuations in the intensity of its light when he observed in different directions, and in so doing argued for a spiral form for the Galaxy. In 1900, for example, Easton presented 'A New Theory of the Milky Way', in the prestigious American-based *Astrophysical Journal* and sketched the Milky Way as a spiral. But when Easton again argued for a spiral structure for the Milky Way in 1913, he not only presented a revised sketch of the Milky Way, but also drew on photographs of the Milky Way.[41]

Around the turn of the twentieth century, a small number of mathematicians and professional astronomers turned to developing models of the galactic system. Among these were a few who employed sophisticated mathematics and emphasized a consistent mathematical account to fashion different versions of the so-called ellipsoidal model. The ellipsoid model came with varying degrees of empirical input. C.V.L. Charlier, at Lund, and Hugo von Seeliger, at Munich, were two of the three chief exponents in the early twentieth century of what Pannekoek would term 'statistical astronomy' in his *A History of Astronomy*.[42]

40 For background on this section, see Smith 2006; and Paul 1993.

41 Easton 1900; and 1913.

42 Pannekoek 1989, 473.

The Dutch astronomer J.C. Kapteyn was recognized as the leading practitioner of statistical astronomy. As we noted earlier, through his skills, drive, and wide range of contacts and collaborators outside the Netherlands, he became an internationally renowned astronomer. He was also, as I have argued elsewhere, 'an astronomer with a grand passion: the solution of the Sidereal Problem',[43] that is, the solution of the problem of 'the present positions and motions of the stars as a stage in the history of a dynamical system (whether in a steady state or not) and the deduction of the presumable history of the system in the past and in the future'.[44] The Astronomical Laboratory at Groningen was designed to allow Kapteyn to tackle the Sidereal Problem.

Kapteyn showed himself to be a new sort of astrophysicist, a practitioner of what we termed earlier as 'Second Stage Astrophysics'. He was not interested in surveys for the sake of surveys or piling-up information for no clear end purpose, in the manner typical of many first-generation astrophysicists. Rather, he sought great masses of accurate data to solve a very major problem. Kapteyn had, therefore, 'married the concern for the diversity of stars, which was so important to astrophysicists, to the traditional values of mathematical astronomers of exactness and rigorous mathematics'.[45] In so doing, Kapteyn had a great influence on Dutch astronomy and astronomers. 'Kapteyn', as Sullivan has argued, 'found a niche. [...] Dutch astronomers were masters at this type of analysis.[46] In Sullivan's view, this sort of research was marked by 'thoroughness, neatness, and precision' and the avoidance of speculation.[47]

Despite his careful approach and concern for errors, Kapteyn's quest to solve 'The Sidereal Problem' ended in a grand failure. Within a decade of Kapteyn's death in 1922, astronomers were agreed that the obscuring matter spread throughout galactic space undermined the reliability of his star-counts and so the distances he derived for distant stars close to the galactic plane.[48] Kapteyn had, of course, been fully aware of the potential seriousness of interstellar absorption for his investigations. He had returned to the question of its existence and nature at various times. Towards the end of his career, however, Kapteyn had been persuaded by Shapley's researches of the colours of stars in remote globular clusters that the effects of a general interstellar absorption are relatively minor. If so, it would have relatively little effect on

43 Smith 2000, 183.

44 Russell 1919, 391. For the context of this paper by Russell, see DeVorkin 2000, 138-152.

45 Smith 2000, 190.

46 Sullivan 2000, 236.

47 Sullivan 2000, 237.

48 van der Kruit 2015; see also Smith 2000, 188.

his derived distances.[49] At the core of Shapley's studies of globular clusters were his estimates of the distances of various sorts of stars within them, and these distances led him to a new and radical view of the stellar system. In early 1918, Shapley told the leading British astrophysicist A.S. Eddington that his determination of the distances to all the known globular clusters had very rapidly settled the 'whole sidereal structure'. The globular clusters, he argued, surrounded and framed the stellar system.[50] Soon after, Shapley was explaining to the Director of the Mount Wilson Observatory that the Galaxy is in effect a collection of star clusters and far bigger than astronomers had believed, some 300,000 light years or so in diameter. The Sun, furthermore, is several tens of thousands of light years away from the centre.[51] Here was a very different picture from the others astronomers discussed in the late 1910s, which always contained a relatively central Sun and which often portrayed the galactic system as lens-shaped and perhaps 30,000 light-years across.

It is misleading, however, to regard the systems of Kapteyn and Shapley as entirely opposed to one another and to assume that astronomers plumped for either Kapteyn's system or Shapley's.[52] As Gingerich remarks, as 'for the divergence between Kapteyn's heliocentric cosmos and Shapley's much vaster galactocentric system, the differences are much more stark in the modern telling than in the historical actuality around 1920.'[53] Both Kapteyn's version of the stellar system and Shapley's Big Galaxy thesis, for example, were larger than the sizes typically quoted (by factors of two and ten). But there were nevertheless major differences in the approach and results that underpinned the two systems. Most significantly, Kapteyn and the statistical astronomers worked outwards from our stellar neighbourhood. Shaply, on the other hand, worked inwards from the Galaxy's outer regions as defined by the globular clusters.[54] Kapteyn's model was well regarded by some astronomers, while many were reluctant, initially at least, to accept Shapley's system as credible. The best known such critic was the Lick Observatory astronomer H.D. Curtis who deployed a range of objections to Shapley's scheme in the so-called Great Debate of April 1920.[55] Pannekoek, however, was one of Shapley's earliest public supporters.

49 Smith 2000, 188; 1982, 57-59; and Paul 1993, 101-106.

50 H. Shapley to A.S. Eddington, 8 January 1918, HS; and Smith 1982, 61.

51 H. Shapley to G.E. Hale, 19 January 1918, HS.

52 Smith 1982, 69.

53 Gingerich 2000, 191.

54 Smith 1982, 68.

55 Hoskin 1976; and Smith 1982, 77-87. Curtis also held strong doubts about Kapteyn's results: Smith 1982, 85.

Pannekoek had already concluded in 1910 that Kapteyn's mathematical approach and assumptions meant an overall, symmetrical ellipsoid shape for the galactic system in which the stars slowly decrease in number as one travels further from the solar neighbourhood. As Tai and van Dongen have argued, Pannekoek instead contended 'that the visual appearance of the Milky Way, with its patchy light structure, completely contradicted such a symmetry. His solution was to focus on specific features of the Milky Way that stood out visually and determine the star distribution function for each of these features individually, while still using Kapteyn's numerical methods.'[56] In examining star clouds in the directions of the constellations of Aquila and Cygnus, Pannekoek decided that instead of a gradual thinning out of stars, there was an increasing number of the fainter stars, thereby contradicting what could be expected from the ellipsoid model.

Pannekoek again emphasized clusters of stars in a paper published in 1919. He noted that the underlying procedure adopted by statistical astronomers was to develop formulae to 'define our star system as a figure of revolution, in which the star density depends on distance and galactic latitude.' Employing this approach, the Sun was represented 'as lying in the midst of a flat star cluster whose densest parts measure some 1000 parsecs.' But Pannekoek, who of course had spent very many hours from his youth studying the intricacies of the Milky Way, emphasized that

> [such a model] is not in accordance [...] with the appearance of the Galaxy. We see the appearance of the Milky Way as a belt of luminous clouds, patches, and drifts, divided by less luminous regions or dark gaps and rifts. If we go in the direction of such a star-cloud, the star-density, after we have left our central cluster, must increase at first on the nearer side of the centre of the cloud, and decrease on the further wide. The aspect of the Milky Way shows that by treating the galactic zone as a whole we intermingle parts of the universe of a great diversity of structure, viz. the aggregation of stars in clouds, separated by regions agreeing perhaps with the galactic poles. In studying the distribution of stars in our universe we must treat the different parts of the Galaxy, especially the great star-clouds and streams, separately.[57]

Pannekoek indeed treated each of the different parts of the Galaxy separately to derive the changes in star density with distance. By this route, he ended-up

56 Tai and van Dongen 2016, 63.
57 Pannekoek 1919, 500.

siding with Shapley. Some of the bright parts of the Milky Way, Pannekoek calculated, were some 40,000 to 60,000 parsecs distant, and 'the starry masses of the Galaxy are spread over space as far as the remotest [globular] clusters, and clearly both belong to one system.' Although Shapley's arguments in favour of an eccentric position of the Sun in the galactic system were, 'contrary to the common view', Pannekoek reckoned that 'Shapley's result is wholly in accordance with the aspect of the Milky Way.'[58]

Shapley, moreover, had not merely expanded the size of the stellar system and placed the Sun in an eccentric position. He had also advanced a dynamic picture of the collection of globular clusters and the Galaxy (what he referred to as a super-system), and Pannekoek looked favourably on this picture too. For Shapley:

> [The] flat form and heterogeneity [of the galactic system], its content of numerous fragmentary systems (open [star] clusters, wide binaries, spectrally-similar groups) of apparently different ages and separate origins, and its control over the motions of the clusters and near-by spirals, have led me for some years to advocate the hypothesis that the Galaxy is a growing composite of disintegrating minor systems.[59]

The globular clusters swing to and fro through the star clouds of the galactic system and on every passage their form changes and their speed is reduced. Over time, the globular clusters are diverted into the galactic regions and 'gradually robbed of their stars' so that they are converted into open star clusters.[60] Globular clusters had not been sighted close to the galactic plane, Shapley pointed out. Maybe the very limited time they spent traversing the galactic plane was the reason.

Shapley's vision of the Galaxy as a growing collection of star clusters was a congenial one for Pannekoek even though it was hardly a well worked-out mathematical model. In what for Pannekoek was a rare moment of speculation, he agreed that perhaps when the globular clusters come from the 'void space into the star-filled galactic regions, [they] are gradually broken up and dispersed into open clusters by the attraction of these stars.'[61] For once,

58 Pannekoek 1919, 507.

59 Shapley 1923, 316. Note that, at the time, Shapley did not believe that spirals were distant galaxies.

60 Shapley 1923, 319.

61 Pannekoek 1919, 500. Pannekoek also noted that Shapley had raised the possibility that the absence of globular clusters from the galactic plane might be explained by the presence of obstructing dark nebulae blocking the view of low-lying globular clusters.

Pannekoek, whose scientific approach can be described as that of a 'close empiricist' very much in the manner of the Dutch school as discussed by Sullivan, agreed with what was an example on Shapley's part of imagination and speculative reasoning.[62]

Other Dutch and Netherlands-based astronomers beside Pannekoek in time also swung behind Shapley's scheme. In May 1922, Shapley met in Leiden with Ejnar Hertzsprung, Pannekoek, and two of Kapteyn's students, including W.J.A. Schouten, who in 1919 had argued that Shaply had overestimated the distances to the globular clusters by a factor of around eight. Shapley, they all decided, was basically correct.[63] Pieter J. van Rhijn, who was a PhD student of Kapteyn's as well as a collaborator of his and his successor at Groningen, however, stuck to his guns. He co-authored 'On the distribution of the stars in space especially at high galactic latitudes' with Kapteyn in 1920, in which they advocated an ellipsoid model for the galactic system. The two of them also argued in 1922 that Shapley had misused the Cepheid variable stars as his main distance indicators and that his distances to them were in fact seven times too big.[64] If that were so, then Shapley's Big Galaxy would have to be shrunk.

In 1922, Kapteyn in some respects pulled together the results of his life's work on the sidereal problem in a paper published in *Astrophysical Journal*. He again argued for an ellipsoid model and concluded that the Sun is close to the centre of the Galaxy, and that the galactic system extends for about 8500 parsecs along the galactic plane and at 1700 parsecs at right angles to the plane before the star density reaches one hundredth of the density in the neighbourhood of the Sun. Even here, after decades of effort to solve the Sidereal Problem Kapteyn still wrote of a 'First attempt at a theory of the arrangement and motion of the sidereal system.'[65]

In 1924, Pannekoek returned to the problem of the distribution of stars within the Galaxy to search for star clusters that, when their light was aggregated, could explain the appearance of the Milky Way. For Pannekoek, the galactic system was to be understood as an accumulation of loose clusters, and so was in line with Shapley's picture, but very different from Kapteyn's ellipsoid.[66]

62 The place of imagination and speculative reasoning in nineteenth-century science has been examined by, among others, Willis 2011. With Shapley's 'Big Galaxy' we see imagination and speculative reasoning in early twentieth-century astronomy.

63 Paul 1981; and van der Kruit 2015, 593. On Schouten's 1919 study, see also Smith 1982, 69.

64 Kapteyn and Rhijn 1922.

65 Kapteyn 1922.

66 Pannekoek 1924. For a commentary on this paper, see Tai and van Dongen 2016, 63-64; and Tai 2017, 242-245.

By the early 1930s, the generally accepted picture of the galactic system was quite different from those in play in the late 1910s and early 1920s. Researches in the second half of the 1920s of a possible galactic rotation, including most importantly by the Swedish astronomer Bertil Lindblad and Dutch astronomer Jan Oort, persuaded astronomers that the galactic system rotates and that the direction of the system's centre is in agreement with the centre of the system of globular clusters as identified by Shapley. Oort placed the centre of the galactic system around 6000 parsecs away, significantly smaller than Shapley's estimate. This difference, most astronomers soon decided in the early 1930s, could be readily explained by the fact that Shapley had taken no account of interstellar absorption in his distance determinations. In 1930, the Lick Observatory astronomer Robert Trumpler advanced arguments in favour of a significant interstellar absorption that other astronomers found convincing.[67] The upshot was that Shapley had overestimated his distances. The Big Galaxy was not as big as he had initially calculated; it needed to be shrunk by about a factor of three. The discrepancy between Oort's and Shapley's estimate of the distance to the centre of the Galaxy now largely disappeared.[68]

In 1932, in his semi-popular book *Kosmos*, Willem de Sitter displayed a diagram of the galactic system provided by Oort. It showed a system with a diameter of around 35,000 parsecs surrounded by globular clusters.[69] This image was strikingly different from those offered by Kapteyn in his two major papers of 1920 and 1922 that, because of his death in 1922, ended his life's work on the sidereal problem. Instead, Kapteyn's model formed just one part of Oort's version of the stellar system.

By the early 1930s and the publication of *Kosmos*, Kapteyn's model was generally reckoned by astronomers to be badly outdated. As Pannekoek put it in his *A History of Astronomy*: 'The main features of the stellar system to which our Sun belongs, its shape and its state of motion, are now established as far different from what had been found in Kapteyn's pioneering investigations.'[70]

The year of the original Dutch publication of Pannekoek's history of astronomy, 1951, was also the year that William W. Morgan, an astronomer at the Yerkes Observatory, and his collaborators provided what astronomers generally agreed was persuasive evidence that the Galaxy has a spiral

67 Trumpler 1930. See also Seeley 1973.
68 Smith 2006, 329.
69 W. de Sitter 1932.
70 Pannekoek 1989, 482.

structure. Numerous astronomers had advocated a spiral structure since the middle of the nineteenth century, including, as we have noted, Cornelis Easton. But it was Morgan's study of the distances to H II regions and bright O and B stars within them that astronomers regarded as decisive and earned him an ovation when he presented a paper on his results at a meeting of the American Astronomical Society in 1951.[71]

At the same meeting, Oort delivered an invited lecture on the 'Problems of Galactic Structure'. Here, he gave a breakdown of what he regarded as the three key phases of the developments in knowledge of the galactic system. The first had been initiated by Kapteyn's researches, though he 'did *not* reach the principal aim he had set out for, because of the unexpected strength of interstellar absorption near the galactic plane.' Shapley had begun the second great development with his investigations of the arrangement of the globular clusters. 'It seems that at present', Oort argued, 'a third phase in the development of galactic research has begun by the successful reception of radiation at radio frequencies. This research is still in its early infancy.'[72] In hindsight, we can now see that Oort was surely correct. The study of galactic structure at radio wavelengths, in which Dutch astronomers were very much in the forefront, *did* open up extremely important new research avenues.[73] But the start of observations of the Milky Way in other wavelength ranges, especially in the infrared, would of course later prove to be crucial additions to the optical and radio. As Pannekoek argued from the vantage point of 1951:

> The establishment of the galactic system is not the end, but rather a beginning of research, specifying a task. Just as many centuries were needed after the establishment of the solar system for the investigation of its contents, structure, and details, its laws and characteristics, so it is now with the stellar system.[74]

## Conclusions

J.C. Kapteyn was the leading Dutch astronomer of the late nineteenth and early twentieth century. When he assumed the professorship of astronomy and theoretical mechanics at Groningen in 1878, many astronomers were

71 Gingerich 1985; and Smith 2006, 331.
72 Oort 1952, 233; emphasis in the original.
73 On the early development of radio astronomy, see Sullivan 2009.
74 Pannekoek 1989, 482.

sceptical of, and some actively hostile towards, astrophysics. But Kapteyn pressed on and in the end helped to establish a new sort of astrophysics, what we described earlier as 'Second Stage Astrophysics', in which he sought to take full account of the range of different sorts of stars as well as analyse rigorously their properties. He had done so despite a lack of resources at Groningen, and had thereby provided one answer to the question of how to be an effective astronomer without a telescope.

Kapteyn died in 1922. By this time, we have seen that a new sort of astrophysics had started to emerge, one that had been given its initial impetus by the researches of Meghnad Saha. We termed this 'Third Stage Astrophysics'. Pannekoek effectively solved the problem of how to be relevant and perform 'competitive' research at the University of Amsterdam despite his lack of resources, including the complete lack of telescopes and an observatory, and in the face of the rise of American astronomy, by rapidly grasping the importance of Saha's path-breaking researches and both developing and applying to actual stars this novel sort of astrophysics. Pannekoek's initial expertise had been in positional astronomy, but he became one of the earliest practitioners of 'Third Stage Astrophysics'.

Pannekoek also positioned himself as a modern astronomer by quickly realizing the importance of Shapley's new picture of the stellar system, advanced publicly in 1918. The next year, Pannekoek became one of the first astronomers to publish additional evidence in support of Shapley, and in so doing underlined the severe limitations of the models developed by the statistical astronomers, including Kapteyn. Pannekoek the astronomer, then, was both very much of, as well as a maker of, his time.

## Archives

HS Papers of Harlow Shapley, 1906-1966, HUG 4773. Harvard University Archives.

## Bibliography

Baneke, David. 2005. '"Als bij toverslag". De renovatie en nieuwe bloei van de Leidse Sterrewacht 1918-1924'. *BMGN – Low Countries Historical Review* 120(2): 207-225.

—. 2010. 'Teach and Travel: Leiden Observatory and the Renaissance of Dutch Astronomy in the Interwar Years'. *Journal for the History of Astronomy* 41(2): 167-198.

—. 2015. *De ontdekkers van de hemel. De Nederlandse sterrenkunde in de twintigste eeuw*. Amsterdam: Prometheus Bert Bakker.

Becker, Barbara J. 2011. *Unravelling Starlight: William and Margaret Huggins and the Rise of the New Astronomy*. Cambridge: Cambridge University Press.

Clerke, Agnes Mary. 1885. *A Popular History of Astronomy during the Nineteenth Century*. Edinburgh: Adam & Charles Black.

DeVorkin, David H. 1998. 'The American Astronomical Community'. Review of *American Astronomy: Community, Careers, and Power, 1859-1940*, by John Lankford. *Journal for the History of Astronomy* 29(4): 389-392.

—. 2000. *Henry Norris Russell: Dean of American Astronomers*. Princeton, NJ: Princeton University Press.

—. 2010. 'Extraordinary Claims Require Extraordinary Evidence: C.H. Payne H.N. Russell and Standards of Evidence in Early Quantitative Spectroscopy'. *Journal of Astronomical History and Heritage* 13(2): 139-144.

DeVorkin, David H., and Ralph Kenat. 1983a. 'Quantum Physics and the Stars I: The Establishment of a Stellar Temperature Scale'. *Journal for the History of Astronomy* 14(2): 102-132.

—. 1983b. 'Quantum Physics and the Stars II: Henry Norris Russell and the Abundances of the Elements in the Atmospheres of the Sun and Stars'. *Journal for the History of Astronomy* 14(3): 180-222.

Dewhirst, David W. 1985. 'Meridian Astronomy in the Private and University Observatories of the United Kingdom: Rise and Fall'. *Vistas in Astronomy* 28(1): 147-158.

Donnelly, Kevin. 2014. 'On the Boredom of Science: Positional Astronomy in the Nineteenth Century'. *British Journal for the History of Science* 47(3): 479-503.

Easton, Cornelis. 1900. 'A New Theory of the Milky Way'. *Astrophysical Journal* 12(2): 136-158.

—. 1913. 'A Photographic Chart of the Milky Way and the Spiral Structure of the Galaxy'. *Astrophysical Journal* 37(2): 105-118.

Gingerich, Owen. 1985. 'The Discovery of the Spiral Arms of the Milky Way'. In *The Milky Way Galaxy: Proceedings of the 106th Symposium, Groningen, Netherlands, May 30-June 3, 1983*, ed. by Hugo van Woerden, Ronald J. Allen, and W. Butler Burton. Dordrecht: D. Reidel, 59-70.

—. 2000. 'Kapteyn, Shapley, and Their Universes'. In *The Legacy of J.C. Kapteyn: Studies on Kapteyn and the Development of Modern Astronomy*, ed. by Pieter C. van der Kruit and Klaas van Berkel. Dordrecht: Kluwer, 191-212.

Helden, Albert van. 1984. 'Building Large Telescopes, 1900-1950'. In *The General History of Astronomy, Volume 4: Astrophysics and Twentieth-Century Astronomy to 1950: Part A*, ed. by Owen Gingerich. Cambridge: Cambridge University Press, 134-152.

Hentschel, Klaus. 2002. *Mapping the Spectrum: Techniques of Visual Representation in Research and Teaching*. Oxford: Oxford University Press.

Herrmann, Dieter B. 1975. 'Zur Vorgeschichte des Astrophysikalischen Observatorium Potsdam (1865 bis 1874)'. *Astronomische Nachrichten* 296(6): 245-259.

Holmberg, Gustav. 1999. *Reaching for the Stars: Studies in the History of Swedish Stellar and Nebular Astronomy, 1860-1940*. Lund: Lund University.

Hoskin, Michael. 1976. 'The "Great Debate": What Really Happened'. *Journal for the History of Astronomy* 7(3): 169-182.

Hufbauer, Karl. 1993. *Exploring the Sun: Solar Science since Galileo*. Baltimore: Johns Hopkins University Press.

Kapteyn, Jacobus C. 1922. 'First Attempt at a Theory of the Arrangement and Motion of the Sidereal System'. *Astrophysical Journal* 55(4): 302-328.

Kapteyn, Jacobus C., and Pieter J. Rhijn. 1922. 'The Proper Motions of δ-Cephei Stars and the Distances of the Globular Clusters'. *Bulletin of the Astronomical Institutes of the Netherlands* 1(8): 37-42.

Kruit, Pieter C. van der. 2015. *Jacobus Cornelius Kapteyn: Born Investigator of the Heavens*. Cham: Springer.

Kruit, Pieter C. van der, and Klaas van Berkel, eds. 2000. *The Legacy of J.C. Kapteyn: Studies on Kapteyn and the Development of Modern Astronomy*. Dordrecht: Kluwer.

Lankford, John. 1984. 'The Impact of Photography on Astronomy'. In *The General History of Astronomy: Volume 4, Astrophysics and Twentieth-Century Astronomy to 1950: Part A*, ed. by Owen Gingerich. Cambridge: Cambridge University Press, 16-39.

—. 1997. *American Astronomy: Community, Careers, and Power, 1859-1940*. Chicago: University of Chicago Press.

Laurie, Philip S., Jacques Lévy, Aleksandr A. Mikhailov, Howard Plotkin, Deborah Warner, Trudy E. Bell, and Dieter B. Herrmann. 1984. 'Astronomical Institutions'. In *The General History of Astronomy: Volume 4, Astrophysics and Twentieth-Century Astronomy to 1950: Part A*, ed. by Owen Gingerich. Cambridge: Cambridge University Press.

Meadows, Arthur Jack. 1973. *Science and Controversy. A Biography of Sir Norman Locky*. Cambridge, MA: MIT Press.

—. 1984a. 'The New Astronomy'. In *The General History of Astronomy: Volume 4, Astrophysics and Twentieth-Century Astronomy to 1950: Part A*, ed. by Owen Gingerich. Cambridge: Cambridge University Press, 59-72.

—. 1984b. 'The Origins of Astrophysics'. In *The General History of Astronomy: Volume 4, Astrophysics and Twentieth-Century Astronomy to 1950: Part A*, ed. by Owen Gingerich. Cambridge: Cambridge University Press, 3-15.

Miller, Howard S. 1970. *Dollars For Research: Science and Its Patrons in Nineteenth Century America*. Seattle: University of Washington Press.

Naik, Pramod V. 2017. *Meghnad Saha, His Life in Science and Politics*. Cham: Springer.

Narrien, John. 1833. *An Historical Account of the Origin and Progress of Astronomy.* London: Baldwin & Cradock.

Oort, Jan H. 1952. 'Problems of Galactic Structure'. *Astrophysical Journal* 116(2): 233-250.

Osterbrock, Donald E. 1999. *Yerkes Observatory, 1892-1950: The Birth, Near Death, and Resurrection of a Scientific Research Institution.* Chicago, IL: University of Chicago Press.

Pannekoek, Anton. 1906. 'The Relation between the Spectra and the Colours of the Stars'. *Proceedings of the Section of Sciences, Koninklijke Akademie van Wetenschappen te Amsterdam* 9(1): 292-302.

—. 1919. 'The Distance of the Milky Way'. *Monthly Notices of the Royal Astronomical Society* 79(7): 500-507.

—. 1922. 'Ionization in Stellar Atmospheres'. *Bulletin of the Astronomical Institutes of the Netherlands* 1(19): 107-118.

—. 1924. *Researches on the Structure of the Universe 1. The Local Star System Deduced from the Durchmusterung Catalogues.* Publications of the Astronomical Institute of the University of Amsterdam 1. Amsterdam: Stadsdrukkerij.

—. 1926. 'The Ionization Formula for Atmospheres Not in Thermodynamic Equilibrium'. *Bulletin of the Astronomical Institutes of the Netherlands* 3(110): 207-209.

—. 1927. 'The Determination of Absolute Line Intensities in Stellar Spectra'. *Bulletin of the Astronomical Institutes of the Netherlands* 4(121): 1-6.

—. 1989. *A History of Astronomy.* New York: Dover.

Paul, E. Robert. 1981. 'The Death of a Research Program: Kapteyn and the Dutch Astronomical Community'. *Journal for the History of Astronomy* 12(2): 77-94.

—. 1993. *The Milky Way Galaxy and Statistical Cosmology, 1890-1924.* Cambridge: Cambridge University Press.

Pyenson, Lewis. 1989. *Empire of Reason: Exact Sciences in Indonesia, 1840-1940.* Leiden: Brill.

Rothenberg, Marc. 1974. 'The Educational and Intellectual Background of American Astronomers, 1825-1875'. PhD Thesis, Bryn Mawr, PA: Bryn Mawr College.

Rue, Warren de la. 1861. 'Proceedings of the Chemical Society'. *Chemical News* 4(92): 130-133.

Rufus, W. Carl. 1921. 'Proposed Periods in the History of Astronomy in America'. *Popular Astronomy* 29(288): 393-404, 468-475.

Russell, Henry Norris. 1919. 'Some Problems of Sidereal Astronomy'. *Proceedings of the National Academy of Science* 5(10): 391-416.

Schaffer, Simon. 1988. 'Astronomers Mark Time: Discipline and the Personal Equation'. *Science in Context* 2(1): 115-145.

Seeley, Daniel H. 1973. 'The Development of Research on the Interstellar Medium c. 1900-1940: Diffuse Nebulae, Interstellar Gas, and Interstellar Extinction'. PhD Thesis, Boston: Boston University.

Shapley, Harlow. 1923. 'The Galactic System'. *Popular Astronomy* 31(305): 316-327.

Sitter, Willem de. 1932. *Kosmos: Course of Six Lectures on the Development of Our Insight into the Structure of the Universe, Delivered for the Lowell Institute in Boston, in November 1931*. Cambridge, MA: Harvard University Press.

Sitter, Wolter Reinhold de. 2000. 'Kapteyn and de Sitter: A Rare and Special Teacher-Student and Coach-Player Relationship'. In *The Legacy of J.C. Kapteyn: Studies on Kapteyn and the Development of Modern Astronomy*, ed. by Pieter C. van der Kruit and Klaas van Berkel. Dordrecht: Kluwer.

Smith, Robert W. 1982. *The Expanding Universe: Astronomy's 'Great Debate,' 1900-1931*. Cambridge: Cambridge University Press.

—. 2000. 'Kapteyn and Cosmology'. In *The Legacy of J.C. Kapteyn: Studies on Kapteyn and the Development of Modern Astronomy*, ed. by Pieter C. van der Kruit and Klaas van Berkel. Dordrecht: Kluwer, 175-190.

—. 2003. 'The Remaking of Astronomy'. In *Cambridge History of Science, Volume 5: Modern Physical and Mathematical Science*, ed. by Mary Jo Nye. Cambridge: Cambridge University Press.

—. 2006. 'Beyond the Big Galaxy: The Structure of the Stellar System 1900-1952'. *Journal for the History of Astronomy* 37(3): 307-342.

Struve, Otto. 1943. 'Fifty Years of Progress in Astronomy'. *Popular Astronomy* 51(509): 469-481.

Sullivan, Woodruff T., III. 2000. 'Kapteyn's Influence on the Style and Content of Twentieth Century Dutch Astronomy'. In *The Legacy of J.C. Kapteyn: Studies on Kapteyn and the Development of Modern Astronomy*, ed. by Pieter C. van der Kruit and Klaas van Berkel. Dordrecht: Kluwer.

—. 2009. *Cosmic Noise: A History of Early Radio Astronomy*. Cambridge: Cambridge University Press.

Tai, Chaokang. 2017. 'Left Radicalism and the Milky Way: Connecting the Scientific and Socialist Virtues of Anton Pannekoek'. *Historical Studies in the Natural Sciences* 47(2): 200-254.

Tai, Chaokang, and Jeroen van Dongen. 2016. 'Anton Pannekoek's Epistemic Virtues in Astronomy and Socialism: Personae and the Practice of Science'. *BMGN – Low Countries Historical Review* 131(4): 55-70.

Trumpler, Robert J. 1930. 'Preliminary Results on the Distances, Dimensions and Space Distribution of Open Star Clusters'. *Lick Observatory Bulletin* 14(420): 154-188.

Werrett, Simon. 2010. 'The Astronomical Capital of the World: Pulkovo Observatory in the Russia of Tsar Nicholas I'. In *The Heavens on Earth: Observatories and Astronomy in Nineteenth-Century Science and Culture*, ed. by David Aubin, Charlotte Bigg, and Heinz Otto Sibum. Durham, NC: Duke University Press.

Whewell, William. 1833. 'Address to the British Association for the Advancement of Science'. In *Report of the Third Meeting of the British Association for the*

*Advancement of Science; Held at Cambridge*. London: British Association for the Advancement of Science, xi-xxxii.

Willis, Martin. 2011. *Vision, Science and Literature, 1870-1920: Ocular Horizons*. London: Pickering & Chatto.

Wright, Helen. 2003. *James Lick's Monument: The Saga of Captain Richard Floyd and the Building of the Lick Observatory*. Cambridge: Cambridge University Press.

## About the Author

**Robert W. Smith** is Professor of History in the Department of History and Classics at the University of Alberta. He has written extensively on the history of nineteenth- and twentieth-century astronomy, and his books include *The Expanding Universe: Astronomy's 'Great Debate' 1900-1931*, *The Space Telescope: A Study of NASA, Science, Technology, and Politics*, and *Hubble Cosmos*. He has been a Fellow of the National Humanities Center and in 2007 was the Lindbergh Chair of Aerospace History at the Smithsonian Institution's National Air and Space Museum.

# 7 'A New Scientific Conception of the Human World'

## Anton Pannekoek's Understanding of Scientific Socialism

*Bart van der Steen*

**Abstract**

This paper sets out to reconstruct Pannekoek's understanding of scientific socialism in order to reconnect Pannekoek's political and astronomical work. It does so through a close reading of Pannekoek's early socialist essays, where he repeatedly referred to socialism's scientific character, explaining it in various ways. From this reading, three different but closely related conceptions of scientific socialism can be abstracted. For Pannekoek, socialism was scientific in that it embraced modern science, in that it supposedly uncovered the laws of societal development, and in that it foretold the advent of socialism. The paper shows how, for Pannekoek, socialism was the only ideology with a true interest in scientific research and findings. This line of reasoning allowed Pannekoek to connect his astronomical and socialist persona.

**Keywords:** Anton Pannekoek, scientific socialism, Isaac Asimov, scientific worldview

## The Role of Science in the Two Lives of Anton Pannekoek

Anton Pannekoek was both a renowned astronomer and a famous socialist, but few attempts have been made to connect his two fields of work. One possible way to do so is by analysing his understanding of the word 'science', because Pannekoek did not only see his astronomical work as a scientific undertaking – he also defined his brand of socialism as 'scientific'. This chapter therefore sets out to reconstruct Pannekoek's understanding of

Tai, Chaokang, Bart van der Steen, and Jeroen van Dongen (eds), *Anton Pannekoek: Ways of Viewing Science and Society*. Taylor & Francis Group, 2019
DOI: 10.5117/9789462984349_CH07

scientific socialism and asks what the term 'science' meant in this context. It does so through a close reading of Pannekoek's early socialist essays, where he repeatedly referred to socialism's scientific character, explaining it in various ways. From this reading, three different but closely related conceptions of scientific socialism can be abstracted, which reinforce each other to a certain extent.

From the early 1900s to the early 1920s, Pannekoek was an internationally renowned socialist.[1] Before the outbreak of World War I, he taught at the cadre school of the Social Democratic Party of Germany (SPD) in Berlin and published in Karl Kautsky's *Die Neue Zeit*, the main theoretical journal of the international socialist movement. There, Pannekoek debated issues of socialist politics and theory with the likes of Kautsky, Rosa Luxemburg, and Eduard Bernstein. Soon after the Russian Revolution, Pannekoek became one of the most prominent radical socialist critics of Leninism, which he denounced as authoritarian and elitist. As an alternative, he developed a strand of revolutionary thinking called council communism, which emphasized workers' self-emancipation and organization in councils that should act independently of trade unions and political parties. After his break with Leninism, Pannekoek was marginalized within the labour movement, his influence limited to the small groups of council communists that remained politically active throughout the interwar period and after.[2] Nevertheless, Pannekoek remained committed to revolutionary politics throughout his life and continued to analyse and comment on the problems of socialist politics and theory until his death in 1960.

While Pannekoek's fame within the socialist labour movement declined, he became an ever more prominent astronomer. In 1919, he gained a position at the University of Amsterdam, after which he became endowed professor in 1919 and full professor in 1932. As a scientist, he not only endeavoured to map the Milky Way and analyse stellar spectra, but also authored various works on the history of astronomy and the philosophy of science.[3]

Pannekoek thus specialized in two disparate fields of work, and his biographers have found it difficult to reconcile the two. The fact that the history of socialism and the history of science have traditionally been remote and distinct fields of research have made this all the more difficult. Most

1 Gerber 1989; van Berkel 2001; Welcker 1986.

2 For Anton Pannekoek and the history of council communism, see: Bock 1993; Bourrinet 2017; Gerber 1988; 1989; Kool 1970; Mergner 1971.

3 van den Heuvel 1982; Tai 2017.

historical research has focused on Pannekoek's political life. As a result, traditional historiography has upheld a divide that was originally introduced by Pannekoek himself. From World War I onwards, he rigidly separated his two fields of work, especially to the outside world. He started to publish his political work under pseudonyms, and in 1944 he authored two different autobiographies; a political and an astronomical one.[4] In explaining Pannekoek's division of political and astronomical persona, various contributors in this volume – most notably Klaas van Berkel, David Baneke, and Edward van den Heuvel – have pointed out that Pannekoek's scientific career was hampered at various moments because of political controversies.[5] Only recently attempts have been made to integrate the two historiographies on Pannekoek and to move beyond the divide. In fact, this volume counts as one of the first attempts to reconnect Pannekoek's two fields of work.

Chaokang Tai, who is currently preparing a new biographical study of Pannekoek, is right in pointing at Pannekoek's theory of knowledge as a way to reconnect Pannekoek's political and astronomical work.[6] Both in his socialist writings and in his writings on (the history of) astronomy, Joseph Dietzgen (1828-1888) and his theory of knowledge acted as a source of inspiration. Dietzgen was a tanner and self-educated worker-philosopher who had corresponded with Marx and coined the term dialectical materialism.[7] Pannekoek was heavily influenced by the works of Dietzgen, who had claimed, mainly in *Das Wesen der menschlichen Kopfarbeit* (1869), that the human mind organized knowledge by abstracting continuously from the particular to the general.[8] This principle not only informed Pannekoek's theory of knowledge, but also his conception of scientific research. Pannekoek even claimed that Dietzgen had transformed philosophy from a speculative endeavour into a 'natural science', stating: 'It is the merit of

4 The two autobiographies were published in 1982 in one volume with two separate, biographical introductions. Pannekoek 1982; Sijes 1982; van den Heuvel 1982. Some of Pannekoek's most important texts were published anonymously or under pseudonyms, for example, as Aartsz 1946. It should not be overlooked, however, that other factors played a role in this process as well. Within the council communist movement, for example, texts were published anonymously as a rule to show that the texts were outcomes of collective ideological labour and discussions. See Brendel 1974.

5 Klaas van Berkel, 'Utopianism in Anton Pannekoek's Socialism and Astronomy', in this volume, 75-86; David Baneke, 'Pannekoek's One Revolution', in this volume, 87-108; Edward P.J. van den Heuvel, 'Anton Pannekoek's Astronomy in Relation to his Political Activities, and the Founding of the Astronomical Institute of the University of Amsterdam', in this volume, 25-50. See also Baneke 2004.

6 Tai 2017; Tai and van Dongen 2016.

7 Burns 2002; Schaaf 1993.

8 Bock 1992; Schaaf 1978.

Dietzgen to have raised philosophy to the position of a natural science, the same as Marx did with history.'[9]

The last paragraph may serve to illustrate that, next to Pannekoek's theory of knowledge, his concept of science offers another way of bridging the gap between the socialist Pannekoek and the astronomer Pannekoek. It has often been overlooked that Pannekoek published his first history of astronomy, 'Die Entwicklung des Weltalls', as a feuilleton in *Die Neue Zeit*.[10] Approaching the topic from another angle, it is hard to miss that Pannekoek's 1951 *History of Astronomy* ended with an overt political statement:

> It is about time that man, by establishing a free, self-governing world community of productive labour and by assuring itself of material prosperity in abundance, liberates all spiritual forces for the perfection of its knowledge of nature and especially the science of the universe.[11]

Both examples suggest that, for Pannekoek, the boundaries between the two fields of work were not as strong as has often been suggested by his biographers.

This chapter analyses and contextualizes Pannekoek's definitions of scientific socialism through a close reading of Pannekoek's early socialist works, with a focus on his writings from before 1918. After 1918, Pannekoek remained politically active, but his conception of scientific socialism did not change significantly. Pannekoek's politics changed, from supporting social democracy to Bolshevism to council communism. His aversion to formal organization grew and he emphasized ever more the need for workers to act and organize independently. Even so, his commitment to orthodox Marxist principles remained and his way of analysing political situations stayed basically the same. In breaking with Lenin, for example, Pannekoek argued that Russia was industrially underdeveloped so that it could not spawn a true workers' revolt, which according to him also explained Lenin's 'backward' ideas of a hierarchically organized vanguard revolutionary party. His political break with Lenin and his emphasis on workers' councils was

9 Pannekoek 1906c, 28. Compared with the German version, Pannekoek 1903. Digitized versions of many publications of Pannekoek can be found on the website of the Association Archives Antonie Pannekoek: www.aaap.be (accessed 16 March 2018).

10 Pannekoek 1908-1909.

11 Pannekoek 1951, 432. This was brought to my attention by Klaas van Berkel, 'Utopianism in Anton Pannekoek's Socialism and Astronomy', in this volume, 75-86.

politically innovative, but rooted in orthodox Marxist ideology.[12] The latter, his conception of socialism as a system, is the focus of this contribution. Pannekoek emphasized socialism's scientific character most strongly in his early works, which is the reason for the focus on these works in this contribution. It needs to be taken into account that his works were published in various languages, and that the word 'science' had different connotations in these languages. This potential problem is corrected by comparing the same texts in different languages.[13]

## Debating Science in Socialism and Academia

When Pannekoek became politically active in the late nineteenth century, what should count as science was very much in flux. This is exemplified by two debates on science that Pannekoek responded to with his texts: one within the labour movement and one among academics. First of all, the term 'science' was used within the socialist labour movement to discern between (orthodox) Marxists and ethical or utopian socialists, with the former accusing the latter of merely 'dreaming up' alternative social models of society instead of truly engaging with contemporary social conflicts. In this context, the Marxists often criticized attempts by 'utopian' socialists to set up self-sustaining and thus isolated socialist communes. According to the Marxists, the utopian socialists thus reduced socialism to blueprints of ideal society, while they themselves based their actions on (scientific interpretations of) current social developments. The main point that Marxists wanted to make was that socialism was, in their view, no longer one social ideal amongst many others, but a future society that would organically grow out of the old, through social developments and conflicts that could be measured, analysed, and anticipated upon.[14] These claims reflected the growing authority of (natural) science in society, and the growing belief that science had a role to play in politics.

12 van der Steen 2006.

13 Such a comparison between versions uncovers interesting differences. In the English translation of Pannekoek's *Marxism and Darwinism* by Nathan Weiser, it is stated: '[T]he teachings of Darwin and of Marx, the one in the domain of the organic world and the other upon the field of human society, raised the theory of evolution to a positive science.' Pannekoek 1912b, 18. The Dutch and German version give the term scientific a different twist, which becomes clear when we re-translate the German passage: 'The scientific meaning of both Marxism and Darwinism lies in their implementation of the principle of evolution, here in the organic world, there in the field of society.' Pannekoek 1909b, 12. Weiser's translation gives the text a bolder character, by speaking of Darwinism and Marxism as a 'positive science' rather than of their 'scientific meaning'.

14 Engels 1880.

The second debate was closely related to the first, but developed in a different setting, namely that of academia. Among social scientists and historians there was a dispute on the works and claims of the positivist philosopher Auguste Comte. An important claim in Comte's philosophy was that the social sciences and history could and should adhere to the same scientific parameters as the natural sciences. In other words, Comte believed that society developed according to certain laws that functioned in the same way as laws of nature – and he felt it was the task of social scientists to uncover these laws. His followers claimed that, ultimately, the social sciences were to develop into a natural science of society, which implied that it could formulate laws of social development, with both the power to explain past developments and forecast future ones. Others, however, heavily criticized this stance. They emphasized the different nature of both fields and claimed that the social sciences could only produce (inherently contested) interpretations and 'images' of society.

In both debates, Pannekoek clearly took his stance, claiming that Marx had transformed both socialism and the social sciences into natural sciences. Even more so, John Gerber notes that Pannekoek 'uses the terms Marxism, spiritual science, social science, and historical materialism interchangeably'.[15] Pannekoek believed that this transformation of the social sciences enabled him and others to abstract 'certain laws and rules' from the past, in order to 'say something about future developments'.[16]

## Scientific Socialism in Pannekoek's Era

The concept of scientific socialism was introduced by Marx and Engels. In the brochure *Socialism: Utopian and Scientific* (1880), Engels claimed that two 'great discoveries' by Marx had turned socialism into a science: Historical materialism and class struggle. As a result, socialism was 'no longer an accidental discovery of this or that ingenious brain'.[17] Instead, historical materialism explained 'the historico-economic succession of events' as a result of class struggle, and even identified 'the means of ending the conflict' – a socialist revolution after which the progress of mankind could unfold unhampered. 'With these discoveries', Engels thus

15 Gerber 1989, 20.

16 Pannekoek 1906a, 25. Compared with the Dutch translation, Pannekoek 1907.

17 Engels 1880.

stated confidently, 'socialism became a science'. Engels saw scientific socialism as 'the theoretical expression of the proletarian movement'. Its task was 'to impart' on the movement that 'universal emancipation is the historical mission of the modern proletariat'. There is a duality in this line of reasoning. Scientific socialism is both a means to understand how historical developments lead to the advent of socialism, and the task of discerning this message to the proletariat. The importance of the latter for the former is emphasized by Pannekoek's claim that the proletariat's knowledge of its coming triumph is crucial for its success in struggle. Pannekoek's writings on scientific socialism are thus characterized by a similar duality.

The extent to which Marx and Engels themselves saw their socialism as scientific and were convinced that socialism was imminent has been a topic of debate. Charles Elkins, for example has vehemently opposed such a view, stating: 'Marx and Engels never claimed for their theories the status of "exact science". They were always careful to describe the "laws" of historical development as "tendencies".'[18] At the same time, Marx and Engels themselves contributed to such confusion, among others by titling a pamphlet *Socialism: Utopian and Scientific*.

Either way, the generation of Marxists that followed *did* see Marxism as (an almost natural) science, stressing the imminence of socialism and its explanatory power. A central aspect of Marxism in Pannekoek's era was the premise that human consciousness was determined by social conditions, and that changes in the latter would ultimately bring about a revolution in the former. Could social and cultural developments thus be explained by changes in the economy alone? Even among prominent Marxists this sometimes led to confusion. In one instance, for example, Engels felt forced to nuance the relationship between social conditions and consciousness. In a letter to Joseph Bloch, Engels wrote: 'According to the materialist conception of history, the *ultimately* determining element in history is the production and reproduction of real life. Other than this neither Marx nor I have ever asserted.'[19]

Still, confusion remained and critics lashed out at Marxism, writing it off as a reductionist and determinist philosophy. Rudolf Stammler, for example, compared the Marxist labour movement of the 1890s to a cult that strove to bring about a lunar eclipse, asking why socialists would labour for a

18 Elkins 1976, 32.

19 Engels to Bloch, 21 September 1890, quoted in: Marx, Engels, and Lenin 1972, 294-296; emphasis in the original. See also: Lukács 1923.

revolution that was bound to unfold either way.[20] Marxists shrugged off these criticisms as reactionary and bourgeois. But even so, some of it must have stuck, for Pannekoek was not entirely convinced by Marxism at first. Only after reading Dietzgen, Pannekoek was fully convinced: 'Here I found for the first time everything that I had been looking for […] I was able to completely clarify my conception of the mutual relationship between Marxism and a theory of knowledge and develop it into a unified whole.'[21] Scholars such as Hans Manfred Bock have claimed that the 'Marx-Dietzgen-Synthesis' of Pannekoek enabled him to overcome the pitfalls of classical Marxism.[22] Pannekoek's early socialist writings, however, show that he remained firmly within the boundaries of orthodox Marxism: He remained committed to the idea that social conditions determined consciousness, that societies pass through certain stages of development, and that socialism would be the ultimate outcome of class struggle.[23]

## Socialism as Scientific Certainty

Pannekoek stressed the scientific character of his socialism in his early works. According to him, Marx had turned socialism into a science that could make claims on social developments in the same way as the natural sciences. In 1912, he described socialism as 'a new scientific conception of the human world',[24] but his equation of socialism with natural sciences reached its apex in 1906, when he stated that scientific socialism was able to 'make some prediction about the future':

> When we speak about the future […] we do not ask: how do we wish to shape the future? Instead, we ask: what will happen in the future. Scientific socialism is the teaching of social development. It has won certain views from the history of society, abstracted certain laws and rules, and these rules and laws allow us to make some prediction about the future, and draw conclusions about how society will be by that time, independent of our desires and wishes.[25]

20 The Russian Marxist Georgi Plekhanov tried to rebuke the critique, but several of the abovementioned issues remained a source of debate well into the 1930s. See Plekhanov 1898.
21 Pannekoek 1982, 94.
22 Bock 1992.
23 van der Steen 2006.
24 Pannekoek 1912a, 4.
25 Pannekoek 1906a, 25.

In another text from the same year, Pannekoek reasserted his views:

> Many have dreamed [...] of a better and brotherly future world [...] without knowing our worldview. The most outstanding quality of the proletarian view of life and [political] struggle lies not in the fact that we want a socialist society, but in the fact that we can see it coming and predict its advent with scientific certainty. Not the wish, but the knowledge that the wish will be fulfilled, and in what way, is the most valuable of our views.[26]

Even so, there was a circular side to his line of reasoning, because Pannekoek believed that 'we can only triumph by fully developing our means of struggle', the most important of which was 'the education [Aufklärung] of the masses'.[27] Scientific socialism taught that the proletariat would bring about revolution, but it could only do so after it had learned that it would ultimately do so. Even more so, Pannekoek at one time seemingly suggested that socialism's scientific character lay exactly in the fact that it could convince the proletariat that their triumph was imminent:

> The proletariat not only needs to long for [herbeisehnen] a better order; historical materialism gives the proletariat the certainty that such an order will come, since the development of the economy contributes to and makes possible its attainment. In this manner, socialism ceases to be a utopia and becomes a science.[28]

For Pannekoek, socialism's scientific character lay not in the certainty that socialism would come, but in its analysis of social developments, which 'showed' that socialism was imminent. Even so, the certainty of socialism's coming and socialism's scientific character were closely connected.

Pannekoek was not unique in voicing such claims. In fact, the idea that historical materialism had enabled socialists to abstract certain laws of social development from history formed the mainstay of Marxist thinking well into the 1930s. But Pannekoek was also an astronomer, who had practical experience with natural science research. This raises the question why he never specified these laws of historical development. His own natural scientific background could have caused him to see the differences between socialism and natural science, rather than their similarities. Asking this question and looking further

26 Pannekoek 1906b, 20-21; Compared with Dutch version, Pannekoek 1905, 31.

27 Pannekoek 1906b, 6. Compared with Dutch version, Pannekoek 1905, 6.

28 Pannekoek 1915.

into Pannekoek's definitions of socialism's scientific character reveals that Pannekoek defined scientific socialism in three different ways: as a science that 'proved' that socialist revolution is imminent, as a method for analysing past and present social developments, and as a worldview that strove for truth in scientific research. It is the first definition of scientific socialism that is most controversial. While Pannekoek's other two definitions emphasize the worldview and method of his socialism, the first one is most ambitious and raises most questions. For if socialism claims the ability to abstract laws from past development and project them onto the future, a future in which socialism reigns supreme, the question inherently comes up why socialists have not been able to determine these laws and predict future developments.

## Marxism as Science: The Concept of Psycho-history

As mentioned, scientific socialism claimed to reveal the laws of social development, but these laws were never precisely articulated, nor were these laws ever operationalized by anyone, including Pannekoek. This divide between the claims of scientific socialism and its practice can be illustrated through the concept of psycho-history, which was introduced by the American science fiction author Isaac Asimov in the early 1940s. Between 1942 and 1950, Asimov wrote a series of stories that would eventually result in the Foundation trilogy, and ultimately into the Foundation saga – 'One of the most staggering achievements in modern SF' according to *The Times*.[29] These stories revolved around the work of Hari Seldon, who had developed an advanced method of mathematical deduction in order to calculate future developments – akin to what we would now call 'big data' science. Dubbing this fictional science 'psycho-history', Asimov explained its essence as follows: 'Psycho-history was the quintessence of sociology, it was the science of human behaviour reduced to mathematical equations'. Its basic principle was that 'the individual human being is unpredictable, but the reactions of human mobs [...] could be treated statistically'.[30] In this

29 The original stories were published in Astounding Science Fiction between 1942 and 1950 and subsequently published in book form as: *Foundation* (1951), *Foundation and Empire* (1952), *Second Foundation* (1953). In the late 1980s and early 1990s, two sequels and two preludes were published: *Foundation's Edge* (1982), *Foundation and Earth* (1986), *Prelude to Foundation* (1988), *Forward the Foundation* (1993). Subsequent quotes stem from the 2010 omnibus version, Asimov 2010. For Asimov's life and writings, see Asimov 2002; and Freedman 2005. For an interesting analysis of the foundation novels, see Käkelä 2016.

30 Asimov 2010, 411.

way, Asimov's character Seldon could predict future developments with a probability of up to 94%. Seldon could do so, because psycho-history 'could forecast reactions to stimuli with something of the accuracy that a lesser science could bring to the forecast of a rebound of a billiard ball'.[31]

Asimov's psycho-history was an allusion to historical materialism, the Marxist notion that social developments are not random, but the result of larger social processes, which unfold according to a certain logic. Because of this, laws of development can be abstracted from the past and projected onto the future. Several authors have claimed this earlier, among them Donald Wollheim, who stated that 'Asimov took the basic premise of Marx and Engels, said to himself that there was a point there – that the movements of human mass must be subject to the laws of motion and interaction, and that a science could be developed based upon mathematics and utilizing all the known data'. For Wollheim, psycho-history therefore was the fictional science that 'Marxism thought it was and never could be'.[32] Still, psycho-history was not so much intended as a parody, but rather as a thought experiment. What if it were actually possible to discern laws of social development from society's history? How would these laws work and how could they be made operational? With his Foundation novels, Asimov provided one possible answer, in which the laws of social development could be ascertained through statistical methods; an answer that continues to fascinate and inspire authors and scientists to this day.[33]

Moving from Asimov's fictional science to the real existing historical notion of historical materialism, we are confronted with a striking paradox. Orthodox Marxists such as Pannekoek claimed that, indeed, laws of social development could be abstracted from the past – and even used to forecast future developments. Friedrich Engels stated at Marx's funeral: 'Marx discovered the law of evolution in human history [...] Marx also discovered the special law of motion governing the present day capitalist method of production and the bourgeois society that this method of production has created.'[34] This, in turn, led Lenin to conclude that 'Marx drew attention and indicated the way to a scientific study of history as a single process which, with all its immense variety and contradictions, is governed by

31 Asimov 2010, 205.

32 David Wollheim, as quoted in Elkins 1976, 32. Elkins' contribution offers an original and thought-provoking critique of Asimov's work. At the same time, however, as he neatly lays out the differences between Asimov's (and Wollheim's) 'crude caricature of Marxism' and the supposedly 'real' ideas of Marx and Engels, his article tends to be rather dogmatic itself.

33 See, for example, *The Economist* 2013.

34 Engels 1883.

definite laws.'[35] Pannekoek went one step further and at one time claimed that these laws 'allow us to make some prediction about the future'.[36] Even so, this scientific method and, more importantly, these laws were never made explicit. This raises the question why Pannekoek did not reflect more explicitly on the differences between social and natural sciences, and the searing divide between the claims of historical materialism – i.e. scientific socialism – and the realization of these claims.

## Nuancing the Scientific Claims of Socialism

In his early works, Pannekoek subscribed to the idea that Marxism could make predictions about the future, especially in forecasting the advent of socialism. As scientific socialism embraced empirical research, it was supposedly able to uncover societal laws of development, which foretold the outbreak of socialist revolution. The last step in this line of reasoning, however, led to tension. In 1906, for example, Pannekoek stated that socialism could provide information on 'what according to our contemporary knowledge *will* be the course of *imminent* social developments'.[37] Yet, at the same time, Pannekoek stated that the forms that the coming revolution would take were 'hard to determine beforehand'.[38] The tension between both these claims was left unresolved in the text. In his 1919 *Historical Materialism*, Pannekoek even explicitly denounced determinist views of how the future would unfold.[39] The other two premises of Pannekoek's socialism remained.

But already by 1909, Pannekoek proposed a different way in which Marxism could be scientific in his well-known treatise *Marxism and Darwinism* (1909).[40] In *Marxism and Darwinism*, Pannekoek held that academic disciplines such as history become scientific when they are able to explain the 'origin and meaning' of phenomena. Along those lines, even ethics could become a science, when this 'science of ethics' would aim to explain and understand the 'origin and essence of ethical phenomena'. In a similar vein, philosophy could become a science, as long as it was premised on the notion that 'the human spirit [religion, art, science, philosophy] is conditioned in

35 Lenin, as quoted in Elkins 1976, 29.
36 Pannekoek 1906a, 25.
37 Pannekoek 1906a, 28.
38 Pannekoek 1906a, 28.
39 Pannekoek [1919] 1972.
40 Pannekoek 1912b. Dutch original and German translation: 1909a; 1909b.

all its expressions by the outside world'. When the human spirit 'simply becomes a part of nature, the humanities turn into natural science.'[41]

This 'social-science-as-a-natural-science', however, did not embody specific well-defined methods or laws such as Asimov suggested in his Foundation novels. Rather, it proposed a general idea of how the social sciences should be conceived. For Pannekoek, the focus should be on explaining social developments. Pannekoek explained this in his comparison between Marxism and Darwinism. Conceding that evolution theory could not be observed directly or be tested in a laboratory, Pannekoek reasoned:

> The best proof for the correctness of this theory would have been to have an actual transformation from one animal kind to another take place before our eyes, so that we could observe it. But this is impossible. How then is it at all possible to prove that animal forms are really changing into new forms? This can be done by showing the cause, the propelling force of such development. This Darwin did.[42]

Pannekoek then continued to claim that Marxism worked along the same principles: 'If we turn to Marxism we immediately see a great conformity with Darwinism. As with Darwin, the scientific importance of Marx's work consists in this, that he discovered the propelling force, the cause of social development.'[43] Scientific socialism thus became a way of understanding and explaining social developments; developments that were moving towards socialism. In this way, scientific socialism as a prediction and as a method were closely linked. When Pannekoek renounced the imminence of socialism, he upheld that Marxism could explain the past and present, but not predict the future.

## Scientific Worldview and Education of the Masses

Pannekoek thus characterized his socialism as scientific because it foretold the advent of socialism, but also because it provided a way to understand social developments. In other works, he defined socialism as a 'scientific conception of the human world'.[44] In 1903, Pannekoek suggested that socialism was not so much one scientific discipline among others, but rather a

41 Pannekoek 1906c, 20.
42 Pannekoek 1912b, 10-11. Compared with 1909a, 10; and 1909b, 7.
43 Pannekoek 1912b, 35.
44 Pannekoek 1912a, 4.

worldview. Dietzgen's 'theory of cognition' was the basis of this 'theory of society and man'. According to Pannekoek, 'anything outside of them is mere fantasy'. In the German version, it even reads: '[O]utside of it, there is only delusion, it forms a satisfying and harmonic worldview.'[45] This third definition implied that scientific socialism was an attitude rather than anything else. This scientific way of looking at the world meant a willingness to engage in scientific endeavours without any class-related prejudices standing in the way, an eagerness to gather and process empirical data, since scientific development was an integral part of the socialist project of human liberation.

In this context, Pannekoek made a sharp distinction between bourgeois science and real science. The former was 'merely the servant of capitalism': 'Not the discovery of truth, but the reassurance of an increasingly superfluous class of parasites is the object of this science. No wonder that it comes into conflict with the truth.'[46] Pannekoek believed that capitalist (or: bourgeois) society was doomed, but that members of the capitalist class were not willing to accept this. As a result, they were not interested in truth, but sought relief in pseudo-scientific endeavours. Pannekoek gave the example of the physician, biologist, and politician Rudolf Virchow, who had supposedly 'assailed the Darwinian theory on the ground that it supported Socialism'. Pannekoek relayed Virchow's response as follows: 'Be careful of this theory'. And Pannekoek concluded: 'What shall be said, however, of the science of a professor who attacks Darwinism with the argument that it is not correct because it is dangerous!'[47]

From this and other examples, Pannekoek concluded: 'Bourgeois thinking cannot solve the mysteries of the world.'[48] The working class, on the other hand, had a different relation towards science. It saw the development of science as a means to further its cause. Pannekoek thus stated:

> Only the physical and natural sciences are admired and honoured by both classes [capitalists and workers]. Their content is identical for both. But science does have a different meaning to different classes. But how different from the attitude of the bourgeois classes, is that of the worker who has recognized these sciences as the basis of his absolute rule over nature and over his destiny in the future socialist society.[49]

45 Pannekoek 1906c, 31; 1903, 24.
46 Pannekoek 1909c, 320.
47 Pannekoek 1912b, 28.
48 Pannekoek 1905, 34.
49 Pannekoek 1912a, 25.

In another text, Pannekoek claimed in a similar manner: 'The Socialist theory restores clearness and scientific exactness by concentrating attention upon the natural divisions of society.'[50] This third line of reasoning is distinct but connected to the other two. The reason that the proletariat strives for truth in science rather than 'reassurance', is that it knows that this truth will help it bring about 'his absolute rule over nature and over his destiny'.[51] Here, the line of reasoning again runs the risk of becoming circular: striving for truth in science leads to a method of explaining past and future developments, which show the advent of socialism, which leads to a dedication to scientific exactness and striving for truth in science.

## Conclusion

This essay has been an attempt at deconstructing Pannekoek's views on scientific socialism, distilling from his early socialist works three distinct but closely linked definitions. For Pannekoek, socialism was scientific in that it embraced modern science, in that it supposedly uncovered the laws of societal development, and in that it foretold the advent of socialism.

Moreover, this essay reveals that the three definitions can strengthen each other in the sense that a scientific worldview could lead to a method of explaining social developments, thus leading to the certainty that socialism is imminent. The three are also linked because the education of the masses takes a central place in all three of them. At the same time however, its weakness and contradictory character is laid bare when Pannekoek is first ambiguous and then denies socialism's predictive powers. If scientific socialism can explain past and present developments in a scientific way, the reason why this method cannot be extrapolated into the future needs to be explained. Therefore, the interlinkage between the three conceptions of scientific socialism falls apart when the last line of reasoning is denounced.

The three definitions of scientific socialism are also linked in a different way. During his whole career as a socialist, Pannekoek emphasized the importance of the 'mental struggles which accompany the social struggles of today'.[52] Thus he stated: 'The material power, which the proletariat possesses due to its size and significance in the production process, would not

50 Pannekoek 1909c, 320-321.
51 Pannekoek 1912a, 25.
52 Pannekoek 1912b, 7.

help it very much, if it were not complemented by its mental superiority.'[53] According to Pannekoek, the proletariat's mental superiority lay in its dedication to truth in scientific endeavours, for the proletariat 'recognized these sciences as the basis of his absolute rule over nature and over his destiny in the future socialist society'.[54] Pannekoek emphasized that this superiority was to be strengthened by 'educating the masses' and instilling on them 'the knowledge that the wish [socialist revolution] will be fulfilled', by 'giving the proletariat the certainty that such an order [socialism] will come'.[55]

For Pannekoek, a divide of his scientific socialism into three separate definitions would not have made a lot of sense. He never spoke of scientific socialism's different meanings, and for him, the three definitions would rather have been aspects of one coherent worldview. Even so, taking the term apart this way reveals the inconsistencies mentioned above.

Pannekoek's views on science can offer a way to connect Pannekoek-the-Astronomer and Pannekoek-the-Socialist. Pannekoek saw both his political work and his astronomical work as scientific endeavours. Although being very different fields, Pannekoek united them by a theory of knowledge, inspired by Dietzgen, that allowed him to approach both fields with a similar attitude and line of reasoning. Even so, scientific socialism adhered to scientific principles only to a certain extent. Deconstructing his lines of argument leads to a number of contradictions. This leaves open the question why it was so important for Pannekoek that his socialism was scientific. First of all, for Pannekoek it showed that socialism was not an 'accidental discovery', but had a different position from other ideologies. Furthermore, for a long time it provided political certainty – the imminence of socialism – and a clear line of political reasoning. But most important was perhaps that for Pannekoek, socialism was the only ideology with a true interest in scientific research and findings. While conservative ideologies, according to Pannekoek, had a functionalist relationship towards science, using its findings when appropriate and denouncing them when they threatened their position, Pannekoek believed that socialism 'restores clearness and scientific exactness'. It was this line of reasoning, above all, that allowed Pannekoek to connect his astronomical and socialist persona.

53 Pannekoek 1906b, 6.
54 Pannekoek 1912a, 25.
55 Pannekoek 1906b, 20-21; 1915.

## Bibliography

Aartsz, P. 1946. *De arbeidersraden*. Amsterdam: De Vlam.

Asimov, Isaac. 2002. *It's Been a Good Life*. Ed. by Janet Jeppson Asimov. Amherst, NY: Prometheus Books.

—. 2010. *Foundation, Foundation and Empire, Second Foundation*. New York: Everyman's Library.

Baneke, David. 2004. '"Hij kan toch moeilijk de sterren in de war schoppen". De afwijzing van Anton Pannekoek als adjunct-directeur van de Leidse Sterrewacht in 1919'. *Gewina* 27(1): 1-13.

Berkel, Klaas van. 2001. 'Pannekoek, Antonie'. In *Biografisch Woordenboek van Nederland*, ed. by Jac Bosmans, Mineke Bosch, Klaas van Berkel, J.B.J. Teeuwisse, F. van Vree, H. van Gelder, Herman de Liagre Böhl, and A.L. Mooij. The Hague: Instituut voor Nederlandse Geschiedenis, 5: 384-387.

Bock, Hans Manfred. 1992. 'Die Marx-Dietzgen-Synthese Anton Pannekoeks und seines Kreises'. In *Die Rezeption marxschen Theorien in den Niederlanden*, ed. by Marcel van der Linden. Trier: Karl-Marx Haus, 106-123.

—. 1993. *Syndikalismus und Linkskommunismus von 1918 bis 1923. Ein Beitrag zur Sozial- und Ideengeschichte der frühen Weimarer Republik*. Darmstadt: Wissenschaftliche Buchgesellschaft.

Bourrinet, Philippe. 2017. *The Dutch and German Communist Left (1900-68): 'Neither Lenin nor Trotsky nor Stalin!' 'All Workers Must Think for Themselves!'* Leiden: Brill.

Brendel, Cajo. 1974. 'Die "Gruppe Internationale Kommunisten" in Holland. Persönliche Erinnerungen aus den Jahren 1934-1939'. In *Marxistische Revolutionstheorien. Außerdem: Korsch-Briefe; Tagungsprotokoll: 'Geschichtsschreibung der Arbeiterbewegung'; Bibliographien; Rezensionen; Hinweise*, ed. by Claudio Pozzoli. Jahrbuch Arbeiterbewegung, Theorie und Geschichte 2. Frankfurt (am Main): Fischer-Taschenbuch-Verlag, 253-263.

Burns, Tony. 2002. 'Joseph Dietzgen and the History of Marxism'. *Science & Society* 66(2): 202-227.

Elkins, Charles. 1976. 'Isaac Asimov's "Foundation" Novels: Historical Materialism Distorted into Cyclical Psycho-History'. *Science Fiction Studies* 3(1): 26-36.

Engels, Friedrich. 1880. 'Socialism: Utopian and Scientific'. https://www.marxists.org/archive/marx/works/1880/soc-utop/index.htm (accessed 16 March 2018).

—. 1883. 'Speech at the Grave of Karl Marx, Highgate Cemetery, London. March 17, 1883'. https://www.marxists.org/archive/marx/works/1883/death/burial.htm (accessed 16 March 2018).

Freedman, Carl Howard, ed. 2005. *Conversations with Isaac Asimov*. Jackson, MS: University Press of Mississippi.

Gerber, John P. 1988. 'From Left Radicalism to Council Communism: Anton Pannekoek and German Revolutionary Marxism'. *Journal of Contemporary History* 23(2): 169-189.

—. 1989. *Anton Pannekoek and the Socialism of Workers' Self-Emancipation, 1873-1960*. Dordrecht: Kluwer.

Heuvel, Edward P.J. van den. 1982. 'Pannekoek als sterrenkundige'. In *Herinneringen: herinneringen uit de arbeidersbeweging; sterrenkundige herinneringen*, by Anton Pannekoek, ed. by Ben A. Sijes, Johanna M. Welcker, and J.R. van der Leeuw. Amsterdam: Van Gennep, 11-69.

Käkelä, Jari. 2016. *The Cowboy Politics of an Enlightened Future: History, Expansionism and Guardianship in Isaac Asimov's Science Fiction*. Helsinki: University of Helsinki.

Kool, Frits. 1970. *Die Linken gegen die Parteiherrschaft*. Olten: Walter Verlag.

Lukács, Georg. 1923. 'Was ist orthodoxer Marxismus'. In *Geschichte und Klassenbewusstsein*. Berlin: Der Malik Verlag, 35-54.

Marx, Karl, Friedrich Engels, and Vladimir Lenin. 1972. *On Historical Materialism*. Moscow: Progress.

Mergner, Gottfried, ed. 1971. *Gruppe Internationale Kommunisten Hollands*. Hamburg: Rowohlt.

Pannekoek, Anton. 1903. 'Die Stellung und Bedeutung von Jozef Dietzgens philosophischen Arbeiten'. In *Das Wesen der menschlichen Kopfarbeit. Ein abermalige Kritik der reinen und praktischen Vernunft*. Stuttgart: J.H.W. Dietz.

—. 1905. *Godsdienst en socialisme. Voordracht op 14 September 1905 te Bremen gehouden*. Leiden: Leidsche Uitgevers-Maatschappij.

—. 1906a. *Ethik und Sozialismus. Umwälzungen im Zukunftsstaat*. Leipzig: Leipziger Buchdruckerei.

—. 1906b. *Religion und Sozialismus. Ein Vortrag*. Bremen: Bildungsausschuss des Gewerkschaftskartells.

—. 1906c. 'The Position and Significance of J. Dietzgen's Philosophical Works'. In *The Positive Outcome of Philosophy*, by Joseph Dietzgen, ed. by Eugene Dietzgen and Joseph Jr. Dietzgen, trans. by Ernest Untermann. Chicago, IL: Charles H. Kerr & Co., 7-37.

—. 1907. *Omwentelingen in den toekomststaat*. Leiden: Leidsche Uitgevers-Maatschappij.

—. 1908-1909. 'Die Entwicklung des Weltalls'. *Die Neue Zeit* 26, 27(2, 1): 37-40, 422-426, 575-577, 966-969, 488-493, 763-767.

—. 1909a. *Darwinisme en Marxisme*. Amsterdam: Brochurehandel S.D.P.

—. 1909b. *Marxismus und Darwinismus. Ein Vortrag*. Leipzig: Leipziger Buchdruckerei.

—. 1909c. 'The New Middle Class'. *International Socialist Review* 10(4): 316-326.

—. 1912a. *Klassenkampf und Nation*. Reichenberg: Verlag von Runge & Co.

—. 1912b. *Marxism and Darwinism*. Trans. by Nathan Weiser. Chicago, IL: Charles H. Kerr & Co.

—. 1915. 'Der Marxismus als Tat'.
https://www.marxists.org/deutsch/archiv/pannekoek/1915/03/tat.htm (accessed 16 March 2018).
—. [1919] 1972. 'Het historisch materialisme'. In *Partij, raden, revolutie*, by Anton Pannekoek, ed. by Jaap Kloosterman. Amsterdam: Van Gennep, 7-21.
—. 1951. *De groei van ons wereldbeeld*. Amsterdam: Wereldbibliotheek.
—. 1982. *Herinneringen: herinneringen uit de arbeidersbeweging; sterrenkundige herinneringen*. Ed. by Ben A. Sijes, Johanna M. Welcker, and J.R. van der Leeuw. Amsterdam: Van Gennep.
Plekhanov, Georgi. 1898. 'On the Role of the Individual in History'. https://www.marxists.org/archive/plekhanov/1898/xx/individual.html (accessed 16 March 2018).
Schaaf, Jasper. 1978. 'De invloed van Dietzgen in de Nederlandse arbeidersbeweging'. Master's thesis, Groningen: University of Groningen.
—. 1993. *De dialectisch-materialistische filosofie van Joseph Dietzgen*. Kampen: Kok Agora.
Sijes, Ben A. 1982. 'Anton Pannekoek 1873-1960'. In *Herinneringen: herinneringen uit de arbeidersbeweging; sterrenkundige herinneringen*, by Anton Pannekoek, ed. by Ben A. Sijes, Johanna M. Welcker, and J.R. van der Leeuw. Amsterdam: Van Gennep, 11-69.
Steen, Bart van der. 2006. 'Anton Pannekoek en het orthodoxe marxisme'. *Vlaams Marxistisch Tijdschrift* 40(2): 73-82.
Tai, Chaokang. 2017. 'Left Radicalism and the Milky Way: Connecting the Scientific and Socialist Virtues of Anton Pannekoek'. *Historical Studies in the Natural Sciences* 47(2): 200-254.
Tai, Chaokang, and Jeroen van Dongen. 2016. 'Anton Pannekoek's Epistemic Virtues in Astronomy and Socialism: Personae and the Practice of Science'. *BMGN – Low Countries Historical Review* 131(4): 55-70.
*The Economist*. 2013. 'Dr Seldon, I Presume', 23 February 2013.
Welcker, Johanna M. 1986. 'Pannekoek, Antonie'. Biografisch Woordenboek van het Socialisme en de Arbeidersbeweging in Nederland. http://hdl.handle.net/10622/ACDE7F4A-5D23-49DE-B973-9F58C1313E1F (accessed 16 March 2018).

## About the Author

**Bart van der Steen** is University Lecturer at Leiden University's Institute of History. His research focuses on interwar labour movements and New Social Movements from 1968 to the present. His most recent publications include: *In Leiden moet het anders. Geschiedenis van een SP-afdeling, 1970-1982* (2019), and *A European Youth Revolt. European Perspectives on Youth Protest and Social Movements in the 1980s* (2016), edited with K. Andresen.

# 8 From Science to Science

## Anton Pannekoek, Willem Bonger, and Scientific Socialism

*Annemarie Rullens*

**Abstract**

Anton Pannekoek was a remarkable man. As a renowned astronomer and equally influential socialist theoretician he set his mark in many ways. As soon as socialism was labelled 'scientific' at the end of the nineteenth century, academics such as Pannekoek started exploring how and why socialism was scientific. In other words, what exactly was scientific socialism? How were science, ideology, and politics related? Pannekoek's particular ideas on scientific socialism were soon contested. His contemporary Willem Bonger can be seen as an interesting counterpart. This article explores Bonger's ideas on socialism as 'applied science', thereby placing Pannekoek's ideas in perspective and demonstrating that there were differing conceptions of the role of science in socialist politics and how, as a science, socialism needed to be practised.

**Keywords:** Anton Pannekoek, scientific socialism, Willem Bonger, technocracy, socialist politics

## Introduction

In 1880, Friedrich Engels stated in his pamphlet *Socialisme utopique et socialisme scientifique* that socialism had become a scientific doctrine. As the ideas of Engels and Karl Marx were embraced by a significant part of the labour movement, they gained a sizable following by the end of the nineteenth century. Soon after, socialism came to be seen by many as a scientific theory that explained society's development through predetermined laws of social evolution. In the Netherlands, too, Marxist intellectuals adopted this worldview. Anton Pannekoek, for example, embraced this

Tai, Chaokang, Bart van der Steen, and Jeroen van Dongen (eds), *Anton Pannekoek: Ways of Viewing Science and Society*. Taylor & Francis Group, 2019
DOI: 10.5117/9789462984349_CH08

conception of scientific socialism to a large degree when he joined the labour movement in 1899.[1]

By defining their socialism as scientific, Marx and Engels made a distinction between their brand of socialism and other, more anarchist-influenced 'utopian', strands of socialism. Even so, it soon became clear that there were different opinions on what it meant for socialism to be scientific. This paper reconstructs the ideas of the Dutch socialist theoretician Willem Bonger on scientific socialism. Bonger envisioned socialism not so much as a reflective science for interpreting social developments, but rather as an applied science. He propagated a socialism that based itself on the newest scientific insights and developed policies in order to build a socialist society. Bonger's ideas thus focused not so much on interpreting and theorizing but on *making* a socialist society. In doing so, he set himself apart from other, more Marxist-inspired socialists like his contemporary Pannekoek. In many ways, Bonger can be seen as a counter example to Pannekoek. This is remarkable since both men had many things in common: both were influential scientists who put their mark on socialism in the Netherlands during the twentieth century. Bonger, for example, co-authored the main interwar policy statements of the Dutch Labour Party and influenced the work of postwar social democratic policymakers.

While Pannekoek and his collaborators published their ideas mainly in the Marxist journal *De Nieuwe Tijd* ('The New Era'), the main vehicle that Bonger used to propagate his views was *De Socialistische Gids* ('The Socialist Guide'), the official scientific journal of the Dutch Social Democratic Workers' Party (Sociaal-Democratische Arbeiderspartij, SDAP). Bonger not only used it to further his own brand of scientific socialism, but also a specific type of socialist intellectual. Bonger explicitly favoured a modern type of scientist; a rational and 'cold-blooded' thinker, as opposed to the other, at times extravert and emotional, kinds of intellectuals of the (radical) left. Both journals represented a distinct intellectual tradition within the SDAP. As such, they form important sources for the intellectual history of the SDAP. *De Nieuwe Tijd* has been the subject of an in-depth study by Henny Buiting, but *De Socialistische Gids* has hardly been of interest for historians so far.[2] This is unfortunate because, as this paper will illustrate, such an exploration demonstrates the broad array of ideas that existed within the party on the relation between science, ideology, and politics.

1 See Gerber 1989.

2 For one of the very few articles on De Socialistische Gids, see: Faassen 1980.

This paper begins by briefly discussing Pannekoek's ideas on scientific socialism. It subsequently analyses how Bonger sought to popularize his vision of scientific socialism as editor of the journal *De Socialistische Gids* and in debates with other Dutch socialists. The paper closes with a brief discussion of Bonger's influence and the differences between him and Pannekoek. In doing so, this paper places Pannekoek in perspective and illustrates the differing ideas in the labour movement on what scientific socialism was and the conflicting conceptions of the role of science in socialist politics.

## Anton Pannekoek on Scientific Socialism

Undoubtedly, Pannekoek saw his writings on socialism as scientific exercises. Being an astronomer by profession, his scientific endeavours were not limited to the observatory.[3] Contributing to the development of Marxism and theorizing about historical materialism, religion, and philosophy provided a similar 'scientific experience' as mapping and analysing the Milky Way. Studying the stars and studying society were equally serious and important undertakings. Even more so, for Pannekoek both activities were related. He considered the study of socialism to be complementary to the natural sciences. Pannekoek made this view explicit during a seminar for physics students in the autumn of 1940, just after the occupation of the Netherlands by Nazi-Germany. There, he claimed that natural sciences had thrived during the nineteenth century and enabled the dominance of men over nature. The natural sciences had however failed to show men how to organize society. According to Pannekoek, socialism sought to do exactly this.[4]

Addressing the students, Pannekoek compared society with a living and growing organism: 'We have seen how steam capacity in machines increased hundredfold, how electricity has grown into an encompassing neural network, how all bodily organs have gained in efficiency. What this organism is still lacking, however, is a conscious mind.' In Pannekoek's view, society had remained a 'headless monster, whose limbs tear itself apart', an explicit reference to the war.[5] While the science of production needed 'no further improvement for the time being', what was now necessary was a collective 'understanding of the social forces' that drove society. This

3 For more information on Anton Pannekoek, see biographical studies: Welcker 1986; Gerber 1989.

4 Anton Pannekoek, Wetenschap en maatschappij, 1940, AP, inv.no. 244.

5 Pannekoek, Wetenschap en maatschappij, 6.

knowledge could not be produced by engineers, but had to 'grow forth from the masses'.[6] According to Pannekoek, scientific socialism did not coincide with the natural sciences. Rather, it was a social science, a science of the human mind. Its goal was to gain an understanding of the human mind in general and of the working-class mind in particular. In doing so, it combined the 'science of society, psychology, philosophy'. Pannekoek argued that an understanding of the human mind through these sciences was a necessary step towards establishing a socialist society. The working class needed to achieve a certain state of consciousness. The mental awakening that Pannekoek envisioned, preceded the formation of a socialist society. He thus concluded: 'This mental development is a scientific development.'[7]

Pannekoek's view of socialism as a science could be seen as representative for the first generation of Marxist intellectuals in the Netherlands. From 1896 onwards, this group publicized their views in a monthly journal devoted to Marxist ideas, politics, and culture, *De Nieuwe Tijd*, which was named after its German counterpart. Among its well-known and influential contributors were the famous poets Henriette Roland-Holst and Herman Gorter, literary critic and co-founder of the Social Democratic Workers' Party, Frank van der Goes, and Anton Pannekoek.[8]

Their ideas, however, did not remain uncontested for long. From the 1910s onwards, the socialist criminologist Willem Bonger challenged Pannekoek's conception of socialism-as-science by formulating a very different idea of how science and socialism were related. Like Pannekoek, Bonger was an academic with a high-standing reputation. Bonger had studied law in Amsterdam and in 1922 became the first professor in criminology in the Netherlands.[9] If Pannekoek had a metaphysical idea of how socialism was a science, Bonger's socialism could be described as an applied science. For Bonger, socialism was not a method for understanding society, but rather a means for literally *making* society. Propagating an explicitly practical approach, Bonger proposed to study society empirically and discover its laws of development in order to assess how scientific insights could be applied to establish a socialist society. His focus was not so much on labour politics per se, or its underlying ideological propositions, but rather on policy development driven by socialist ideals. Unlike Pannekoek, Bonger did not believe that a socialist society would result from a socialist revolution, but

6 Pannekoek, Wetenschap en maatschappij, 7.

7 Pannekoek, Wetenschap en maatschappij, 10.

8 For more information on *De Nieuwe Tijd*, see: Buiting 2003.

9 For more information on Bonger see: van Heerikhuizen 1987.

rather believed that it had to be made or built by the right policy decisions. In his view, engineers, mathematicians, doctors, criminologists, economists played a key role in the forthcoming of a socialist society; not 'the masses', nor the philosophers, poets, or literary critics trying to unravel the working-class mind as Pannekoek believed. For Bonger, practising socialism as a science meant practising disciplines like economics, criminology, mathematics, and medical studies. He saw socialism as an 'applied' science.[10]

## The Development of Bonger's socialism

Willem Bonger was born on 6 September 1876 in Amsterdam in a large liberal family. At the University of Amsterdam, he became acquainted with socialism through the student circle *Clio*. Bonger was attracted to socialism for its humanistic appeal and he vividly discussed its premises, both orally and in written form. For the student newspaper *Propria Cures*, he wrote an article on 'socialism and being a student'. Soon after, he joined the more overtly political Socialist Reading Circle (Socialistisch Leesgezelschap, SL).

The SL consisted of young students, most of whom had a bourgeois background. The group was close to the SDAP – although SDAP membership was optional – functioning as a bridge between student life and the party.[11] In 1900, Bonger became president and under his leadership, the society became an active organization with over a hundred members. It circulated a portfolio of socialist magazines such as *Die Neue Zeit*, *Sozialistische Monatshefte*, *Le Mouvement Socialiste* and *De Nieuwe Tijd*. The international outlook of SL was corroborated by invitations to Emile Vandervelde, Edward Anseele, and Karly Kautsky. Besides prominent Marxists, Bonger also invited liberal thinkers to speak for the students – a move that was controversial among the SL members. Bonger, however, claimed that only debate could strengthen the arguments in favour of socialism and its theoretical basis. Moving away from his early views, Bonger started to emphasize that the strength of socialism lay in its scientific foundation, rather than its humanist appeal to empathy and solidarity.

Bonger's outlook was strongly influenced by Karl Kautsky. Since the late 1890s, Kautsky was considered to be the most influential Marxist thinker on the continent, to the point he became known as the 'pope of Marxism'.

10 Bonger 1925.
11 van Veldhuizen 2015, 175.

Kautsky popularized Marxism as a positivist 'scientific' worldview that proclaimed that the development of capitalism would inherently lead to revolution through the accumulation of wealth in the hands of a few. At the same time, it placed this development in a far future, thus legitimizing moderate and reformist politics. Because of this, 'Kautskyan Marxism' was open to multiple interpretations.[12] Both radical and moderate socialists could endorse it.[13] It informed the influential *Erfurter Programm* of the Social Democratic Party of Germany (SPD) as well as the party programme of the SDAP. Bonger spoke out ever more clearly in favour of Kautsky's Marxism; at first in the student newspaper *Proria Cures*, and later in his dissertation. Focusing on the social dimension of crime, Bonger's 1905 dissertation *Criminalité et conditions économiques* was clearly inspired by Marxist ideas. In it, Bonger analysed how social relations determined crime rates. Capitalism increased poverty but at the same time the desire for material wealth. This way, capitalism not only affected the working class but also the bourgeoisie and created the conditions for criminal behaviour.[14]

Rather than exploring social relations philosophically, Bonger used statistical methods to uncover causal relations between economic conditions and human behaviour. Being aware of the limitations of this approach, he also did complementary qualitative research. His conclusion, however, was utopian: in a socialist society, Bonger claimed, there would be no crime. It was typical for Bonger's view that human behaviour could be reduced to societal conditions. Years later, in 1932, Bonger weakened his conclusions, admitting that even in a socialist society criminal behaviour would continue to exist, since some people were simply 'bad'. For this, however, Bonger had another, quite practical, solution: eugenics. Sifting out criminal genes would end criminality once and for all. Such ideas were not uncommon within the socialist movement at the time and in Bonger's case, the turn to eugenics again underlines his continued belief in the power of policy measures to make a socialist society.[15]

It was exactly this belief – that a socialist society could be *made* – that motivated Bonger to accept the position of editor of *De Socialistische Gids* at the end of 1915. This journal had been founded as the official scientific journal of the SDAP by its executive board following various failed attempts

12 For more information on Kautsky, see: Hünlich 1992; Salvadori 1979; Steenson 1978.

13 Buiting 1989, 629.

14 Bonger 1905.

15 Lucassen 2010.

to take over *De Nieuwe Tijd*, and the turn of several of the journal's members to the communist party. In their decision, the SDAP leaders were driven by the ambition to tie critical intellectuals to the party and neutralize criticism from independent intellectuals within the party, who called for a more principled politics. In various ways, the new journal was to function as a 'safety valve', allowing for debate but primarily under the control of the party leadership.[16] Seen from a grassroots perspective, it is hard to miss that the foundation of the party's first official scientific journal also responded to urgent calls from local party branches, who felt that such a journal was crucial in their efforts to educate the working class. They hoped the journal would provide 'popular scientific leadership and education'.[17] Various actors thus shared a belief that a scientific journal was necessary for the party's development, but each of them had their own arguments and held different ideas on what the journal was supposed to look like.

Bonger was generally regarded as the ideal candidate to lead the new journal. He was a renowned academic and an experienced writer and editor. Furthermore, he had concerned himself with several educative initiatives within the party. Most importantly, however, he was a moderate socialist. After the left Marxists had split from the SDAP in 1909, Bonger had taken an explicitly moderate stance and downplayed his original Marxist views.[18] He no longer believed in a socialist revolution, but remained faithful to the idea of establishing a socialist society. He knew several critical intellectuals within the party, but had not been part of any oppositional group himself. This made him an ideal figure to reconcile the more critical and more moderate wings within the party. With nine votes to one, the executive board of the party voted in favour of his candidacy. Bonger agreed but sought to negotiate the terms of his appointment.[19] His request for a higher salary was met with hesitation by several board members, who claimed that the editor would only have to review proposed texts, while others supported Bonger and his ambition to be a more proactive editor. Ultimately, his salary was raised to a mere 750 guilders a year. After the issue was settled, Bonger feverishly started working on the new journal. From the very start, it became clear that he would not simply act as an editor but that he had a clear agenda for *De Socialistische Gids*.

16 Notulen partijbestuur en dagelijks bestuur, 17 April 1915, SDAP, inv.no. 27.

17 Congresverslag 1914, SDAP, inv.no. 263.

18 For more information on the SDAP during this period, see: Buiting 1989.

19 Notulen partijbestuur, 23 October 1915, SDAP, inv.no. 27. Not all members of the SDAP board where present during the vote: those present: Vliegen, Schaper, Loopuit, Hermans, Bergmeyer, De Roode, and Matthysen; not present: Hoejenbos, van Kuykhof, and Troelstra.

## Different Expectations of Science

In the very first issue of *De Socialistische Gids*, Bonger immediately made his vision of the journal explicit. He published an article written by the well-known engineer Theo van der Waerden on the new production model developed by Winslow Taylor. Aiming to increase efficiency, Taylor proposed a system of production the main feature of which was the production line, reducing complex work to a series of simple tasks for each worker. The choice for this article was typical for Bonger. It focused on an economic issue, was written by an engineer and contributed to the understanding of the labour process. At the same time, the mathematical models and economic laws applied by Taylor, and explained by Van der Waerden, were not written for workers, even those who were educated. Rather, the text was academic in content and style. It was exactly what Bonger wanted for *De Socialistische Gids*.

Bonger was the dominant voice in the editorial board of this journal, which generally consisted of five members. He fully dedicated his time and energy to the journal and as a result singlehandedly formed it into an intellectual forum for the rational and 'cold-blooded' thinkers, who he believed were essential for the development of the socialist movement. Amongst Bonger's favourites were the engineer Van der Waerden, the economic historian N.W. Posthumus, and the economist J. van den Tempel. They wrote on economy, physics, and statistics. But even medical studies were discussed in *De Socialistische Gids*. Bonger explicitly sought to promote the exact and social sciences as opposed to the humanities as he believed socialist policies needed to be based on the former.[20]

Soon, however, Bonger's choice of topics and the format of the journal led to criticism. One critic, for example, judged that however 'interesting' Van der Waerden's paper was, 'now' (i.e. 1915) was not the time to 'spend hours studying this kind of literature'.[21] According to many, the ongoing world war had put science on a second tier, a sentiment that was even shared by some of the editors of *De Socialistische Gids*. Meanwhile, SDAP board members expected the journal to cover more popular and politically acute

20 This did not mean that articles on arts and humanities were completely missing from *De Socialistische Gids*. Under pressure of the publishers short entertaining pieces were published in 1925 and 1926. From 1931 onwards, a column called 'Film, Music, and Architecture' appeared in the magazine.

21 Opwaarts, Orgaan van den Bond van Christen-Socialisten, 28 January 1916, DSG, no. F2.

topics. SDAP president Pieter Jelles Troelstra hoped to provide answers to specific political controversies related to the war and socialist politics by means of the journal. Rudolf Kuyper, another party leader, on the other hand, preferred the journal to reflect on Marxism and socialist theory. Contrary to what Marxist theory had predicted, labour movements in all European countries supported their governments' decision to go to war. For Kuyper, theoretical reflection was now necessary in order to rethink socialism. When Bonger appeared unwilling to take 'his' journal in either of these directions, this resulted in fierce debates. Two of which will be shortly discussed here, to illuminate not only Bonger's views, but also those of his co-editors.

By the end of 1917, Troelstra argued that *De Socialistische Gids* had failed to discuss 'the greatest problems of the imminent future'.[22] He specifically referred to the 'ministerial question'; the question of whether the SDAP should join a liberal-led coalition government. Since universal male suffrage was to be granted to the Dutch population in 1917 (women would gain the vote two years later), the SDAP expected a sizable increase of votes. A few years earlier, in 1913, it had been decided that only in the case of 'utter necessity', the party would join a coalition government with liberal parties.[23] According to Troelstra, such a situation had developed by 1917 but others disagreed. The following year, the situation became more complex when Troelstra attempted to start a socialist revolution and failed. In the resulting debate on parliamentary versus revolutionary politics, Troelstra expressed his views in *De Socialistische Gids* in an article where he defended his failed revolution.

The article provoked strong reactions. Bonger attacked Troelstra's politics as well as similar politics propagated by Henriette Roland Holst. Typical of his line of reasoning was his dismissal of the arguments of Roland Holst as unscientific. Bonger called her text a 'volcano of emotions' and the expression of 'a restless and impatient artist-temperament, that wished to skip some phases in the development of society'.[24] Denouncing a socialist revolution, Bonger wrote: 'The home that will one day house a prosperous mankind will be large and strong. Like everything man-made, however, it will arise stone by stone and not suddenly, like a castle in the air'.[25] Troelstra's response was sarcastic. He called Bonger an 'extremist of legality, of

22 Notulen redactievergadering, 17 December 1917, DSG, no. A1.
23 Congresbesluiten 1913, SDAP, no. 262.
24 Bonger 1919, 333.
25 Bonger 1919, 361.

"moderation" and "sobriety"'.[26] Troelstra defended Roland Holst's emotional appeal, claiming that 'by reducing the results of world events to mere mathematics, the soul and inner foundation of the revolution is dismissed and its actual legitimation discarded.'[27] According to Troelstra the coming of a socialist society required something more than scientific insights, namely emotional involvement.[28] Science clearly had a different meaning for Troelstra than for Bonger. For Troelstra, it was a trump card, to be played every once in a while in favour of his own political position. The label 'scientific' gave him prestige and lent a certain weight to his party-political manoeuvres, not only within the SDAP but also in other arenas such as parliament. Being part of the editorial board of *De Socialistische Gids* served his position within the SDAP, and at the same time the journal was one of his instruments of power.

The second example discussed here is an argument in 1926 between Bonger and Kuyper, who was also a member of the editorial board. Their argument revolved around the role of intellectuals within the socialist movement. In a speech for socialist students in 1925, Bonger had claimed that the SDAP faced a shortage of intellectuals. Since he believed the SDAP to be a 'constructive party' destined to one day rule the country, he regarded this as a serious problem.[29] Bonger's explanation for the lack of intellectuals in the SDAP was that the party did not appreciate intellectual work enough. He therefore argued in favour of better compensation for contribution to journals and other publications. Kuyper disagreed with Bonger and stated that not a lack of appreciation, but a lack of emotional appeal kept intellectuals away from the party. Moreover, Kuyper held *De Socialistische Gids* responsible, which in his view had become a beacon of 'one-sided, intellectualist dryness and scholasticism'.[30] The journal lacked any emotional appeal, and thus failed to attract younger intellectuals and artists.[31]

Kuyper subsequently stated that *De Socialistische Gids* should take as its example the new socialist youth organization, the *Arbeiders Jeugd Centrale*, which explicitly cultivated a socialist culture and mentality.[32] Bonger was not impressed. In his view, it was nothing more than a 'German import product'

26 Troelstra 1919, 513.

27 Troelstra 1919, 577.

28 Hagen 2010.

29 Bonger 1919, 361.

30 Kuyper 1926a, 367.

31 Kuyper 1926a, 365-366.

32 For more information on the *Arbeiders Jeugd Centrale* and its most prominent leader Koos Vorrink see: Hartveld, de Jong Edz., and Kuperus 1982; Wiedijk 1986; Wiedijk and Harmsen 1988.

and 'a romantic sect'.[33] Again, Bonger opposed a strong emotional appeal. He wrote: '[T]hose who wish to experience emotion should not join the labour movement [...] but go to the theatre or concert hall instead.'[34] What the socialist movement needed, according to Bonger, was professionals who were able to hold their emotions in check. He even went so far as to equate Kuyper's position with that of utopians, searching for an ideal society in a faraway place. On such journeys 'skilful seamanship and hard labour are required', not 'aestheticians' who soon lose their interest. 'Emotions', Bonger claimed, 'are not entirely resistant to seasickness'.[35]

> Devising the New is not as easy as one would think, since food and shelter must be secured. Imaginary manna cannot be eaten and people cannot live in castles in the air. Those who seek emotions soon had enough and wanted to 'go home'. But the workers kept at it, because they wish to realize what they had in mind, if not for themselves, then at least for the next generations.[36]

A furious Kuyper wrote a reply which was so aggressive in tone that the editors publicly denounced his style.[37] Kuyper argued that the youth movement at least secured a 'minimum level of ideology', which counterbalanced the matter-of-fact tendency stimulated by *De Socialistische Gids*.[38] What the party really suffered from was a rigid and uninspired atmosphere and for this, Kuyper blamed Bonger. From the start, Kuyper had sought to use the journal as a platform for the theoretical development of Marxism, just as *De Nieuwe Tijd* had done before. Constantly frustrated in his efforts, Kuyper decided to leave his position as editor of *De Socialistische Gids* in July 1926.

For Kuyper, practising socialism as a science meant discussing and popularizing Marxism. It had nothing to do with statistics, economic analysis or even medical topics; issues that were central in Bonger's approach. Bonger focused on politics and policy proposals, while Kuyper wished to prolong an intellectual tradition introduced and embodied by *De Nieuwe Tijd*-group.

33 Bonger 1926, 378.
34 Bonger 1926, 381-382.
35 Bonger 1926, 382.
36 Bonger 1926, 382.
37 Van der Waerden to Johan Frederik Ankersmit and Willem Vliegen, 23 April 1916, PJT, no. 756.
38 Kuyper 1926b, 479.

## Engineer of Technocracy

With *De Socialistische Gids*, Bonger introduced a new way of doing science in the SDAP. Exact and social sciences were explored and moderate professionals given a stage. It was the result of the way Bonger viewed science and its relation to socialism and the SDAP. When Bonger spoke of science, he thought of applied science. For Bonger, the goal of science was intervening in and shaping society. It was explicitly practical and pragmatic. He did not embrace the philosophy of science that had been dominant in the socialist movement up to that moment, which had been strongly influenced by Marx and Hegel. Hegel had presented philosophy, together with religion and art, as the highest branches of science. These sciences enabled self-reflection and were therefore considered crucial for human progress. According to Bonger, on the other hand, the applied sciences were the means of social progress.

In 1925 Bonger thus asked a group of students:

> What can socialism be for idealistic intellectuals? [...] It is a beautiful ideal in its own right [...] but the meaning of socialism goes much further. It is not the ideal of placing one class over another. Rather, it is the making of a society of cooperating forces of manual and collar workers.[39]

For Bonger, socialism not merely expressed a political aspiration; it was the *making* of a socialist society. Thus, science stood at the core of Bonger's political beliefs. Because of this, socialism was an amalgamation of political worldview and scientific knowledge.

As such, Bonger remained convinced of the necessity of actively manufacturing a socialist society. He dismissed revolutionary politics and arousing working-class spirits as a means towards that goal. Instead, he laboured to develop policy measures that could change society in a socialist direction. He held a firm belief that in this way, not only society, but even human nature could be changed. Although he firmly believed in socialism, he was neither a supporter of Kautsky nor the 'revisionist' Eduard Bernstein. With his applied-science approach, Bonger developed his own brand of scientific socialism. He believed in the malleability of society and envisioned a key role for professionals, making his socialism close to technocratic. Bonger used *De Socialistische Gids* to express his views within the SDAP, since he believed that the party was destined to govern and able to realize his vision. Publishing *De Socialistische Gids* was his contribution to the socialist cause.

39 Bonger 1925, 1011.

However, Bonger not only expressed his views in *De Socialistische Gids*. There were other opportunities for him to further his programme, of which the Report on Socialization (*Socialisatie-rapport*, 1920) is the best example. Next to Bonger, the engineers Theo van der Waerden and Jan Goudriaan and several party board members contributed to the programmatic text on how to socialize the means of production through (parliamentary) democracy. The Report proposed a gradual overtaking of key industries by the state. Remarkably, however, the arguments in favour of such a move were not political but focused on supposed gains in economic efficiency.

In the 1930s, Bonger's views were partly adopted by a new generation of socialist intellectuals. Two of his former students, Jan Tinbergen and Hein Vos, wrote the key text, *Plan van de Arbeid* ('Planned Labour', 1935), where social inequalities and tensions were approached from an economic and statistical perspective, closely related to the way Bonger worked. Economic planning was proposed as a way to counter the economic crises and improve the welfare of the working class. However, distinct from Bonger's Report on Socialization, the goal was now to find a new balance within a capitalist system, rather than the establishment of a socialist society.

A similar mixture between admiration and criticism can be found in another of Bonger's students, Hilda Verwey-Jonker. She had graduated on a thesis supervised by Bonger and once prompted her fellow socialist students to 'work hard and think of Bonger'.[40] Nevertheless, in 1931, Verwey-Jonker and her friends, the so-called 'kenteringsocialisten', wrote a letter to the party board, complaining about Bonger's leadership of *De Socialistische Gids*. They argued that the journal had failed to develop 'a generally accepted, all-encompassing and well-developed socialist "ethics", which could guide our actions'.[41] The letter echoed the earlier complaints of Kuyper, claiming that *De Socialistische Gids* was too down to earth and lacked a sense of idealism. One of the reasons for their criticism was the fact that the new generation did not share Bonger's irrefutable belief in a socialist society. They embraced Bonger's notion of making and shaping society through policy measures, but dismissed his premise that this would lead to a socialist society. Because of it, rethinking socialist ideals and ethics was more important than it had been for Bonger. The difference in worldview and mentality caused the younger group to start its own journal: *De Sociaaldemocraat*. By doing so, they took a new direction, just as Bonger had done in response to *De Nieuwe Tijd* group.

40 Quoted in van der Steen 2011, 88.

41 Kenteringsocialisten to the SDAP board, 21 April 1931, SDAP, no. 2739.

## Conclusion

In the 1890s, Anton Pannekoek and his collaborators introduced classical scientific socialism in the Netherlands, seeking to contribute to the socialist struggle through theoretical explorations and popularizations of Kautskyan Marxism. This intellectual tradition was taken in a different direction by Pannekoek's contemporary Willem Bonger. Originally, Bonger was inspired by Kautskyan Marxism, but soon he dismissed the idea of a socialist revolution and developed the idea of socialism as an applied science. Instead of approaching socialism as a reflective study of society, he promoted a socialism that was practical and pragmatic. Based on statistical and technical knowledge, policies were devised which would further the socialist cause. Bonger's firm belief in his own programme explains the dedication and fierceness with which he did his job as editor of *De Socialistische Gids* and the fights he picked with co-editors, party board members, and other critics. The closed and exclusive nature of his programme disgruntled many, but because he had a clear agenda for the journal, Bonger was unwilling to compromise. For him, the journal was a means to express his own interpretation of socialism and – by advancing the exact and social sciences – of quite literally manufacturing a socialist society. This coincided with his core ideological belief that, above anything, socialism was the science of making a socialist society. This redefinition of socialism-as-science subsequently influenced a younger generation of socialist intellectuals on how science, ideology, and politics were related. This younger generation developed its own intellectual tradition, just as Bonger had done before them. Bonger's thinking in terms of malleability and his focus on policy design became central aspects of post-war social democratic thinking, even if his strong belief in manufacturing a socialist society was dismissed.

Bonger poses an interesting and illuminating example of the ways in which socialist intellectuals thought about science, ideology, and politics. In many ways, he can be seen as a counter example to Pannekoek. Bonger put science at the core of his ideological and political programme. Applied sciences were not just a *means* to carry out socialist politics, they formed an integral part of Bonger's socialism. In comparison, the natural sciences, which Pannekoek mastered as part of his academic position, did not form a part of Pannekoek's Marxism. Rather, Pannekoek considered natural sciences and socialism to be distinct but complementary. The differences between Bonger and Pannekoek are remarkable since the two men had many things in common. They were scientists, contemporaries, both embraced Marxism in their younger years, and had become members of the SDAP because of it. Nevertheless, Bonger was far from the Marxist that Pannekoek was. As a result, he did not share Pannekoek's international prestige within the labour movement. While Pannekoek in many ways remained an orthodox Marxist philosopher, Bonger developed himself into an engineer of technocracy.

## Archives

| | |
|---|---|
| AP | Archief Anton Pannekoek. International Institute of Social History, Amsterdam. |
| DSG | Archief De Socialistische Gids, Maandschrift der Sociaal-Democratische Arbeiderspartij. International Institute of Social History, Amsterdam. |
| PJT | Archief Pieter Jelles Troelstra. International Institute of Social History, Amsterdam. |
| SDAP | Archief SDAP. International Institute of Social History, Amsterdam. |

## Bibliography

Bonger, Willem Adriaan. 1905. 'Criminalité et conditions économiques'. PhD Thesis, Amsterdam: University of Amsterdam.

—. 1919. 'Evolutie en revolutie'. *De Socialistische Gids* 4(4-5): 321-361.

—. 1925. 'Intellectueelen en socialisme'. *De Socialistische Gids* 10(12): 993-1012.

—. 1926. 'Over een paladijn en over nog meer. Een repliek'. *De Socialistische Gids* 11(4): 369-383.

Buiting, Henny. 1989. *Richtingen- en partijstrijd in de SDAP. Het onstaan van de Sociaal-Democratische Partij in Nederland (SDP)*. Amsterdam: Stichting Beheer IISG.

—. 2003. *De Nieuwe Tijd: sociaaldemokratisch maandschrift, 1896-1921. Spiegel van socialisme en vroeg communisme in Nederland*. Amsterdam: Aksant.

Faassen, Sjoerd van. 1980. 'Ten koste van de helderheid. De overgang van De Socialistische Gids tot Socialisme en Democratie, 1938'. In *Het tweede jaarboek voor het democratisch socialisme*, ed. by Jan Bank, Martin Ros, and Bart Tromp. Amsterdam: De Arbeiderspers, 149-162.

Gerber, John P. 1989. *Anton Pannekoek and the Socialism of Workers' Self-Emancipation, 1873-1960*. Dordrecht: Kluwer.

Hagen, Piet. 2010. *Politicus uit hartstocht: biografie van Pieter Jelles Troelstra*. Amsterdam: De Arbeiderspers.

Hartveld, Leo, Frits de Jong Edz., and Dries Kuperus. 1982. *De Arbeiders Jeugd Centrale AJC: 1918-1940/1945-1959*. Amsterdam: Van Gennep.

Heerikhuizen, Bart van. 1987. *W.A. Bonger, socioloog en socialist*. Groningen: Wolters-Noordhoff.

Hünlich, Reinhold. 1992. 'Gab es einen Kautskyanismus in der Epoche der II. Internationale?' In *Marxismus und Demokratie: Karl Kautskys Bedeutung in der sozialistischen Arbeiterbewegung*, ed. by Jürgen Rojahn, Till Schelz-Brandenburg, and Hans-Josef Steinberg. Frankfurt am Main: Campus Verlag.

Kuyper, Rudolf. 1926a. 'Het socialisme, de hoofdarbeiders en de komende cultuur'. *De Socialistische Gids* 11(3): 344-368.

—. 1926b. 'Nogmaals: het socialisme, de hoofdarbeiders en de komende cultuur'. *De Socialistische Gids* 11(5): 464-480.

Lucassen, Leo. 2010. 'A Brave New World: The Left, Social Engineering, and Eugenics in Twentieth-Century Europe'. *International Review of Social History* 55(2): 265-296.

Salvadori, Massimo L. 1979. *Karl Kautsky and the Socialist Revolution, 1880-1938*. London: NLB.

Steen, Margit van der. 2011. *Drift en koers. De levens van Hilda Verwey-Jonker (1908-2004)*. Amsterdam: Bert Bakker.

Steenson, Gary P. 1978. *Karl Kautsky, 1854-1938: Marxism in the Classical Years*. Pittsburgh, PA: University of Pittsburgh Press.

Troelstra, Pieter Jelles. 1919. 'De revolutie en de S.D.A.P'. *De Socialistische Gids* 4(3-9): 201-222, 377-401, 512-524, 566-579, 681-703.

Veldhuizen, Adriaan van. 2015. *De partij: over het politieke leven in de vroege S.D.A.P.* Amsterdam: Prometheus Bert Bakker.

Welcker, Johanna M. 1986. 'Pannekoek, Antonie'. Biografisch Woordenboek van het Socialisme en de Arbeidersbeweging in Nederland. http://hdl.handle.net/10622/ACDE7F4A-5D23-49DE-B973-9F58C1313E1F (accessed 28 March 2017).

Wiedijk, C.H. 1986. *Koos Vorrink: gezindheid, veralgemening, integratie: een biografische studie (1891-1940)*. Groningen: Wolters-Noordhoff.

Wiedijk, C.H., and Ger Harmsen. 1988. 'Vorrink, Jacobus Jan'. Biografisch Woordenboek van het Socialisme en de Arbeidersbeweging in Nederland. http://hdl.handle.net/10622/77504969-F0C4-40D1-ABFA-91115F59FFC8 (accessed 28 March 2017).

## About the Author

**Annemarie Rullens** studied Political Culture and National Identities at Leiden University. Her research focused on the relationship between socialist politics, science, and ideology in the Netherlands in the first half of the twentieth century. By analysing the writings of a group of intellectuals within the Social Democratic Workers' Party (Sociaal-Democratische Arbeiderspartij, SDAP) she researched the meaning of science for the party and its influence on political thought. Currently, Rullens works as a consultant for both the public and private sector.

# 9 Popularizing the Cosmos

## Pedagogies of Science and Society in Anton Pannekoek's Life and Work

*Jennifer Tucker*

**Abstract**

New expectations about the role of the astronomer in modern life emerged in the early twentieth century. This chapter sketches Anton Pannekoek's role in fostering new forms of public and political engagements with astronomy. Through his scientific writings and photography, Pannekoek did more than foster the wonders of nature and science. He also presented astronomy as a field that instilled large-scale visions of society and human progress. After considering Pannekoek's efforts to build stronger connections between science and polity in both galactic astronomy and council communism, it concludes with thoughts about Pannekoek as a key early twentieth-century figure in a new tradition of historical writing about scientific instruments and practice.

**Keywords:** Anton Pannekoek, photography, science popularization, historiography of science, Marxism, British Astronomical Association

> What really is the Milky Way? Exactly speaking, it is a phantom; but a phantom of so wonderful a wealth of structures and forms, of bright and dark shapes, that, seen on dark summer nights, it belongs to the most beautiful scenes which nature offers to man's eyes.
>
> Anton Pannekoek, *History of Astronomy*, 474.

In his 1961 historical account of the origin and development of astronomy, Pannekoek touched only lightly on the two subjects that had brought him greatest scientific fame: his investigations of the Milky Way and stellar spectra. Instead, he mapped the unfolding of what he referred to as 'the

Tai, Chaokang, Bart van der Steen, and Jeroen van Dongen (eds), *Anton Pannekoek: Ways of Viewing Science and Society.* Taylor & Francis Group, 2019
DOI: 10.5117/9789462984349_CH09

concept of the universe', which he defined as a 'new concept of the world' that had opened 'new ways of thinking'. Driven by 'a strong social development', he stated, astronomy since the sixteenth century had unsettled beliefs and certainties, disclosing 'that what seemed the most certain knowledge of the foundation of our life' (that is, the immobility of the sun) 'was merely an appearance'. The replacement of a fixed earth to an idea of 'endless space' had been a 'revolution', forcing the reorientation of humankind to the world. In those centuries of revolution, the contest over astronomical truth was 'an important element in the spiritual struggle accompanying the great social upheavals', he wrote, stirring the minds of 'mankind'. While other sciences (physics, chemistry, and biology were the ones he named) had gradually surpassed astronomy in their practical applications, he wrote, astronomy stood apart as a leading index of the "transformation of the human race" from the ancient world to the present, an 'essential part in the history of human culture."[1]

Studies on relations between Pannekoek's career as a scientist and his political ideals necessarily must weigh the tremendous growth of public prestige for astronomy in the first half of the twentieth century. Across his long career, involvement with astronomy and especially the scientific study of the Milky Way, Anton Pannekoek combined scientific discovery with a passion for fostering public understanding and an awareness of the leading astronomical debates of his day. Although he is recognized as a leading twentieth-century astronomer, Pannekoek's creativity and capacity to think beyond the boundaries of existing paradigms were not confined to astronomy. As a theoretical leader of the radical left wing of the communist movement, he also tried to imagine a radically new social order on earth. Given his prominence, it is tempting to see him mainly in terms of his unique characteristics. By taking a closer look at his navigations through the diverse contexts of early-twentieth-century scientific and political circles, however, we come to understand him not as an idiosyncratic lone individual, but as a figure who was greatly concerned with new forms of public engagement in both radical politics and the new astronomy and was fully immersed in their public networks.

This paper considers three spheres in which Anton Pannekoek worked to build stronger connections between science and polity in both galactic astronomy and council communism: his early associations with the British astronomical community, his involvements in photography, and his work as an author and leading popularizer of astronomical science. Across his life

1 Pannekoek 1961, 14-15.

and career, Pannekoek's ideas about the role of institutions in the changing conditions of knowledge, the preeminence of machines and instruments in his concepts of science, and *techne* as a kind of labour were shaped within a wider international astronomical community that was itself focused on fostering stronger relationships between astronomy and society. These wider ideas and associated practices informed both his astronomical practice and his approach to writing the history of science.[2]

The paper concludes with thoughts about Pannekoek as a figure who put the telling of historical narratives about astronomy in the foreground of his efforts to bridge practice and theory. From his earliest contact with British amateur astronomical associations in the 1890s to his later writings about the history of science Pannekoek always placed science *in* society: the changing conditions of knowledge, and the place of the scientist in those changes, were at the centre of his thinking and writings. He wrote about the conditions of science in a world that he genuinely believed was evolving into a socialist system. In both astronomy and radical politics, Pannekoek publicly articulated a strong connection between astronomy and its publics, from the scientific gathering of data on one hand, to the dissemination of research findings, on the other. In this reading of his work, astronomy is not seen as something separate from the wider public sphere (the sphere in which politics is supposed to operate). Instead, it is evident that although it is challenging to trace the connections between his 'scientific' and 'political' views, what connected them was his approach to a conception of the proper relation of science and the publics with which they were concerned. More than most astronomers of his day, Pannekoek was interested in bringing science to the public (through public lectures, scientific education, popularization, and dissemination of research findings in ways that were generally accessible, e.g. through his historical writings). At the same time, as even his earliest work in amateur astronomy shows, he recognized the value of members of the public to the construction of a culture of astronomy (e.g. through creation of norms and even, in some cases, as with the BAA, data).

## The Hand of Social Institutions in Regulating Astronomy

Pannekoek's ideas about science as a productive tool strongly informed his practical approach to astronomy networks. This may be seen, for example,

2 See especially: Tai 2017.

right from the beginning of his astronomical career, when he was in his early twenties and engaged with British astronomers about the need for techniques for synchronizing the work of many individual observers. Pannekoek's early interest in the British astronomical community's programme of astronomical research demonstrates an early commitment to widen public participation in astronomy.

The British amateur and professional astronomical community had high ambitions for a vast international network of scattered observers around the world. The British Astronomical Association (BAA), for example, was founded in 1890 to encourage amateur astronomers, but its members also included many professional astronomers, and the society was unusual in including women – making it more progressive than many other scientific organizations at the time.[3] The BAA became a central site for the collection of empirical data from disparate members scattered across the globe, who sent their astronomical observations and drawings to Burlington House in London, where they were discussed and often incorporated in maps and charts of astronomical objects.[4] Recognizing the challenge of coordinating the work of multiple observers, the BAA devised strategies and provided what might be termed 'epistemic advice' to prospective observers around the world.

These ideas resonated with the young Pannekoek. He especially was drawn to the work of Nathaniel Green, the president of the BAA. Green was also a painter and taught astronomers how to sketch physical landscapes showing surface details and cloud patterns. He was a long-time advocate of artistic and subjective skills in planetary drawing and engaged in discussions with other amateur astronomers about how to create maps and charts using data from far-flung individuals with different seeing and drawing styles.[5]

Like the British amateur astronomers, Pannekoek regarded the study of the Milky Way as a useful field open to young enthusiasts like himself, whose data-gathering contributions he both encouraged and actively sought to channel. In 1897, he published a paper titled 'On the existing Drawings of the Milky Way and the Necessity of Further Researches' in the *Journal of the British Astronomical Association*, in which he noted that for observations of the Milky Way '*no instrument whatever is required*; nothing, indeed,

3 Elizabeth Brown, possibly the only woman in England at the time to own her own observatory, became head of the Solar Section.

4 The work of British amateur astronomical observers is discussed in Tucker 2005, chapter 5.

5 On the role of drawings and photography in astronomy, c. 1870-1930, see Nasim 2013; Pang 1997; Tai 2017; Tucker 2005, 209-211; and Chaokang Tai, 'The Milky Way as Optical Phenomenon', in this volume, 219-247.

except a pair of rather sharp eyes and a sky of tolerable clearness'.[6] He praised standardization, social networks, utility, training and discipline, mechanical reproduction in the new amateur astronomy, seeing them as markers of progress. The following year, Pannekoek related to Green in a letter that he welcomed the formation of a section dedicated to the study of the Milky Way.[7]

Pannekoek's ideals of observation were not unique to Pannekoek, but were instead shared by many contemporary planetary astronomers in British amateur astronomical networks. Yet how were these ideals to be achieved in practice? In astronomy, there were widespread differences in various representations of the same object; as Pannekoek explained the familiar problem: 'In certain parts the two drawings seem to represent wholly different objects'.[8] These differences between the representations of different observers, Pannekoek explained, were caused by errors and problems of 'subjectivity'. Visual subjectivity thrived beside invocations of 'mechanical objectivity' in nineteenth-century science. The British astronomer Walter Maunders had even coined the term 'artistic personal equation' to describe the discrepancy between observers' planetary drawings – referring to the technical term, 'personal equation', that astronomers used to denote differences in the reaction times of observers who recorded the transit of stars across a telescopic meridian.[9] Although Pannekoek agreed that the multiplication of observers introduced subjective errors, he thought that they could be overcome. As he put it, 'the influence of all of these causes of error may be greatly diminished by the co-operation of many persons with a certain ability of observing, and much good will accrue to the cause of science'.[10]

Pannekoek saw the mechanical reproduction and distribution of charts as a possible solution for subjective errors. Most star charts, however, were not adapted for use for the Milky Way: Pannekoek thought their scale was too small, and left little space for drawing minute differences of brightness. He also noted that his fellow Dutch astronomer and leading describer and interpreter of the Milky Way, Cornelis Easton (1864-1929) had published a catalogue that made it possible to draw a star chart in cylindrical projection

6 Pannekoek 1897b, 39; emphasis added.

7 Anton Pannekoek to Nathaniel Green, 1 December 1898, RAS.

8 Pannekoek 1897b, 40.

9 For more on the 'artistic personal equation', see Tucker 2005. On the problem of the personal equation in Victorian astronomy, see esp. Schaffer 1988.

10 Pannekoek 1897b. For more on this, see Chaokang Tai, 'The Milky Way as Optical Phenomenon', in this volume, 219-247.

for use in Milky Way studies. Pannekoek took it upon himself to make these charts and had them lithographed by Easton, 'and now they may be had by anyone studying this subject at very slight expense'.[11]

Pannekoek's engagement with the BAA offers an early glimpse of what would become a long career of public engagement with astronomy. It also prefigured what would become a lively and enduring correspondence with leading British thinkers in both science and politics for the duration of his lifetime. Astronomy was well suited to public engagement, since it needed a large pool of observers, equipped with basic instruments. From his earliest days as an astronomer to the end of his career, Pannekoek showed a dedication to the value of contributions from extended networks of disparate observers – as well as to the practical challenges involved in building them.[12] This was no fleeting interest, for even at the end of his career, he expounded on how amateur astronomers could contribute to the observations of the Milky Way without any astronomical instrument, especially in the Southern subtropical zones known for their especially clear skies.[13]

## Theories of Science and Marxism, and the Pre-eminence of Machines

Pannekoek's vision of the role of the astronomical observer in visualizing the natural order of the galaxies coincided with his strenuous and devoted efforts, in other contexts, to visualize a new social order. As is discussed in other parts of this volume and in an important recent paper by historian of science Chaokang Tai, this can be seen in his attempts to clarify the relationship between science and Marxism in his writings beginning as early as the early 1900s.[14] Yet, it is in how he saw the relationship between technology, knowledge and labour, and his larger conception of the world that he shared much in common with his astronomical and political contemporaries, especially those who reached out to new public audiences

11 Pannekoek 1897a, 80.

12 Pannekoek 1961, 422. He credited amateurs with opening up new fields of astronomy, such as observations of shooting stars.

13 Pannekoek 1957.

14 Tai 2017. Pannekoek's influence in the field of Marxism continues to receive attention, following an upsurge of interest among a new generation of socialists and historians in the late 1960s. For more background on his politics see esp. Gerber 1989; Roth 2015; Boggs 1995; and Hoffrogge 2015. Reviews of Pannekoek's political writings are also useful, see, e.g. Lane 2005; Schurer 1963; Malandrino 1984; Souyri 1979.

beyond professionals (in the case of astronomy) and party leaders (in the case of politics).[15] For Pannekoek, as for other contemporary scientists, the essence of historical change was in technology:

> The basis of society – productive power – is formed chiefly through technology [...] Technology does not merely involve material factors such as machines, factories, coal mines and railroads but also the ability to make them and the science which creates this ability. Natural science, our knowledge of the forces of nature, our ability to reason and cooperate are all important as factors of production. Technology rests not only on material elements alone, but also on strong spiritual elements.[16]

Some glimpses of his thinking on this point can already be identified in the 1904 article 'Klassenwissenschaft und Philosophie', in which he called into question some tenets of orthodox Marxism. This article addressed two levels: 'an examination of the methods, meanings and objects of inquiry behind scientific knowledge; and an analysis of the position of science in human social and mental activity'.[17] Among the various forms of what the socialist philosopher Joseph Dietzgen (1828-1888) described as the 'thought activity' of a historical epoch, none had more importance for him than science – which, Pannekoek suggested, 'stands as a mental tool next to the material tools and, itself a productive power, constitutes the basis of technology and so is an essential part of the productive apparatus'.[18]

One outlet where Pannekoek's expressed his ideas about technology and knowledge was his writing on photography. Furthermore from 1908 to 1914, he wrote several articles for socialist papers with the intention of developing a body of popularized theory addressed to the average worker. As a Marxist, he faced the challenge of reconciling the need for a revolution with the idea of scientific progress: scientists believed that science was cumulative, at odds with the notion of revolution. In his writings from the first two decades of the century he held that the scientific disciplines of the nineteenth century were the 'spiritual basis of capitalism' yet at the same time that 'a certain form of science

15 On the professionalization of nineteenth- and twentieth-century astronomy in particular see, among others, Andrews and Siddiqi 2011; DeVorkin and Smith 2004; Heyck 1983; Pang 2002.

16 Pannekoek as in Gerber 1978, 9-10, n. 24.

17 Gerber 1978, 9.

18 Pannekoek 1948a, 19. Pannekoek played a major role in assuming Dietzgen's currency among rank-and-file working-class militants, see Gerber 1978, 4.

can be both an object and a weapon of class struggle'.[19] Historically, he thought, science had been subordinated to the requirements of class relations within a given social system. The science and technology of the socialist future would necessarily develop out of the foundations laid by previous scientific and social developments, but he lamented the fact that scientists in the twentieth century seemed isolated in their specialties or bearers of reactionary ideas.[20] By 1937, Pannekoek rejected determinism, and he believed that the key to socialist victory lay in a mental revolution amongst the working class, freeing them from their 'spiritual dependence' on the ruling class through the latter's control over the press, science, schools, and the church and the persistence of traditional ways of thinking, handed down 'in the form of prevailing beliefs and ideologies, and transferred to future generations in books, in literature, in art and in education'.[21]

## 'The Application of Photography was a Revelation'

Pannekoek regarded photography as a social and material technology – *techne* – and work with machines as a kind of labour.[22] Astronomy, in particular, he thought, rested on broad social and artisanal foundations, which became a major theme of his later writings in the history of astronomy. In *History of Astronomy*, for example, he explained:

> Astronomy profited from the increasing perfection of optical techniques developed on behalf of practical photography since its discovery in 1839. Laborious theoretical computations [by scientists] combined with the practical inventiveness of constructors [...] gradually creating a number of increasingly more perfect types of optical systems. [...] The demands for great brightness, an extended field, and faultless depicting could

19 Gerber 1978, 10.

20 Gerber 1978, 11.

21 Pannekoek as in Gerber 1978, 19.

22 For more on Pannekoek's photography, see Tai 2017, 226-230. Tai shows that, despite his contributions to astronomical photography, Pannekoek believed that photography could never replace human-eye observations (p. 226). By 1919, he had developed a method that would make photographic representation of the Milky Way possible through a technique of extrafocal photography. However, as Tai demonstrates (p. 228), unlike other photographs, his attempt was not to depict the Milky Way exactly: 'Instead, its purpose was for photometry, to provide numerical values for the brightness measurement of the Milky Way, which in turn could be used to construct isophotic maps.'

> not be met at the same time; [...] thus a wealth of different types has been invented and constructed. [...] Portrait objectives of larger angular aperture for the use of amateur photographers are found all over the world in thousands of cameras. [...] [I]n larger dimensions, they have been made especially for [astronomical] observatories, providing a new type of instrument that offered new aspects of celestial objects.[23]

While Pannekoek was not unusual in advocating the use of photography and other instruments in astronomy, he stood out from other scientists in his vocal and public advocacy of photography and in his effort to construct historical narratives about the role of machines and instruments in astronomical practice that included artisanal labour and collective achievement. According to Pannekoek, 'the application of photography was a revelation',[24] even if he also acknowledged the role of experience and subjectivity in photography, and its limitations compared to human-eye observation.

Pannekoek's extensive writings about photography of the Milky Way span from his earliest papers in the late 1890s, to his latest at the end of his career.[25] Although he praised visual observations of the Milky Way, he declared that the photographic method offered 'far greater wealth of detail', describing the difference between visual and photographic methods this way:

> We might describe the aspect the Milky Way would present to eyes that were far more sensitive to faint glares of light than ours and at the same time able to distinguish smaller details. A comparison with the focal photographs of Barnard and Ross shows a smoothing out of all sharp detail, thus gaining a true representation of the surface intensity which is lacking there.[26]

Photographs, Pannekoek stated, revealed 'the most picturesque aspect' of the Milky Way, 'the dark features, empty spaces almost without stars, often sharply defined.[27]

Ultimately, Pannekoek would devote around forty pages to the role of photography and photometry in astronomy in his *History of Astronomy*. He

23 Pannekoek 1961, 337.
24 Pannekoek 1961, 475.
25 See, e.g. Pannekoek 1897b; 1923; 1925; 1940; Pannekoek and Koelbloed 1949.
26 Pannekoek and Koelbloed 1949, 28.
27 Pannekoek 1897b, 39.

wrote that 'For a body so rich in detail as the moon, photography meant invaluable progress. A single photograph picturing the entire disc at once replaced hundreds of drawings that would have taken months and years at the telescope; moreover it was trustworthy as a document'.[28] Nevertheless, Pannekoek also recognized the failures and frustrations of photographic work:[29] 'A photographic atlas of the moon differs from a visual atlas, in that it gives the direct aspect of the moment with all its shadows; it is not a topographic map constructed by the astronomer out of a number of drawings at different phases.' For this reason, he thought, 'visual work should not be abandoned: many amateurs with good telescopes [...] continued their study of the details of special objects, chiefly to check the occurrence of small changes.'[30]

Pannekoek's drawings of the Milky Way – widely regarded as among the most accurate in the world at the time – were more than discovery tools, for exchanges among astronomers, however. Their reproducibility made them a critical part of the public's very image of astronomy and of science. In his lifetime, and partly under his influence, photography began taking on broader pedagogical, social, and conceptual aims: goals that in turn envisaged a much greater social role for science and scientists. On this reading, the 'revelations' of astronomy through the reproduction of photographs could be made available for wider projects (both socialist and humanist).[31]

For more than simply instructing people on the wonders of science and nature, Pannekoek and others also recognized that photographs of the Milky Way and other astronomical phenomena presented new opportunities for instilling visions of society and opening new possibilities for social progress. Over the course of his career his writings and lectures were filled with discussions about the productive value of labour in relation to the progress of knowledge.

28 Pannekoek 1961, 373.

29 He said that 'the work was never finished, because every succeeding year brought a new opposition, with new observations'; if this work were neglected, however, 'the predicted result would be more and more in error'. Pannekoek 1961, 354.

30 Pannekoek 1961, 374.

31 Beyond instructing people on the wonders of science and nature, Pannekoek recognized that depictions of the Milky Way and other astronomical phenomena presented new opportunities for instilling visions of society and opening new possibilities for social progress. Across Europe and the Soviet Union, exhibitions and public displays of astronomical phenomena, including planetaria, were being made to serve broader socialist and humanist goals. For more on the wider context, see esp. Benjamin 2006; Smolkin-Rothrock 2011; Strickland 2015.

## Pedagogies of Science as Historical Narratives

Pannekoek quit the Communist International in 1921 and went into a self-imposed six-year long break from revolutionary politics. In 1921, he built an astronomical laboratory dedicated to teaching and research. Including the measurement and reduction of photographic plates of the Milky Way taken by other observatories throughout the world.[32] This research remained an important topic at the astronomical institute for the next couple of decades, even as Pannekoek redirected his attention to the newly emerging field of astrophysics of stellar atmospheres. Meanwhile, the political movement that he had led remained relatively small, and according to some historians, struggled to gain any support or relevancy.[33]

In his astronomical work and popular-science writings, as in his ongoing political work with workers' councils and left-wing politics, Pannekoek tackled questions of truth but also of visibility and epistemology: what could be known, and through what means. To Pannekoek, what mattered *was the integrity of the process itself – the scientific method.* From the middle years through to the end of his career, Pannekoek's continued correspondence with Communists and other labour activists provided an outlet for his critical reflections, as did his later widely read popular history of science writings.

Across the spheres in which he worked, his epistemic values were engaging with both Marxist theories of science and standard disciplinary norms within astronomy; albeit often in unpredictable and uneven ways. During the course of his life, a clear connection is exhibited between his ideas and values about science and the civic ideals that he upheld. It is manifest in his philosophical understanding of the public role of science and, even more important perhaps, in the way that he practised astronomy and wrote its history with a keen eye to its social conditions. Yet Pannekoek's ideas about science and Marxism, far from being idiosyncratic, must be seen as representative of a wider response to a wider set of rising ideas in Europe in the early twentieth century about practical astronomy, mass culture, and Marxism that others also shared.

32 By founding an astronomical laboratory, Pannekoek explicitly followed in the footsteps of Groningen professor J.C. Kapteyn. See Edward P.J. van den Heuvel, 'Anton Pannekoek's Astronomy in Relation to his Political Activities and the Founding of the Astronomical Institute of the University of Amsterdam', in this volume, 25-50.

33 Discussed in Gerber 1978; Hoffrogge 2015, among others.

## History of Astronomy as a Vision of Politics and Society

Pannekoek's writings on the history of science clearly attest to his idea about the role of technology and astronomy in human history. History of science writing was a site of significant political debate in the 1930s and 1940s. Pannekoek's commitment to popularizing the scientific method in the wider public sphere and seeking to change the terms in which its role in society was understood, was a goal shared by many fellow Marxist scientists who also turned to history at around the same time.

In 1931, at a landmark event in many origin stories of the history of science, the Second International Congress of the History of Science and Technology took place in London, where the Soviets were represented by a delegation led by Nikolai Bukharin. It was here that Boris Hessen, the Soviet physicist and philosopher, presented his now famous paper, 'The Socio-Economic Roots of Newton's Principia', which became foundational in the history of science, opening the door to modern studies of scientific revolutions and sociology of science. In the paper, Hessen argued that Newton's work was not the disinterested study of the natural world, but was motivated by an attempt to solve the problems of the day.[34]

Hessen's focus on the relationship between society and science attracted significant attention, yet similar views had been expressed a year earlier by Pannekoek, in a paper that he published titled 'Astrology and its Influence on the Development of Astronomy' in *the Journal of the Royal Astronomical Society of Canada*.[35] There, he argued that scientific knowledge was embedded in the social and economic conditions of its time, especially in the case of astronomy. Indeed, he explained, 'For the other sciences, history is confined almost entirely to the last three or four centuries; their development took place within the walls of universities and laboratories, far from the convulsions of social and political life.' Their practitioners were always modern men whose traditions were similar to own. 'With astronomy matters are different', he said 'its history accompanies the development of mankind from its first beginnings.'

> Our predecessors were Babylonian priests and magicians, Greek philosophers, Arabian princes, medieval monks, Renaissance noblemen

34 On Hessen, see Freudenthal and McLaughlin 2009; Graham 1985.

35 Pannekoek 1930; this lecture was originally delivered in Dutch in 1916 as inaugural lecture in Leiden when Pannekoek started as unpaid lecturer in the history of astronomy, see 1916. I am grateful to Chaokang Tai for the reference.

> before they turned into modern university professors. For them the science of the stars stood not apart from their other opinions but was intimately interwoven with their philosophical and religious conception of life. In the sixteenth century the contest about astronomical truth was part of a struggle between world conceptions, and was deeply connected with the social struggles of that time. In this strife astronomy cleared the way for the freedom of scientific research more generally.[36]

He continued, saying that as astronomy was more closely shaped by practical necessities of life such as commerce and time reckoning:

> What was the reason that these primitive people turned their eyes to the stars and began to observe them regularly? Was it admiration of the beauty of the heavens, was it the dawning impulse of study to find out the cause of the phenomena? No, it was the hard necessity of life that induced them to look at the sky, the practical needs connected with their labour and intercourse with other people.[37]

He concluded by gently rebuking historians who dismissed the idea that astrology was founded on 'reasons from experience'. In fact, he retorted, 'modern writers are in the habit of speaking of old astrology as a regrettable aberration of the human mind, and of trying to wash the famous astronomers of history clean from the stain of having believed this superstition'. Yet 'nothing could be more false than this standpoint'. 'In our textbooks only that part of ancient astronomy is reproduced which coincides with the beginnings of our modern science. But in reality, the ancient science of the stars was in the first place astrology'.[38]

As a young astronomer, Pannekoek had struggled with what he saw as lack of the social relevance of his work. As he once wrote:

> Why can't I participate and find my place in the struggle? While everyone (?) contributes to improving the [social] conditions, I'm here, reducing meridian positions. Science is surely the only lasting and progressive factor in the changing of people and conditions. It must prepare for a

36 Pannekoek 1930, 159.
37 Pannekoek 1930, 160.
38 Pannekoek 1930, 169.

> better future: it is the reason why humanity has come to the point where it can enable her to become free and happy. But what mental gymnastics is required to follow the connection, in all its twists, between social happiness and reducing stellar positions.[39]

Pannekoek wrote this in a period before he moved to Germany, when he was a young aspiring astronomer and struggling with his political ideas. Now, he found history as a resource for answering this question. Pannekoek dedicated himself to writing popular histories of astronomy from 1930 onwards, becoming one of the discipline's leading chroniclers at a time when history of science itself was on the rise. In writing popular histories of astronomy, he sought to promote the status and progressive values of science in modern society. He also wanted to correct the popular image of astronomy as a labour-neutral endeavour; far from being otherworldly, astronomy was a science that was embedded in the rough-and-tumble of life, in contrast, he thought, to other sciences.

Pannekoek expressed the view that science was not an individual pursuit of knowledge, but a social activity; one in which the main idea was producing results through practice. He also emphasized the social-economic roots of science with early astronomy as a primary example. Pannekoek's popular writings about astronomy should be placed in the context of other contemporary Marxist writers on the history of science, including British counterparts like J.D. Bernal (1901-1971) as well as the Austrian philosopher of science Edgar Zilsel (1891-1944).[40] As a historical materialist, he affirmed the social origins of ideas, rather than stressing only the genius of individual men. The works of Ptolemy, Copernicus, Kepler, Newton, and Laplace were not epiphanies but 'consecutive steps in our knowledge of world-structure'.[41]

Advancing what historians might now recognize as a rejection of the 'Whig interpretation of history', Pannekoek contended that it was necessary for the historian to inhabit the worldview of the times, rather than assessing its strength or weakness from the standpoint of present understanding. This may be seen in his historical treatment of the importance of medieval astrology in the historical development of modern science. Pannekoek

39 See Anton Pannekoek, 'Wijsbegeerte en Politiek, Notebook 1898-11', June 1899, API, on 12. Translation by Chaokang Tai.

40 See Zilsel 1942. Although Hessen is generally brought up to do with Marxist accounts of the history of science, Zilsel gets discussed far less. Long 2011 has a section on Zilsel. I am grateful to Robert W. Smith for pointing out the relevance of Zilsel's work in this context.

41 Pannekoek 1961, 422.

argued that astrology and astronomy shared their basis in socioeconomic conditions: He asked:

> If we place ourselves a moment into the ideas of those times then what more sublime aim could there be imagined than to investigate and to discover the most intimate connection between men and the world, between the course of the stars and happenings on earth? [...] From this standpoint we will not be astonished to find most of the great astronomers of antiquity also believing in astrology.[42]

On Pannekoek's account, astrology, far from being a false vision, helped lay the foundations for the later development of modern astronomy with its careful and assiduous observation of irregularities with celestial bodies, including those of no use for calendar and travel. Instead of saying that Renaissance Europe was in the grip of false superstitions like astrology, limited to a few outlier magicians, he declared that the entire 'world concept' of medieval Europe was 'pervaded and dominated by astrology', a development that he linked with the facilitation of a spirit of 'wonder' that he connected with the rise of inductive science.[43]

In Pannekoek's later life, then, science and socialism finally came together in a direct manner. In his history of science publications he actively attempted to relate the study of astronomy with the practical needs and technological advances of the societies in which it was developed. He argued, for example, that the discovery of Neptune was much better received in France than in England because in France the bourgeoisie was still struggling for power and could make use of such a remarkable scientific achievement, while in England, the battle had already been won.[44]

For Pannekoek, the materialist conception of history was neither a fixed system nor a certain theory; it was a method of research that searched for the plausible causes of social developments.[45] Yet often, the focus of Marxist historians of the 1930s and 1940s was on economic factors such as ownership of the means of production, labour relations, and the distribution of capital. Pannekoek's research method, on the other hand, was to explain how ideas emerge as the result of economic, social, and ideological conditions. In

42 Pannekoek 1930, 170.

43 Pannekoek 1961, 176-177.

44 On the Neptune discovery and its interpretation, see Smith 1989; Hubbell and Smith 1992; and Bart Karstens, 'Anton Pannekoek as a Pioneer in the Sociology of Knowledge', in this volume, 197-217.

45 Tai 2017, 247.

particular, he put a strong emphasis *on the role of the mind in interpreting these material factors.* The role of the mind, he argued, differentiated historical materialism from mechanical materialism, which, according to him, reduced the entire world to the deterministic movement of particles.[46] Historical materialism was not limited to physical matter alone; instead, it was expanded to include anything that could have an objectively observable effect. As he wrote:

> The human mind is entirely determined by the surrounding real world. […] [T]his world is not restricted to physical matter only, but comprises everything that is objectively observable. The thoughts and ideas of our fellow men, which we observe by means of their conversation or by our reading are included in this real world. Although fanciful objects of these thoughts such as angels, spirits, or an Absolute Idea do not belong to it, the belief in such ideas is a real phenomenon, and may have a notable influence on historical events.[47]

Pannekoek emphasized what his contemporaries referred to as the 'mental' factors of materialism (e.g. tradition and religion). Only by removing the strict demarcation of mind and matter, did he think that society could be researched with scientific methods.

Pannekoek's *History of Astronomy* was published in Dutch in 1951, and later translated and published in English in 1961. It stands out from other contemporary works about the history of science in several respects. While it praised scientific achievements and instrumental advances, it did not focus on single individuals or great discoveries; instead, its central themes were the refinement of practice and theory, the struggles between 'world systems' of knowledge, and the relevance of geography, climate, and trade, and sociopolitical conditions as historical forces that shaped astronomical science and instrumentation. In his discussion of science in antiquity, he writes, for example, that 'What constituted the character strength of the Romans, their sense of social-political organization, created a mode of time-reckoning destined to dominate the entire future civilized world'.[48] In the book, he carried forward his ideas about the history of science beyond his papers on Neptune and astrology, and made his views accessible to a broader reading public.

46 Tai and van Dongen 2016, 66.

47 Pannekoek 1937, 451. See also Tai 2017, 248.

48 Pannekoek 1961, 146.

In a further nod to the importance of social and material practice in science, he also pursued a novel approach in his practice as a historian. This is reflected in his inclusion of images, a relatively novel departure from other general histories that were being published (on any subject) in the 1950s and 1960s: Pannekoek's popular history was unusually well illustrated with 24 historic maps, photographs and drawings. These images were not merely included for decorative purposes, they reinforced a central theme of the book: the centrality of visual observation and representation in the history of scientific practice.[49] Pannekoek wrote extensively about the importance of the practice of visual representation in the history of science. His *History* arguably contains some of the key historical analysis at mid-century of the role of observation, photography, and drawing in astronomy, including extensive discussions about the benefits of the photographic method in astronomy since 1839, from the first photographs of the sun and moon, to the 'first usable photographs of the starry heavens' in 1864, to the measurement of the position of stars in atlases and beyond.[50] He did not gloss over the practical challenges of photography, or discuss only its virtues. He emphasized the material practices that were associated with stellar photography and put the development of photographic methods in a wider historical context of societal changes, including the transformation of scientific *labour*. He wrote that 'Technical precision in electrical control of gigantic instruments is the material basis of modern astronomy', adding that 'Modern development of astronomical instruments would not have been possible without the growth of techniques in nineteenth-century industry, which revolutionized the entire aspect of society'. The astronomer's labour was like the driver of other colossal machines: 'the small brain of the huge steel organism'.[51]

Just as images were central to his practice of science, they were also central to his practice of history. Similarly, his emphasis on the importance of amateurs in his history of astronomy mirrored his discussion about the significance of amateur observation for opening up new fields in contemporary astronomy.[52] *History of Astronomy*, then, represented a return, in his historical writings, to some of the core epistemic concerns of his scientific writings.

49 Pannekoek's 1951 *History of Astronomy* resonates with British left-wing scientist Joseph Needham's later volumes of *Science and Civilisation in China* (1954-2015).

50 Pannekoek 1961, 337; see on photography, 345, 373-4, 405-6, 434-5, and 485-6; and on photometry, 385-6, 438, 440-1, 446, among others.

51 Pannekoek 1961, 338.

52 Pannekoek 1961, 422.

## 'The Scientific Worker in Overalls'

A second aspect that differentiates Pannekoek from other contemporary scientists who were writing about the history of science in the post-war period was his insistence on referring to scientists as *workers*. The theme of scientific research as a form of work, with its own relation to changing economic and social conditions, reverberates in his historical writings. In contrast to philosophers and historians who argued for science's internal logic of development (and although he did refer at times to the role of 'genius' in discovery) he refused to describe scientific enterprise as the mere unfolding of truth to superior intellects. The timing was crucial, for Pannekoek's thesis was in sharp contrast to C.C. Gillispie's historical narrative in *Edge of Objectivity: An Essay in the History of Scientific Ideas*, published in 1960, a year before the English edition of Pannekoek's *History of Astronomy*. Gillispie portrayed the rise of modern science as the development of objectivity through the study of nature. From Galileo's analysis of motion to the theories of evolution and relativity, Gillispie evoked personalities over instruments, and individual genius over collective labour.[53]

Pannekoek, on the other hand, complained that in publications of important scientific researches, the labour that had been involved was obscured. In a 1948 essay on Kepler's planetary theory, he praised Kepler for being different from others in this respect:

> Usually in the publication of new important researches only the results with the data and arguments are given; the discoverers keep to themselves how they arrived at them, their fruitless endeavours, their detours, their failures, and exhibit the result as a well-rounded harmonious structure, as a work sometimes of art, constructed straightforwardly, where all traces of the difficult searching have been effaced. Thus Copernicus, Newton, Laplace, Gauss. This is fine for study and admiration. But in this way outsiders get a wrong idea of the making of science; they do not suspect, what every scientific worker knows through his own practice, how many painful failures and long detours one must go through before finally the direct way is found which then afterwards is easily seen as the obvious truth.[54]

Kepler, by contrast, did not hide that, when he did science, he was *working*.

53 Gillispie 1960.

54 Pannekoek 1948b, 63.

> Kepler, differently, exposes his entire course of research, his errors, his false suppositions, and their disclosure, his perplexities and new endeavours, till the simple truth springs forward; all is laid open before the reader.[55]

And, perhaps in a more autobiographical mode, Pannekoek went on to discuss the fact that for years Kepler faced political battles and did not receive a salary. According to Pannekoek, Kepler's *Astronomia Nova*, showed 'a special character different from most of the great works of science'.[56] Kepler's book, Pannekoek argued, offered 'a true image of the growth of scientific discovery; here we see, as it were, the *scientific worker in overalls*'.[57] Pannekoek was talking here about the need for visualizing *astronomical work* – not simply celestial *phenomena*.

Pannekoek's distinctive position was noticed. The Harvard historian of science George Sarton (1884-1956) followed Pannekoek admiringly in his own foundational book in the history of science, *Introduction to the History of Science* (1927-1948).[58] Much later, British sociologist Barry Barnes, one of the founders of the strong programme in the Sociology of Scientific Knowledge, used Pannekoek's analysis of the discovery of Neptune in his account of Kuhn and scientific discovery.[59]

## Conclusion

This paper has suggested that historical appraisals of Pannekoek require consideration of broad changes that transformed European astronomy in the early twentieth century, when wider public discourses and political debates focused intently on the role of science in society. Efforts to understand his life and work and to take account of both his scientific and political activities have faced the obstacle that Pannekoek himself tried to keep his two careers strictly separated. Another impediment to thinking about connections across his spheres of work may be that scholarship about the history of politics and history of science themselves both tend to engage

55 Pannekoek 1948b, 63-64.

56 Pannekoek 1948b, 63.

57 Pannekoek 1948b, 64; emphasis added.

58 Sarton 1927-1948.

59 Barnes 1982. See also Barnes 1974; Shapin 1982. For an assessment of these references to Pannekoek, see Bart Karstens 'Anton Pannekoek as a Pioneer in the Sociology of Knowledge', in this volume, 197-217.

separate literatures and assign importance to different facets of his life and career. Recently, however, new attempts have been made to discover links between Pannekoek's science and political ideology through a focus on his biography, including individual methodological decisions and epistemic beliefs or virtues.[60]

This work sees Pannekoek as embedded in the wider fabric of both radical politics and twentieth-century astronomy. Pannekoek engaged the publics of astronomy on many levels. From his early associations with the British amateur astronomers, through his contacts with British political figures such as Sylvia Pankhurst and other supporters of workers' councils in Britain, to his largely neglected prominence in the creation of the field of history and sociology of science in the UK, Pannekoek's explorations of the social conditions of science both shaped and reflected a large public vision of astronomy in a new age of contradictions and political tensions.

Anton Pannekoek straddled both science and social criticism: as a scientist, he was concerned with how we can learn about galaxies beyond our capacity to observe; as a socialist, he wondered how we can imagine and bring into being a better future society. Early in his astronomical career, he concluded that the involvement of a mass of amateur observers was one of the best methods of gathering data to plot the structure of the Milky Way. Similarly, as a socialist he later concluded that the involvement of large numbers of people through workers' councils was the best way to organize social decision-making – an approach that would leave him at odds with many of the centralized solutions favoured by the Bolsheviks who seized power in Russia. To better understand the relationship between Pannekoek's science and his politics, we need to reflect on the period in which he worked. This was a world, as these circumstances clearly illustrate, in which science and astronomy were seen as critical to the elucidation, not just of astronomical discoveries, but also of public understandings of its essence.[61] In his left-wing politics, as in his scientific research, Pannekoek saw history (in his words) as 'only a preamble to the future'. What occupied him and others in his political circle, was not so much the 'past in retrospect' but 'the outlook towards the future' when, in the study of the universe, as in the tumultuous political history of the early twentieth century, technology was the driving force of revolutionary change.[62] Only proper attention to the worker, in astronomy and elsewhere, could capture all developments.

60 Tai and van Dongen 2016; Tai 2017.

61 Pannekoek 1961, 496.

62 Pannekoek 1961, 483.

## Archives

API Archive of the Anton Pannekoek Institute, University of Amsterdam.
RAS RAS Letters. Royal Astronomical Society.

## Bibliography

Andrews, James T., and Asif A. Siddiqi, eds. 2011. *Into the Cosmos: Space Exploration and Soviet Culture*. Pittsburgh, PA: University of Pittsburgh Press.

Barnes, Barry. 1974. *Scientific Knowledge and Sociological Theory*. London: Routledge.

—. 1982. *T.S. Kuhn and Social Science*. London: Macmillan.

Benjamin, Walter. 2006. 'Zum Planetarium'. In *Gesammelte Schriften IV.1: Kleine Prosa. Baudelaire-Übertragungen*, ed. by Tillman Rexroth. Frankfurt am Main: Suhrkamp, 146-148.

Boggs, Carl. 1995. *The Socialist Tradition: From Crisis to Decline*. New York: Routledge.

DeVorkin, David H., and Robert W. Smith. 2004. *The Hubble Space Telescope: Imaging the Universe*. Washington, DC: National Geographic Society.

Freudenthal, Gideon, and Peter McLaughlin. 2009. 'Classical Marxist Historiography of Science: The Hessen-Grossmann-Thesis'. In *The Social and Economic Roots of the Scientific Revolution*, ed. by Gideon Freudenthal and Peter McLaughlin. Dordrecht: Springer Netherlands, 1-40.

Gerber, John P. 1978. 'The Formation of Pannekoek's Marxism'. In *Pannekoek and the Workers' Councils*, by Serge Bricianer. Saint Louis: Telos Press, 1-30.

—. 1989. *Anton Pannekoek and the Socialism of Workers' Self-Emancipation, 1873-1960*. Dordrecht: Kluwer.

Gillispie, Charles Coulston. 1960. *The Edge of Objectivity: An Essay in the History of Scientific Ideas*. Princeton, NJ: Princeton University Press.

Graham, Loren R. 1985. 'The Socio-Political Roots of Boris Hessen: Soviet Marxism and the History of Science'. *Social Studies of Science* 15(4): 705-722.

Heyck, Thomas William. 1983. *The Transformation of Intellectual Life in Victorian England*. New York: Palgrave Macmillan.

Hoffrogge, Ralf. 2015. *Working-Class Politics in the German Revolution: Richard Müller, the Revolutionary Shop Stewards and the Origins of the Council Movement*. Ed. by Radhika Desai. Trans. by Joseph B Keady. Chicago, IL: Haymarket Books.

Hubbell, John G., and Robert W. Smith. 1992. 'Neptune in America: Negotiating a Discovery'. *Journal for the History of Astronomy* 23(4): 261-291.

Lane, Fintan. 2005. 'Contradicting the Bolsheviks: Anton Pannekoek and European Marxism'. Review of *Workers Councils*, by Anton Pannekoek. *Saothar* 30: 103-113.

Long, Pamela O. 2011. *Artisan/Practitioners and the Rise of the New Sciences, 1400-1600*. Corvallis, OR: Oregon State University Press.

Malandrino, Corrado. 1984. 'Il socialismo internazionalista di Anton Pannekoek tra guerra e rivoluzione'. *Studi Storici* 25(2): 405-429.

Nasim, Omar W. 2013. *Observing by Hand: Sketching the Nebulae in the Nineteenth Century*. Chicago, IL: University of Chicago Press.

Pang, Alex Soojung-Kim. 1997. 'Visual Representation and Post-Constructivist History of Science'. *Historical Studies in the Physical and Biological Sciences* 28(1): 139-171.

—. 2002. *Empire and the Sun: Victorian Solar Eclipse Expeditions*. Stanford, CA: Stanford University Press.

Pannekoek, Anton. 1897a. 'New Charts for Inserting the Milky Way'. *Journal of the British Astronomical Association* 8(2): 80-82.

—. 1897b. 'On the Existing Drawings of the Milky Way and the Necessity of Further Researches'. *Journal of the British Astronomical Association* 8(1): 39-42.

—. 1916. *De astrologie en hare beteekenis voor de ontwikkeling der sterrekunde*. Leiden: Brill.

—. 1923. 'Photographic Photometry of the Milky Way and the Colour of the Scutum Cloud'. *Bulletin of the Astronomical Institutes of the Netherlands* 2(44): 19-24.

—. 1925. 'On Photographing the Corona Spectrum during Total Solar Eclipses'. *Bulletin of the Astronomical Institutes of the Netherlands* 3(86): 43-44.

—. 1930. 'Astrology and Its Influence upon the Development of Astronomy'. *Journal of the Royal Astronomical Society of Canada* 24(4): 159-176.

—. 1937. 'Society and Mind in Marxian Philosophy'. *Science & Society* 1(4): 445-453.

—. 1940. 'Two New Instruments for the Reduction of Stellar Spectra'. *Bulletin of the Astronomical Institutes of the Netherlands* 9(335): 155-157.

—. 1948a. *Lenin as Philosopher: A Critical Examination of the Philosophical Basis of Leninism*. New York: New Essays.

—. 1948b. 'The Planetary Theory of Kepler'. *Popular Astronomy* 56(552): 63-75.

—. 1957. 'Colour Differences in the Milky Way?' *The Observatory* 77(901): 241-242.

—. 1961. *A History of Astronomy*. London: George Allen and Unwin.

Pannekoek, Anton, and David Koelbloed. 1949. *Photographic Photometry of the Southern Milky Way*. Publications of the Astronomical Institute of the University of Amsterdam 9. Amsterdam: Stadsdrukkerij.

Roth, Gary. 2015. *Marxism in a Lost Century: A Biography of Paul Mattick*. Chicago, IL: Haymarket Books.

Sarton, George. 1927-1948. *Introduction to the History of Science*. 3 vols. Baltimore, MD: Williams and Wilkins.

Schaffer, Simon. 1988. 'Astronomers Mark Time: Discipline and the Personal Equation'. *Science in Context* 2(1): 115-145.

Schurer, Heinz. 1963. 'Anton Pannekoek and the Origins of Leninism'. *The Slavonic and East European Review* 41(97): 327-344.
Shapin, Steven. 1982. 'History of Science and Its Sociological Reconstructions'. *History of Science* 20(3): 157-211.
Smith, Robert W. 1989. 'The Cambridge Network in Action: The Discovery of Neptune'. *Isis* 80(3): 395-422.
Smolkin-Rothrock, Victoria. 2011. 'The Contested Skies: The Battle of Science and Religion in the Soviet Planetarium'. In *Soviet Space Culture: Cosmic Enthusiasm in Socialist Societies*, ed. by Eva Maurer, Julia Richers, and Monica Ruthers. New York: Palgrave Macmillan, 57-80.
Souyri, Pierre. 1979. Review of *Les Conseils Ouvriers*, by Anton Pannekoek. *Annales. Histoire, Sciences Sociales* 34(4): 881-884.
Strickland, Manuela Pallotto. 2015. 'Under a Starry Vault. Warburg, Jung and the Renaissance of Ancient Paganisms at the Beginning of the 20th Century'. *Aisthesis. Pratiche, Linguaggi e Saperi Dell'estetico* 8(2): 41-57.
Tai, Chaokang. 2017. 'Left Radicalism and the Milky Way: Connecting the Scientific and Socialist Virtues of Anton Pannekoek'. *Historical Studies in the Natural Sciences* 47(2): 200-254.
Tai, Chaokang, and Jeroen van Dongen. 2016. 'Anton Pannekoek's Epistemic Virtues in Astronomy and Socialism: Personae and the Practice of Science'. *BMGN – Low Countries Historical Review* 131(4): 55-70.
Tucker, Jennifer. 2005. *Nature Exposed: Photography as Eyewitness in Victorian Science*. Baltimore, MD: Johns Hopkins University Press.
Zilsel, Edgar. 1942. 'The Sociological Roots of Science'. *American Journal of Sociology* 47(4): 544-562.

## About the Author

**Jennifer Tucker** is a cultural historian of science specializing in the study of nineteenth-century technology, law, photography, and visual culture. The author of *Nature Exposed: Photography as Eyewitness in Victorian Science*, she recently completed a book manuscript about photography and facial likeness in the courtroom. The editor of an issue of *History and Theory* about photography and history, she is co-editor of the forthcoming *Photography and Law Reader*, Image Editor of *History and Technology*, and co-chair of *Radical History Review* journal. She is Associate Professor of History at Wesleyan University.

# 10 Anton Pannekoek as a Pioneer in the Sociology of Knowledge

*Bart Karstens*

**Abstract**

Thirty years after the publication of Pannekoek's 1953 paper on the history of the discovery of Neptune, it was cited as an early forerunner of the strong programme in the Sociology of Scientific Knowledge. This recognition, however, was achieved by laying Pannekoek's paper on a Procrustean bed. On close inspection we find that Pannekoek's approach to history accords best with Merton's sociology of knowledge. Thus, Pannekoek gained a reputation as an important innovator in historiography of science for the wrong reasons. This paper offers a much-needed correction, which facilitates a more precise evaluation of the innovative aspects of Pannekoek's historical work, especially with respect to the effect of external factors on the course of science.

**Keywords:** Anton Pannekoek, sociology of science, strong programme, Robert K. Merton, discovery of Neptune

## Introduction

The activities of Anton Pannekoek in the fields of social theory and astronomy are well known and the subject of most contributions to this volume. Perhaps less well known are his efforts in the history of science. Pannekoek devoted two books to the history of astronomy. Early in his career he published *De Wonderbouw der Wereld: De grondslagen van ons sterrekundig wereldbeeld populair uiteengezet* (1916) and after his retirement he wrote *De Groei van ons Wereldbeeld: Een geschiedenis van de sterrekunde* (1951), which was translated into English as *A History of Astronomy* (1961).

Tai, Chaokang, Bart van der Steen, and Jeroen van Dongen (eds), *Anton Pannekoek: Ways of Viewing Science and Society*. Taylor & Francis Group, 2019
DOI: 10.5117/9789462984349_CH10

While *Wonderbouw* contains an exposition of the state of the art of astronomical knowledge at the time when it was published, offering only occasional historical perspectives, *De Groei van Ons Wereldbeeld* is a truly historical work, tracing the history of astronomy from Antiquity to the present. From this book, Pannekoek lifted a paper on the discovery of Neptune, which was published in *Centaurus* in 1953. Interestingly this paper was taken up from the 1980s onwards by scholars in the Sociology of Scientific Knowledge (henceforth referred to as SSK) as a splendid early example of how to properly explain the history of science. Pannekoek's interpretation of the discovery of Neptune even made it into a 2004 *Handbook of Epistemology*, as David Bloor in his contribution on SSK cited Pannekoek's study as one of the prime examples in support of the SSK approach.[1] This is a remarkable feat, given that Pannekoek was not a professional scholar in the History and Philosophy of Science.

It is even more remarkable that Pannekoek's paper was taken up by a group of scholars that defend a radical position in the sociology of knowledge, which is identified as the *strong programme*. The leading idea of the *strong programme* is that the rejection and acceptance of all claims to knowledge (i.e. what we come to hold as true or false about the world) is *always* ultimately determined by social factors. This idea dates from the mid-1970s and represented a further step in the development of epistemological relativism.[2] Thomas Kuhn and others had already rejected the idea that the structure of the world is just there, waiting to be discovered, and hence that nature itself is the sole referee of the correctness of our theories about nature.

Yet, to let the human element play a decisive role in all of science was a radical step to make. Other positions in the sociology of knowledge exist as well, granting an important role to the notion of science as a social process, because science after all is a human endeavour, but at the same time acknowledging the importance of nature in our theories of the world too. One can, for example, argue that society to a large extent decides which topics scientists will investigate. The direction of research is then determined by social factors, but the content of it does not have to be, since one can take the view that scientific research roughly proceeds in the same manner

1 Bloor 2004. The *Handbook of Epistemology*, together with *The Oxford Handbook of Epistemology*, is one of two of the most recent handbooks of epistemology. Its goal is to provide an overview of the historical roots and systematic development of the theory of knowledge. It contains 28 entries written by a variety of scholars. Bloor's 'Sociology of Scientific Knowledge' is one of these entries.

2 A key text is Bloor 1976.

everywhere. In such approaches the social only *weakly* determines scientific outcomes.[3]

Overall, Pannekoek's historiography is marked by optimism and progressivism. He makes it abundantly clear that the historical development of science should be interpreted as a logical succession of stages towards uncovering the truth. Next stages build on earlier ones and in this way our understanding of the world gradually grows. Now this perspective on the history of science cannot be squared with SSK's epistemological relativism at all, given that the traditional notion of truth no longer holds in the *strong programme*. To find Pannekoek hailed as a pioneer of SSK is thus surprising. Moreover, while there were *weak programmes* in the sociology of science in the 1950s, no one was already pursuing something as radical as SSK scholars later started to do. So, what is going on here? Has Pannekoek's account of the discovery of Neptune perhaps been richly interpreted by SSK scholars? Has he simply been misread? Or did Pannekoek's paper indeed contain the contours of a new approach to the history of science, perhaps rooted in the special brand of Marxism he had developed?

The present paper aims to provide an answer to these questions. It is structured as follows. In the first section I summarize the interpretation offered by Pannekoek in his 1953 paper. Then I show how this paper was cited by SSK scholars, respectively Barry Barnes (1982), Steven Shapin (1982), Simon Schaffer (1986), and finally David Bloor (2004). While these authors are very sympathetic towards Pannekoek, I will argue that they nonetheless do not do full justice to his views. Because Pannekoek clearly grants an important role to social factors in the discovery process of Neptune I then proceed in section three to consider whether his account befits other sociological approaches. I will argue that this is indeed the case and that his historical work best matches Merton's sociology of science. In section four I conclude by maintaining that Pannekoek can be seen as a forerunner, not of SSK, but of other forms of contextualism in the historiography of science. In 1953, it was still rare to grant such an important role to external factors in historical explanations. More specifically, Pannekoek's treatment of competition between countries was innovative because he was one of the first to take the discussion of such competition beyond mere priority issues, allowing for a richer and more complex treatment of the subject.

3 Historians of science may have remained agnostic about the *strong programme* but Jan Golinski (2005) shows that it has nonetheless set the agenda in terms of research topics of much historical work in the past few decades. For the agnosticism of historians, see Shapin 1992.

## 'The Discovery of Neptune' (1953)

In 1781, William Herschel discovered the planet Uranus. In the decades after this discovery, observations registered perturbations to the expected orbit of Uranus, as predicted by Newton's theory of gravity. This was not taken as a serious problem for Newton's theory but led to the idea that there must be another mass present causing the perturbations, most likely another planet nearby. In the 1840s, both Urbain Le Verrier in France and John Couch Adams in England started to calculate the position of the planet. Independently, they arrived at the same result. This then still required confirmation by observation. The French turned out to come first here as a Berlin observatory confirmed Le Verrier's calculation in September 1845. As a consequence, the credit for the discovery of Neptune went to France.

The *first* part of Pannekoek's account of this episode offers an explanation for the fact that the English lost the priority dispute. After all, Herschel had also discovered Uranus and the British were better equipped with observatories than the French. Pannekoek explains the difference in pace, in making the discovery 'complete', with respect to differences in the social structure of England and France. In the 1840s, the rising bourgeoisie in France still had to fight the authority of Church and nobility. Science, according to Pannekoek, was an important asset in this fight because it showed people that another world order, governed by scientific principles, was possible. The fact that one could predict the structure of the universe (i.e. the existence of a planet) based on a scientific theory (Newton's Law of Universal Gravitation) was a splendid proof of the power of science. This explains the efforts of the French to make the discovery known to the public. Le Verrier also published a book of more than 200 pages, simultaneously showing the difficulty of the problem and the skill required to overcome it. According to Pannekoek, this way of publishing the result of the discovery of the new planet also had the purpose of enhancing the prestige of science.[4]

In contrast, Adams only wrote a small paper of 31 pages in which he presented his calculations of the orbit of Neptune. In this paper, Adams clearly indicated a number of uncertainties in his calculations. Pannekoek saw this as an expression of Adams' modesty (which he praises), but also of a lack of sense of urgency to work on these uncertainties and match the calculations with observation. Pannekoek explains this as follows: in England no battle of the bourgeoisie against Church and nobility was needed

4 In a recent article, James McAllister (2015) called this 'the rhetoric of effort': painstaking labour (extreme carefulness, avoidance of error, etc.) lends credibility to a scientific result.

anymore. The reign of Cromwell had broken the self-evident government of the royals. Gradually, this led to legal fortification of the position of the bourgeoisie: 'after many stages of advance, the bourgeoisie had already reached a situation in which the Reform Act of 1832 and the repeal of the Corn Laws in 1846 left it as virtually the supreme power in the land'.[5] In the Netherlands and France this only happened in the revolutionary year 1848, and in Italy and Germany even later because these countries still needed to become unified states.

Because no social battle was needed anymore in England, using the weapon of science was not an urgent matter. This, according to Pannekoek, explains the open expression of uncertainties and the slower pace of doing research in England compared to France. Social factors thus played a role in determining the speed of scientific development. Pannekoek was forced to positively evaluate the role of bourgeoisie in this period. Although the bourgeoisie came to oppress the working class later in the nineteenth century, and should thus be overthrown, it had also played a crucial role in overthrowing the traditional institutions of power. For a Marxist this was not a common line of reasoning.[6] It also led Pannekoek to defend the, at first sight, odd position that science in the more advanced England (in terms of social structure) prospered less than in France. The *second* part of the story, in which the United States enters the picture, will make this more understandable.

The first part of the story has not gone uncontested in historiography of science. Robert W. Smith questions Pannekoek's idea of the 'calm indifference' with which Adams's calculations and the loss of the priority dispute were met in Britain. Smith shows that the events surrounding the discovery of Neptune did arouse huge passions. The English had thought that the discovery of Neptune should naturally be awarded to them and regretted missing it very much. This even led to serious damage in the prestige of a number of English scientists, most notably the Astronomer Royal, George Biddell Airy. Smith also attributes the French winning of the priority dispute to social factors, but in his account *internal* social struggles slowed the British down. There were fights over financial budget between universities (and other research institutes). On the one hand, this made people cautious in making claims of scientific breakthroughs. One had to be certain, because wrongful claims would jeopardize the reputation of a scientist as well as

5 Pannekoek 1953, 130.

6 More on Pannekoek's anomalous Marxism and the history of science follows in the section below.

the institution to which he was affiliated, and this could lead to a decrease in financial resources. On the other hand, competition also meant a slower exchange of ideas, in the case of the discovery of Neptune for example between Cambridge University and the Greenwich observatory.[7]

Others have sought more personal reasons for the hampering communication. According to William Sheehan and Steven Turber, the modesty of Adams and his habit of keeping things for himself have to be attributed to an autistic mental disorder. The discovery of the new planet required the ability to fully concentrate on a problem, paired with the mathematical skill to solve it, *and* social skills to discuss calculations and cooperate with astronomic observers. Adams had the first ability but lacked to second, and that is why England failed to win the priority dispute.[8]

The latter interpretation is hard to check against the historical record. Smith, however, has at least convincingly proved that the discovery of Neptune aroused much passion in England, which is already enough to seriously question Pannekoek's explanatory scheme. Pannekoek can perhaps be credited for opening the door to more intricate explanations of priority disputes between nation states, because he was a pioneer in paying attention to differences in social structure.[9] Yet, it is highly unlikely that his account can stand up to scrutiny. For the present purposes this does not matter very much, because I am investigating how Pannekoek's scheme of explanation orbited into the SSK publication trail. If his interpretation of the discovery of Neptune does indeed contain major flaws, this even more begs an explanation of the later positive citation of Pannekoek's paper by SSK scholars.

For these purposes the *second* part of the story is actually the most interesting one. Le Verrier and Adams had calculated roughly the same orbit for Neptune and the observation in Berlin matched this calculation. However, two Americans, Sears Cook Walker and Benjamin Peirce, noticed that earlier observations of presumably the same heavenly body could

7 Smith 1989.

8 Sheehan and Thurber 2007.

9 Pannekoek also considers cooperation between England and France, which makes the story even more complex. Le Verrier and Adams became friends and in 1848 they were honoured together by the Royal Astronomical Society when Le Verrier visited England. According to Pannekoek (1953, 133) their simultaneous calculation had demonstrated the power of science because it was proof that the discovery was not just a chance happening but something more profound. He again praised Adams's modesty and lack of ambition, which according to him made a friendship with Le Verrier possible 'in a situation that contained all the elements necessary for a fierce personal conflict'.

not be fitted into the orbit calculated by Le Verrier and Adams. A later observation, after 1850, also diverged from the calculated pattern. Walker and Peirce proceeded to calculate another orbit that had to be empirically adequate with respect to all observed positions. This orbit turned out to be significantly smaller than the one calculated by Le Verrier and Adams.

Now, there were thus two ways of accounting for the same data. Perhaps the observations dating from earlier days did not refer to the same planet? Peirce initially suggested that this might be possible and hence that the two orbits referred to two different planets. The French however rejected this possibility. According to Pannekoek, this was because their whole publication strategy, including outreach to the general public, would then fail: the predictive power of science allowed for only one planet to be discovered. The English joined the French. Both Herschel and Struve initially saw the competing American calculation as an unjustified, wild attack. This attitude led Europeans to ignore for a while the further calculations produced by the Americans. Yet, these calculations demonstrated beyond reasonable doubt that data of Neptune obtained between 1800 and 1850 fitted both orbit calculations (Figure 10.1). However, only the smaller orbit could account for the additional data. Because of the similarity during the period 1800-1850 the existence of two planets had to be ruled out. The Europeans could only save their one planet theory by admitting their calculations had been wrong.[10]

This was not an easy matter. Pannekoek again explains the lack of openness to the American critique with reference to differences in social structure. Both in England and in France, the bourgeoisie could not yet deal with a blow to their reputation. Social embarrassment in his view blocked an open discussion. Pannekoek approvingly quotes the Leiden astronomer Frederik Kaiser who complained in 1851 that in the United States people did not have to claim so hard that they had discovered something and

10 Walker and Peirce calculated what we now think is the correct orbit of Neptune. Thus Le Verrier and Adams had calculated *the same* erroneous orbit independently from each other. This resembles another occasion in the history of science in which both Galileo and Descartes (together with Isaac Beeckman) independently arrived at a mistaken formula of free fall, namely with speed as a function of traversed distance. Only Galileo managed to change his mind and calculate the speed of an object in free fall as a function of time elapsed. Alexandre Koyré ([1939] 1978) attributes the occurrence of this double error to the reigning 'thinking cap' of impetus physics. The same type of explanation can be applied to the erroneous orbit calculation. Both Le Verrier and Adams used the Titius-Bode Law in order to reduce the number of variables, as this Law at least fixed the average distance to the Sun of the 'new' planet. The Titius-Bode Law approximately states that each planet is twice as far from the Sun as the one before but this only gives a rough estimation of the position of the planets and simply fails with Neptune.

**Figure 10.1 The observed orbit of Neptune compared with the calculated orbit of Neptune, as drawn by Pannekoek**

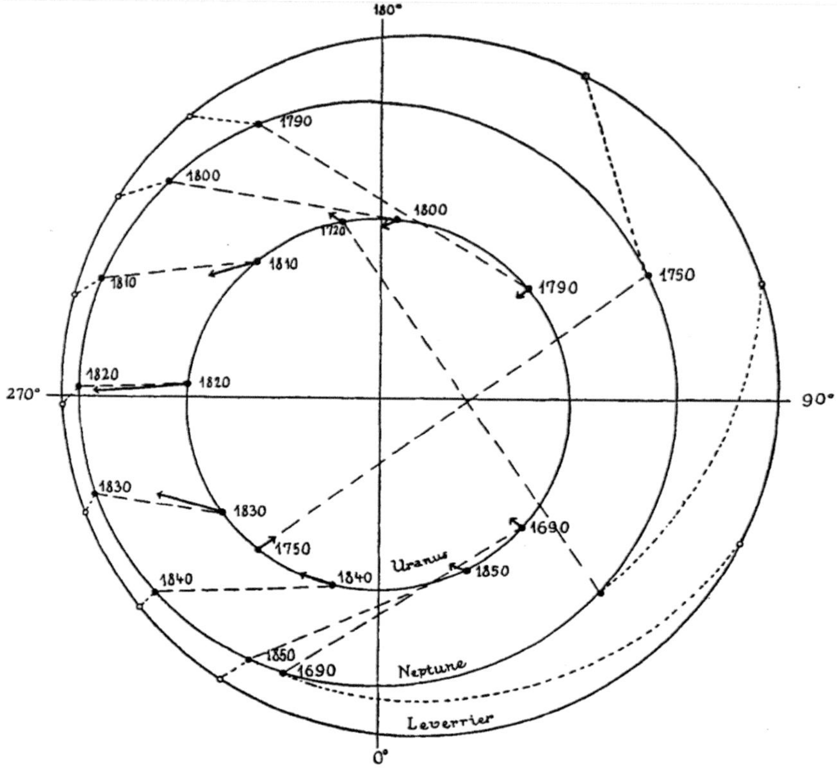


Source: Anton Pannekoek, *De groei van ons wereldbeeld* (Amsterdam: Wereldbibliotheek, 1951)

hence in that climate there was much more room for critical discourse. A hypothesis could be falsified without directly undermining the whole authority of science. Pannekoek therefore concludes: 'it appears clearly that not science itself but its social function ruled the attitude of scientists in Europe'.[11] When a prediction is falsified, this is a step forward for science but (in this case) a step backward for the use of science in social struggle. The implication is that, when all social struggles are behind us, free thinking and a maximum of critical discourse is possible. Hence in a society free of social classes there will be no obstructions to 'science itself'.[12]

11 Pannekoek 1953, 136.

12 As John G. Hubbell and Robert W. Smith have shown (1992), Pannekoek's account of the American part of the story has been sketchy and narrow. There were more scientists involved

This puts the earlier discussed 'England had a better social structure than France' evaluation in perspective. England was far from perfect either, because it was still plagued with social hindrances to scientific pursuit. The distinction that Pannekoek draws between Europe and the United States is the key to understanding how social factors play a role in his explanatory framework. Social factors can enhance *and* temper scientific development, but they never touch the heart of science. There is an autonomous realm in which scientific development, marked by imagination, hard work and openness to critical discourse, can flourish, if freed from unwelcome social forces. This viewpoint is important to keep in mind as it crucially differs from the SSK approach, as I will demonstrate in the next section.

## References to Pannekoek in SSK

The first two references to Pannekoek's paper on Neptune by SSK scholars focus on the second part of the story, because here two different methods of accounting for the data were possible and it was not directly apparent at first which one of them was correct. Both Barry Barnes and Steven Shapin argue that social factors forced a decision here. Thus Barnes writes about Pannekoek: 'What is interesting is his explanatory strategy. He perceives that at least two methods of accounting were possible, and he explains their association with two distinct contexts by references to goals and interests. Many sociologists would recognize this as a sound procedure'.[13] According to Barnes, the Europeans were busy with ideological struggles in which science played an important role and the Americans were busy with demonstrating technical competence. The data could not decide between the two interpretations: 'If we are capable of recognizing that the data do

in the orbit calculations than just Walker and Peirce and the discovery of Neptune also received considerable popular interest. Hubbell and Smith argue that the American response to the discovery has to be understood as typical with respect to the relationship between American and European science. The Americans, and especially the so-called Lazzaroni group, were involved in raising the standards of scientific education and research in their country, and in this context exposing 'European arrogance' was very much welcome and actively promoted as such. Thus, the Americans too protected a national interest and used competition with other countries to further the case of science in their own country. Still, one could maintain that openness to critical discourse was what made American science at the time stronger than European science, as Pannekoek had argued. In any case, as I have already said above, the aim of this paper is not to correct Pannekoek's interpretations but rather to correct the way in which these interpretations were portrayed and used by SSK scholars three decades later.

13 Barnes 1982, 98.

not settle matters in favor of either the opposed accounts by the scientists, then we should be able to recognize also that any interpretation we prefer will have no special status in relation to that "data". Our preferences will be a matter of how we are, more than how reality is'.[14] Indeed, *if* this would be the case, the SSK view – which places the human element, or 'the social', at science's core – would follow. However, the uncertainty about the data only lasted for a very short period of time as the Americans quickly demonstrated that both calculations could account for the data observed in the period 1800-1850, but only theirs for data observed before and after that period. As a final decision over the correct orbit fell only *after* the Americans had shown this, there certainly was a sense in which the data settled matters.

Therefore, it is also hard to follow Shapin's conclusion: 'if Pannekoek is right, one of the most fundamental acts of cognitive judgement (are natural objects the same or not the same?) was in this case structured by interests in the professional status and social standing of the scientific community'.[15] Again this explanation can only work when we consider a very small time frame and can certainly not be attributed to Pannekoek. With respect to the interest in professional status of Walker and Peirce, we can ask whether they were really chiefly busy with demonstrating 'technical competence'. According to Pannekoek, they were busy furthering the cause of science. In his view getting at the truth was the sole purpose of the American investigations. Reading Barnes and Shapin, it is as if the two scientific communities were epistemically equal. This does not square with Pannekoek's paper, because such a reading completely neglects his point that with less social impediments we get better science and hence that in the 1850s the climate for science was *better* in the United States compared to Europe.

How can this selective reading of Pannekoek by both Barnes and Shapin be explained? I believe the answer lies in SSK's attempts to obtain a respectable place within science studies. After all, the *strong programme* is quite radical and may have been hard to swallow, even for those willing to grant an important role to social factors in determining the course of science. Barnes's book has the clear purpose of positioning the SSK approach with respect to other approaches, such as Kuhn's model of alternating paradigms and Garfinkel's ethnomethodology. At the same time, his aim was to demonstrate the viability of the SSK approach and 'allies' were welcome to serve

14 Barnes 1982, 96.

15 Shapin 1982, 175. Shapin refers to Barnes and it appears that he relied on Barnes' reading of Pannekoek's paper. To be fair, he also indicates that much more research is needed to establish the correct interpretation of the episode.

this purpose. Showing that central ideas of SSK already featured in earlier historical papers made the approach less out of bounds and hence more credible. This rhetorical strategy possibly led to the crude presentation of Pannekoek's argument.[16]

Simon Schaffer also wrote a paper with clear programmatic overtones in 1985. His citation of Pannekoek is even more curious than the ones by Barnes and Shapin. Schaffer does not even refer to these earlier published papers but only directly to Pannekoek. He focuses exclusively on the *first* part of the story, which he represents in an odd way. According to Schaffer, completely different research programmes were pursued in France and Britain, both based on different 'techniques of observation'. We should therefore not speak of a multiple discovery of Neptune but of two *distinct* discoveries, both pertaining to the respective research programmes.[17] This reading of the first part of the episode strikes me as nonsensical. Le Verrier and Adams together received honour for the discovery of Neptune by the Royal Astronomical Society in 1848. Before that, French and English scientists were well aware of each other's work and corresponded through letters.[18] What, then, were these different 'techniques of observation' causing a split in distinct research programmes? There is no hint to such differences in Pannekoek's paper at all. I cannot find a serious argument against the logical conclusion that on both sides of the Channel roughly the same methods of calculation and observation were used.

In my view, Schaffer's citation of Pannekoek exemplifies one of the main weaknesses of SSK, which is to create strong oppositions between conflicting parties. This is done to leave no room to settle the conflict by argument or by experimental data. Closure of scientific controversies can then only be explained with reference to social factors, as these are the only tiebreakers left. What makes Schaffer's application of this mode of explanation to the *first* part of the story odd is the fact that there was no epistemic conflict at all. At least in the second part of the story there was such as conflict. In the first part of Pannekoek's paper (1953), competition relates only to gaining priority for the discovery. Again, if Schaffer wants to maintain that there was a major epistemic conflict between Le Verrier and Adams, he cannot rely on Pannekoek, but strangely enough he does.

16 Various papers in Velody and Williams 1998 address SSK's rhetorical strategies. Contributions to Meister et al. 2006 investigate the 'political' strategies of SSK proponents to obtain a respectable place in science studies. See also Richards and Ashmore 1996.

17 Schaffer 1986.

18 For their correspondence, see Smith 1989.

With David Bloor, in 2004, we are back on track with a focus on the second part of the story where the Americans join in. He reproduces the interest theory already proposed by Barnes and Shapin in the early 1980s. Thus, Bloor writes: 'Wittgenstein was right; sameness is problematic, even when we are dealing with huge pieces of matter like planets'.[19] He argues that astronomers in Europe and the USA pursued different interests and this led to different predictions and discoveries. If this reading of the episode can be maintained at all, and this is in my view highly doubtful, it can surely not be attributed to Pannekoek (which is what is at stake here).

Social factors play an important role in Pannekoek's account of the discovery of Neptune and the estimation of its orbit. His mode of explanation can, however, not be squared with SSK. SSK scholars take the impact of social factors a major step further and let it touch the heart of science, that is, social factors directly affect our epistemological commitments. This major step, however, cannot be found in Pannekoek's paper. Yet, he is cited as if this can be done. In the papers discussed here by Barnes, Shapin, and Schaffer, we explained this 'rich interpretation' as a result of rhetorical strategy with the purpose of trying to make the SSK approach *salonfähig*. Bloor's contribution to the *Handbook of Epistemology* (2004) is merely a repetition of earlier work by Barnes and Shapin. This contribution shows that SSK has indeed become socially accepted, and also makes Pannekoek's paper an important point of reference in science studies. These feats have been achieved at the cost of misrepresenting Pannekoek's perspective on the efficacy of social factors in the history of science. Therefore, to do him full justice, I investigate with which other major approach in the sociology of knowledge Pannekoek's mode of explanation can better be aligned in the final section of this paper.

## Pannekoek and Other Sociologists: Marx, Elias, and Merton

Perhaps the most natural place to look for likeminded souls is among Marxist historians of science. After all, Pannekoek's paper was published in a special issue of *Centaurus* devoted to Marxist historiography to which almost all leading Marxist historians of science of the day contributed. As sociologist Norbert Elias once contended, we owe the perspective of a strong connection between human psyche and social structure to Marx and the Marxists.[20]

19 Bloor 2004, 927.
20 Cf. Burke 2012.

'Human psyche' can be broadly conceived to include science. Notwithstanding the practical work that goes into scientific research, gaining scientific knowledge is foremost a mental activity. For a classical Marxist, the psychological is always fully determined by the social. From this, it follows that scientific thinking is a function of social forces. For Pannekoek, the relation between society and human psyche was an important theme, but following Joseph Dietzgen's *Das Wesen der menschlichen Kopfarbeit* (1869), he defended the unorthodox position that mental change could be the *cause* of social change. In other words, consciousness could determine (social) being.[21] In this respect, Pannekoek differed from other Marxist historians such as Boris Hessen or J.D. Bernal, who were inclined to follow the traditional idea that culture and mental life formed a superstructure on the basic socioeconomic premises.

Pannekoek opted for the less rigid interpretation of this relation because he saw in science an important driving force of social change. We have seen a clear example of this in the first part of the discovery of Neptune story. According to Pannekoek, modern science served as an important weapon for the bourgeoisie in France in the 1840s. The idea that science has the capacity to refute untenable authority (including social and political power) gives an important reason for Marxists to be interested in studying the history of science. History can show how science helped to liberate us from the rule of superstition, church, nobility, absolutism, etc. At least this led Pannekoek to another adjustment of orthodox Marxism as he recognized a positive force in the rise of the bourgeoisie as part of the whole process of the liberation of humankind. While the bourgeoisie oppressed the working class during the twentieth century, and hence needed to be defeated, the bourgeoisie was actually very effective in breaking traditional power structures during the nineteenth century.[22]

The general idea, to consider all mental phenomena in relation to social structures, could thus be interpreted in multiple ways. The lack of unanimity among Marxist historians of science was actually one of the reasons why Léon Rosenfeld took the initiative for a themed issue of *Centaurus* in 1953. It was his hope that discussions surrounding this project would lead to consensus on the matter how to conceive of the relation between science and society.[23] This led to interesting contributions, which addressed a wide variety of topics, such as the relation between internal and external factors,

21 See Pannekoek 1913.

22 This point of view of the bourgeoisie as a temporary beneficial force led to a bitter dispute between Pannekoek and Lenin. See Pannekoek [1938] 1973.

23 Jacobsen 2008.

the relation between the individual and society, and the proper interpretation of the succession of scientific theories.

This set of papers could have been the start of fruitful discussions on these topics. The example of Pannekoek shows that not all Marxists were inflexible dogmatists. It is very well conceivable that reference to 'Marx' would drop out entirely with increasing sophistication of discussions on the role of social factors in science.[24] The message of Marxist historians, however, was badly received in the Cold War climate of the 1950s. First of all, this climate was hostile to Marxism for obvious political reasons. Secondly, intellectual historians for the most part continued to ignore the effect of social and economic circumstances on science. It did not help that 'externalism' came to be associated with Marxism. It also did not help that instead of having an open critical discourse, Marxist historians tended to fight over disagreements in a rather harsh way. In the development of thinking about science and exploring approaches to the past, we thus find something of a break-off. In a sense, SSK has picked up the trail of earlier Marxists historians of science because in SSK, social factors gain central importance again.[25] From this perspective, their positive reception of Pannekoek's account of the discovery of Neptune is not surprising. The problem, however, is that Pannekoek can be aligned neither to the epistemological relativism of SSK, nor to the unilateral determinism of orthodox Marxism. Pannekoek's thought better fits with contemporary modernist sociologists such as Norbert Elias and Robert K. Merton.

Links between Elias and Pannekoek have been explored by others.[26] There are indeed many similarities between the two. In both we find the connection between the social and psychological realms of human existence, including the notion that this connection is reciprocal. Both tend to study short-term events in the context of long-term processes. With respect to the latter, they have both granted an important role to the theory of evolution. In the development of science, they have also both called for attention to the study of competition between countries.[27]

Elias, however, following Karl Mannheim, has taken this notion of competition very far in his sociology of knowledge: 'The whole figuration

24 For example, Merton (1938) acknowledged his indebtedness to Hessen for thinking in terms of external factors, but at the same time distancing himself from Hessen's materialism.

25 For an example, see Schaffer 1984, in which he calls attention to the contributions of Hessen to the history and sociology of science. A more recent attempt at rehabilitation of Hessen can be found in Freudenthal 2005.

26 Westbroek 2012.

27 Elias 2009; see also Burke 2012.

is animated by a continuous competitive struggle for preservation, avoidance of loss or rise of status and power chances. There is also competition for economic resources'.[28] It appears that this would go too far for Pannekoek. In his account of the discovery of Neptune, the struggle to be the first to claim the discovery does not occupy central stage. Pannekoek's main concern is the role science plays in society and how to improve social structure, in order for scientific research to optimally blossom. In this optimal picture, competition is not at the heart of science but cooperation is. It is not without reason that Pannekoek praises Adams's modesty as an important scholarly virtue and that he stresses the good relations between Le Verrier and Adams after the priority dispute was over.

We can see something similar in Pannekoek's evaluation of Darwinism, when applied to the study of the evolution of human societies. Where Elias was inclined to take the analogy of selection and survival of the fittest very far, Pannekoek clearly saw limits to the analogy.[29] Again, he makes the point that Darwinism had been a useful weapon for the bourgeoisie in the nineteenth century to use against the 'feudal classes'. Yet, in the twentieth century the bourgeoisie could also use the theory against the proletariat in order to maintain the correctness of capitalism. In drawing an analogy between capitalist competition and the fight for survival in the animal kingdom, the bourgeoisie could justify the inequalities in wealth and power in twentieth-century Western societies. With reference to Darwin, the bourgeoisie could even present their coming out on top as a law of nature. Darwinism could be seen as complementary to Marxism in stressing continuous evolution towards improvement, but it also failed to set a good example, because it underpinned social inequality whereas socialism strives for equal access to wealth and power for everyone.

So, for Pannekoek, not all Darwinian principles could be applied to human society. On the one hand, man is a social animal and only full participation of all members of the group would make the group strongest. On the other hand, the analogy with nature breaks down when we consider the speed of change in human societies. While evolution works slowly, society changes at a rapid pace and this means that it qualitatively differs from nature. Pannekoek stressed that cultural innovation such as tools, techniques, and language acted as social cement. In order to improve living conditions, it was essential to recognize this profoundly social aspect of human culture.[30]

28 Elias 2009, 137.
29 Pannekoek 1909.
30 See again Westbroek 2012.

Notwithstanding the many points of overlap, Pannekoek's thought differed from Elias in crucial respects. With Robert K. Merton, another prominent early sociologist of knowledge, we do not find such discrepancies. To Merton we owe a number of sociological concepts, such as 'role model', 'self-fulfilling prophecy', and 'unintended consequences', which have become so ubiquitous, that we often do not even realize he invented them. With respect to the study of science, Merton was a pioneer in the discussion about the relation between internal factors and external factors.[31] He thought that socio-cultural circumstances influence the course of science to a considerable degree. However, he also thought that these external factors could never touch the real heart of science. The exact shape of this internal realm differed of course from field of study to field of study, but the institution of science as a whole could be characterized as being governed by a set of invariable norms, namely: communalism, universality, disinterestedness, and organized scepticism. Good science (or real science) is marked by cooperating scientists, who are after the truth without other agendas (note how this contrasts with SSK's interest theory) and with an open mind to criticism of others (note how this corresponds to Pannekoek's evaluation of the science climate in the United States in the 1850s).

Still, the direction and pace of research could be determined by external factors. Merton's famous example comes from a comparison he made between science in protestant countries and science in catholic countries.[32] Merton argued that Protestantism exhibited a set of dominant values, such as ascetism, self-reliance, ethic of hard work, discipline, etc., that created a beneficial climate for scientific pursuit. This made science move forward more rapidly in protestant countries than in catholic countries. Without the Reformation, science would thus have progressed at a much slower pace, but it would eventually have led to the same discoveries and scientific theories, most likely even in the same *sequence*. This Merton thesis has been met with a lot of criticism, but a discussion of the merits of the Merton thesis is not the point of this paper.[33] Here, it is enough to see similarities between Merton's thesis and Pannekoek's account of the discovery of Neptune. Pannekoek argues that the French proceeded at a quicker pace than the English for social reasons, but eventually observations in England would lead to a very similar match with the calculated orbit that was already present.

31 A collection of Merton's most important papers in the sociology of knowledge can be found in Merton 1973.

32 Merton 1938.

33 Such a discussion can be found in Shapin 1988.

That Pannekoek recognized a realm of pure science can also be inferred from remarks he makes in his two books on the history of astronomy. In *De groei van ons wereldbeeld* he writes, for example: 'While the other sciences celebrate their triumph in the complete transformation of the material world, astronomy has become an adventure of the mind, a piece of pure culture. Thus, her history remains now the same as in the past: a part of humanity's cultural history'.[34] From this, we can clearly infer that without practical applications, there is a realm of pure science, here called 'pure culture'. Clearly, the term 'pure culture' is used as bearing no relation to social (or external) factors.

Dick Pels once made a distinction between two main traditions in sociology of knowledge.[35] On the one hand, there is the Mannheim-Merton tradition in which external factors matter, but there is also a special internal realm of science, which safeguards an epistemological realism. On the other hand, there is the Wittgenstein-Kuhn tradition, to which SSK also belongs. In this tradition, an internal realm of science, marking a continuous development of the scientific enterprise, is not recognized, which leads to a high degree of epistemological relativism. Elias can be placed somewhere in the middle with his stress on competition and Darwinism.[36] As Pannekoek differs exactly on these points from Elias, it draws him towards the Mannheim-Merton tradition. Orthodox Marxists may also be part of this tradition, but it is especially interesting for our purposes that Merton, despite complimenting Hessen for drawing attention to the role of external factors in science studies, distanced himself from Hessen because he reduced all explanation to economic forces.[37] Just like Pannekoek, Merton thought such a 'vulgar materialistic' scheme of explanation was too unidirectional.

For a number of reasons, then, I believe Pannekoek's account of the discovery of Neptune accords with Merton's sociology of institutions. Apparently Pannekoek was unaware of this because we find no references to Merton, neither in the 1953 paper nor in *De groei van ons wereldbeeld*. Merton, on the other hand, devoted a paper to priority disputes in which he even mentions the one between Le Verrier and Adams. Although it was published after Pannekoek's paper, no reference to Pannekoek can be found in Merton's paper. Merton's conclusion there reiterates the key elements of his sociology of knowledge:

34 Pannekoek 1951, 9.

35 Pels 1996.

36 See Jardine 2001; Burke 2012.

37 Merton 1938.

> The pursuit of science is culturally defined as being primarily a disinterested search for truth and only secondarily, a means of earning a livelihood. In line with the value-emphasis, rewards are to be meted out in accord with the measure of accomplishment. When the institution operates effectively, the augmenting of knowledge and the augmenting of personal fame go hand in hand; the institutional goal and the personal reward are tied together. But these institutional values have the defects of their qualities. The institution can get partly out of control, as the emphasis upon originality and its recognition is stepped up.[38]

For SSK scholars, getting 'out of control' can just be part of the game of science. For Merton, and also for Pannekoek, it is the point where we start to transgress the boundaries of science.

## Conclusion

In this paper, I have indicated how and why Pannekoek's account of the discovery of Neptune has been hailed as a pioneer of the radical SSK approach to the study of science. However, Pannekoek can only be seen as a pioneer of the *strong programme* if crucial aspects of his account are skipped over. I have shown that this has happened and have suggested that this has occurred mainly for rhetorical reasons. As these are easily forgotten, especially as late as 2004, when Pannekoek's paper found its way into a handbook of epistemology, the present paper provides a necessary correction on the now apparently state-of-the-art reading of Pannekoek's interpretation of the history of science.

Pannekoek can still be seen as a pioneer when we consider that in his time historians of science paid almost no attention to the social structure of countries as an important causal factor in past priority disputes. Adding the dimension of external factors to the story was thus an innovative step towards more complex accounts of achievements of scientific discoveries. While Marxist historians of science, especially Hessen, can be credited for introducing external factors to the study of science, Pannekoek, despite being a Marxist, cannot be aligned with those historians either. He developed his own anomalous version of Marxism, which was less materialistic than orthodox versions. In Pannekoek's version, room was left for the mental to act causally upon the social and this formed an important building block of

38 Merton 1957, 659.

his argumentation in his 1953 paper on the discovery of Neptune. Comparing his interpretation of this episode in the history of science with other sociologists of knowledge, we found that it accords best with Merton's sociology of institutions. While some have claimed that since the advent of SSK, explaining science with a model that is based on (some interpretation of) the distinction between internal and external factors has become obsolete,[39] others have recently started to return from SSK's radicalism because of the negative effects of embracing an omnibus epistemological relativism. One of SSK's early proponents, Harry Collins, has for example called for a return to Mertonian values, which he now calls 'elective modernism'.[40] Should this revaluation of Merton find more support in the future, Pannekoek can again be hailed as a pioneer, as his 1953 paper can be read as an early exemplification of the Mertonian approach to the study of past science. Should this identification occur, this paper demonstrates that it would at least be on justificatory grounds.

## Bibliography

Barnes, Barry. 1982. *T.S. Kuhn and Social Science*. New York: Columbia University Press.

Bloor, David. 1976. *Knowledge and Social Imagery*. London: Routledge & Kegan Paul.

—. 2004. 'Sociology of Scientific Knowledge'. In *Handbook of Epistemology*, ed. by Ilkka Niiniluoto, Matti Sintonen, and Jan Woleński. Dordrecht: Kluwer, 919-962.

Burke, Peter. 2012. 'Norbert Elias and the Social History of Knowledge'. *Human Figurations* 1(1). http://hdl.handle.net/2027/spo.11217607.0001.102 (accessed 16 September 2016).

Collins, Harry M. 2010. *Gravity's Ghost: Scientific Discovery in the Twenty-First Century*. Chicago, IL: University of Chicago Press.

Collins, Harry M., and Robert J. Evans. 2002. 'The Third Wave of Science Studies: Studies of Expertise and Experience'. *Social Studies of Science* 32(2): 235-296.

Elias, Norbert. 2009. *Essays I: On the Sociology of Knowledge and the Sciences*. Ed. by Richard Kilminster and Stephen Mennell. Collected Works of Norbert Elias 14. Dublin: University College Dublin Press.

Freudenthal, Gideon. 2005. 'The Hessen-Grossman Thesis: An Attempt at Rehabilitation'. *Perspectives on Science* 13(2): 166-193.

39 Shapin (1992).

40 Collins 2010. For his ideas to move beyond SSK and follow a third wave of science studies dating back as far as 2002, see Collins and Evans 2002.

Golinski, Jan. 2005. *Making Natural Knowledge: Constructivism and the History of Science*. Chicago, IL: University of Chicago Press.

Hubbell, John G., and Robert W. Smith. 1992. 'Neptune in America: Negotiating a Discovery'. *Journal for the History of Astronomy* 23(4): 261-291.

Jacobsen, Anja Skaar. 2008. 'The Complementarity Between the Collective and the Individual: Rosenfeld and Cold War History of Science'. *Minerva* 46(2): 195-214.

Jardine, Nicholas. 2001. 'Sammlung, Wissenschaft, Kulturgeschichte'. In *Sammeln als Wissen: das Sammeln und seine wissenschaftsgeschichtliche Bedeutung*, ed. by Anke te Heesen and Emma C. Spary. Göttingen: Wallstein.

Koyré, Alexandre. (1939) 1978. *Galileo Studies*. Hassocks, UK: Harvester Press.

McAllister, James W. 2015. 'Rhetoric of Effortlessness in Science'. *Perspectives on Science* 24(2): 145-166.

Meister, Martin, Ingo Schulz-Schaeffer, Stefan Böschen, Jochen Glässer, and Jörg Strübing, eds. 2006. 'What Comes after Constructivism in Science and Technology Studies'. *Science, Technology & Innovation Studies*, Special issue 1.

Merton, Robert K. 1938. 'Science, Technology and Society in Seventeenth Century England'. *Osiris* 4: 360-632.

—. 1957. 'Priorities in Scientific Discovery: A Chapter in the Sociology of Science'. *American Sociological Review* 22(6): 635-659.

—. 1973. *The Sociology of Science: Theoretical and Empirical Investigations*. Chicago, IL: University of Chicago Press.

Pannekoek, Anton. 1909. *Marxismus und Darwinismus. Ein Vortrag*. Leipzig: Leipziger Buchdruckerei.

—. 1913. 'Dietzgens Werk'. *Die Neue Zeit* 31(28): 37-47.

—. 1951. *De groei van ons wereldbeeld*. Amsterdam: Wereldbibliotheek.

—. 1953. 'The Discovery of Neptune'. *Centaurus* 3(1): 126-137.

—. 1961. *A History of Astronomy*. London: George Allen and Unwin.

—. (1938) 1973. *Lenin als filosoof. Een kritische beschouwing over de filosofische grondslagen van het leninisme*. Amsterdam: De Vlam.

Pels, Dick. 1996. 'Karl Mannheim and the Sociology of Scientific Knowledge: Toward a New Agenda'. *Sociological Theory* 14(1): 30-48.

Richards, Evelleen, and Malcolm Ashmore. 1996. 'More Sauce Please! The Politics of SSK: Neutrality, Commitment and Beyond'. *Social Studies of Science* 26(2): 219-228.

Schaffer, Simon. 1984. 'Newton at the Crossroads'. *Radical Philosophy* 37: 23-38.

—. 1986. 'Scientific Discoveries and the End of Natural Philosophy'. *Social Studies of Science* 16(3): 387-420.

Shapin, Steven. 1982. 'History of Science and Its Sociological Reconstructions'. *History of Science* 20(3): 157-211.

—. 1988. 'Understanding the Merton Thesis'. *Isis* 79(4): 594-605.

—. 1992. 'Discipline and Bounding: The History and Sociology of Science as Seen through the Externalism-Internalism Debate'. *History of Science* 30(4): 333-369.
Sheehan, William, and Steven Thurber. 2007. 'John Couch Adams's Asperger Syndrome and the British Non-Discovery of Neptune'. *Notes and Records of the Royal Society* 61(3): 285-299.
Smith, Robert W. 1989. 'The Cambridge Network in Action: The Discovery of Neptune'. *Isis* 80(3): 395-422.
Velody, Irving, and Robin Williams, eds. 1998. *The Politics of Constructionism.* London: Sage.
Westbroek, Peter. 2012. 'Civilizing Earth'. *Human Figurations* 1(1). http://hdl.handle.net/2027/spo.11217607.0001.108 (accessed 3 February 2016).

## About the Author

**Bart Karstens** is Postdoctoral Research Fellow at the Vrije Universiteit Amsterdam and the Vossius Center of the University of Amsterdam. In 2015, he defended his PhD thesis 'Pluralism within Parameters', on the conceptualization of error in historiography of science, at Leiden University. He is interested in discipline formation, interaction between disciplines, assessments of past science, historical epistemology of 'structure' and 'structuralism', comparativism, digital humanities, and digital culture.

# 11 The Milky Way as Optical Phenomenon

## Perception and Photography in the Drawings of Anton Pannekoek

*Chaokang Tai*[*]

**Abstract**

One of Anton Pannekoek's main scientific projects was to provide a representation of the appearance of the Milky Way – an object he believed to be an optical illusion. This paper elucidates how Pannekoek thought the Milky Way appearance was formed by a combination of human psychology and physiology, and why he attributed such significance to it. In doing so, it explores the connections between Pannekoek's scientific methodology and his socialist epistemology. The paper also outlines the various techniques Pannekoek employed in his research. To observe the Milky Way, he used both extrafocal photography and visual observations by himself and others. To represent the results, he combined naturalistic drawings with verbal descriptions, numerical tables, and isophotic diagrams.

**Keywords:** Anton Pannekoek, Milky Way, Marxism, astronomical drawing, perception, photography

Ever since they were first published in the 1920s, the Milky Way images created by Anton Pannekoek have captured the imagination of astronomers

* This research has been funded by the Institute of Physics and the Anton Pannekoek Institute for Astronomy, University of Amsterdam, the Descartes Centre, Utrecht University, and Stichting Pieter Zeeman Fonds. I would like to thank Omar W. Nasim for our conversations on astrophotography, which helped me to further develop and clarify many of the ideas present in the paper, and Emma Mojet, Sjang ten Hagen, Robert W. Smith, Bart van der Steen, and Jeroen van Dongen for their valuable comments on earlier drafts of this paper.

Tai, Chaokang, Bart van der Steen, and Jeroen van Dongen (eds), *Anton Pannekoek: Ways of Viewing Science and Society*. Taylor & Francis Group, 2019
DOI: 10.5117/9789462984349_CH11

and the public alike. Astronomers have used them as a definitive source for the distribution of galactic light, while the public got to know them through their inclusion in Zeiss planetaria and the Lund Panorama of the Milky Way.[1] More recently, they inspired visual artist Jeronimo Voss in the creation of his exhibition 'Inverted Night Sky', which was displayed at the Stedelijk Museum Bureau Amsterdam.[2] Joseph Ashbrook, editor of *Sky and Telescope*, even considered Pannekoek to be the '[g]reatest of all naked-eye observers of the galaxy'.[3] A striking feature of Pannekoek's Milky Way research was that he used both visual observations and photographical methods to determine the distribution of galactic light, which he then represented using many different techniques, including naturalist drawings, verbal descriptions, isophotic diagrams, and numerical tables. In this chapter, I focus on how these various representations were made and why they were made in the first place. Revealing how and why Pannekoek employed such wide-ranging methods for observing and representing the visual aspect of the Milky Way provides crucial insight into the development of early twentieth-century astronomy. It illustrates the complex relation between naked-eye observations and photography during this period, reveals how astronomers coped with the characteristics of human psychology and physiology, and deepens our understanding of the connections between political philosophy and scientific epistemology.[4]

To explain the coexistence of various representational methods in Pannekoek's research, we must first examine the role he attributed to astronomers in observing the Milky Way. In particular, how he thought certain characteristics and limitations of human physiology and psychology combined to create the image of the Milky Way. On this issue, it is informative to draw a parallel with late-nineteenth-century epistemic debates concerning the inherent differences between astronomical observers. Following the realization that well-skilled observers recorded different coordinates for the same star even when using the same instruments and diligently abiding to the same methods, astronomers had to reconsider the role of human perception in visual observation and develop strategies to either minimalize or stabilize these differences.[5] This reflexive inward look of astronomers was part of a greater 'reflexive turn' in

1 For the Zeiss Planetarium, see King 1958; for the Lund Panorama, see Lundmark 1957.

2 See SMBA 2016; Alena J. Williams, 'A Galaxy of Appearances', in this volume, 305-318; and Johan Hartle and Jeronimo Voss, 'Cross-Fading the Milky Way', in this volume, 285-303.

3 Ashbrook 1984, 375.

4 This chapter expands on earlier research presented in Tai 2017, 218-230.

5 See Schaffer 1988; Canales 2001; Hoffmann 2007. As Hoffmann indicates, the term 'constant differences' was used in the early nineteenth century before the concept of a personal equation

observational science during the mid-nineteenth century and it caused several astronomers to venture beyond their own field and participate in a cross-disciplinary exchange of ideas.[6] More than half a century later, Pannekoek too was deeply concerned with the anatomy of the human eye and the psychology of the human brain when developing his method for visual photometry of the Milky Way. He too ventured beyond astronomy to develop his ideas. In his case, however, it was not experimental psychology, but Marxism he turned to.

There are clear advantages to actively considering Pannekoek's Marxism when discussing his scientific methodology, even if he himself tried to keep his socialist and astronomical careers separate from one another. It is in his Marxist writings that Pannekoek developed his philosophy of the human mind: that humans have an innate ability to analyse and synthesize sense perceptions, but that this ability is implicitly influenced by prior experience. Historians of science Lorraine Daston and Peter Galison have argued that scientific epistemology is inextricably linked to conceptions of the self, as scientists seek to counteract the weaknesses of the self while emphasizing its strengths.[7] In Pannekoek's Milky Way research we find that he wanted to utilize the intuitive analytical character of the human mind while eliminating the effects of implicit bias.[8] In doing so, he concurred with contemporary ideas on scientific collaboration. The late nineteenth century saw the emergence of large-scale scientific collaborations taking on grand transnational projects. The organizers of these projects considered it vital for participants to show self-restraint and follow predetermined methods; for contributions to be mutually compatible, individual discrepancies had to be minimalized.[9] Although Pannekoek's Milky Way research was conceived on a much smaller scale, he advocated a similar ethos in the hope of eliminating individual subjectivity while preserving collective subjectivity.

The question of how to observe and represent the Milky Way inevitably leads to a discussion on the role of photography in early twentieth-century astronomy. When discussing the development of astrophotography, it is tempting to list vivid and increasingly more detailed photographic images of visually striking astronomical objects, like nebulae, clusters, or the moon,

that was tied with individual physiology and psychology emerged in Greenwich in the second half of the nineteenth century.

6 Canales 2001; for a similar reflexive turn in microscopy, see Schickore 2007.

7 Daston and Galison 2007; Galison 2004. See Daston 2008 for the importance of the visual in bridging psychology and epistemology.

8 For a detailed discussion on Pannekoek's epistemic virtues in astronomy and socialism, see Tai and van Dongen 2016; and Tai 2017.

9 Galison and Daston 2008.

being produced by the latest technological innovations. Such a listing, however, ignores the fact that the acceptance of photography in astronomy was far from straightforward: it was accompanied by genuine epistemic concerns about the usefulness and trustworthiness of photography.[10] Historical research on this topic has mostly focused on the second half of the nineteenth century, but these concerns persisted well into the twentieth century. When we look at Pannekoek's Milky Way research, we find that drawing and visual observation still played a prominent role in his work precisely because he believed contemporary photographic images of the Milky Way were inadequate for his purposes. Moreover, it was rare for photographs depicting astronomical objects to find their way into professional publications at all. Rather, photography was used as a tool for gathering, storing, sharing, and measuring large amounts of observations without needing constant access to a telescope and clear skies.[11] The information they contained was then usually presented in the form of large tables of numbers. Pannekoek's use of astrophotography fits in this profile. He was not interested in the way the Milky Way was depicted by photographic images, but in the measurement of its light intensity on photographic plates.

This chapter will begin by investigating Pannekoek's ideas on what the Milky Way actually was; how, as a phenomenon, it was related to human physiology; and how astronomers could best take advantage of this physiology while counteracting its flaws. In doing so, it is vital to look beyond his scientific writings and consider his Marxist philosophy. The next section will illustrate how these epistemic concerns were then translated into astronomical practice. It explores Pannekoek's method of photometry through visual observations, how he combined observation from various observers, and the various ways in which he represented the final results. The final section will discuss his method of photographic photometry as a way of replacing visual observations and address the striking continuity between his visual and photographic programme.

## The Milky Way as Optical Phenomenon

To understand what Pannekoek wanted to achieve by researching and representing the Milky Way, it is necessary to first establish what he

10 See, e.g. Lankford 1987; Rothermel 1993; Pang 1997; 2002; Canales 2002; Tucker 2005, chapter 5; Nasim 2018.

11 See, e.g. Bigg 2000; Ratcliff 2008, 60-74; Wilder 2009a, 34-38; Hoel 2016; Daston 2017.

believed the Milky Way was. In the introduction to his 1920 publication on the northern Milky Way, Pannekoek goes into this explicitly: 'The Milky Way image that we observe is an optical phenomenon on whose creation various optical, physiological and psychological conditions work together. [Cornelis] Easton once referred to the Milky Way as an optical illusion; this expression may be even more true than the author himself had intended'.[12] The Milky Way, according to Pannekoek, was not a real entity that existed in the external world; it was the result of the combined light of countless faint stars, as processed by the human eyes and brain. Even so, he still believed it was valuable to investigate and represent this optical illusion. To understand why, we must turn to his Marxist philosophy, where he examined both the essence of scientific laws and the nature of the human mind.

According to Pannekoek, the task of the human mind was to analyse and abstract the information it received from the sense organs. This intuitive abstraction was required to make sense of the external world, which was a constant flow of infinitely varied and ever-changing phenomena. The mind turned these phenomena into stable objects and causal effects that we could understand. In his own words: 'The mind is the faculty of generalization. It forms out of concrete realities, which are a continuous and unbounded stream in perpetual motion, abstract conceptions that are essentially rigid, bounded, stable, and unchangeable'.[13] For Pannekoek, this reasoning extended to natural laws uncovered by science. These had no existence outside of the human mind, but were, in their essence, abstract rules extracted from our sense perceptions, formulated to bring structure and understanding to our observation of the external world of appearances. The aim of scientific research then should not be to search for the true structure of reality, but to summarize knowledge and provide economy of thought. By organizing and systematizing natural phenomena into laws and models, it became possible to comprehend them.[14] In light of this conceptualization of natural law, one can begin to understand why Pannekoek thought it worthwhile to investigate and represent the Milky Way. Even if it was not a real physical object, it was still valuable as a *scientific* object. As an intuitively created abstraction of the distribution of stars in the galaxy, it allowed astronomers to use it as a comparison for statistical astronomy and to track changes in the general distribution of stars.

12 Pannekoek 1920, 14.
13 Pannekoek 1906.
14 Pannekoek 1917; 1932.

What makes the Milky Way phenomenon especially interesting in the context of Pannekoek's philosophy of science, is that he explicitly discussed the various conditions that played a role in transforming the light of countless faint stars into the Milky Way as perceived by our eyes. He divided these conditions into three classes: the optical-anatomical, the psychological-physiological, and the purely psychological. Optical-anatomical conditions referred to such properties as the size and number of photosensitive nerves on the retina. The limited number of these retinal elements meant that the light of multiple stars, which otherwise would have been too faint to be detected individually, combined onto a single nerve. At the same time, the light of each star was not detected by just one nerve but was spread out over multiple. The combination of these two effects obscured the individuality of stars in rich agglomerations and made their light appear to human eyes as a flat image of gradually changing surface brightness. This flat image, Pannekoek identified as 'the theoretical Milky Way'.[15]

The theoretical Milky Way was not how one actually perceived the Milky Way, however, as this image was further altered by psychological-physiological conditions. An example of such a condition was the visual stimulus threshold, which was a function of both the size and brightness of an observed object. The smaller the object, the brighter they had to be to still be detectable.[16] Additionally, small bright features were also blurred over a larger area, making them appear less distinct. Crucially, both optical-anatomical and physiological-psychological conditions were tied to individual personal properties – like the number of retinal elements, visual acuity, or sensitivity to faint light – which meant that the Milky Way appeared differently to each observer.[17]

It was impossible to discern the extent to which personal differences in physiology and anatomy affected the appearance of the Milky Way, however, as the effect was drowned out by a much more significant effect. As Pannekoek explained it: 'The personal Milky Way image is not objectively determined by the earlier mentioned conditions, but is subject to still other influences, which can best be described as purely psychological'.[18] Due to the elusive faintness of the Milky Way light, the brain inevitably created patterns where there were none. Unlike the other two classes of conditions, purely

15 Pannekoek 1920, 15.

16 Here, Pannekoek explicitly referred to the work of physiologist Hans Edmund Piper, which later became known as Piper's law. Pannekoek 1920, 15, n. 1.

17 Pannekoek 1920, 14-16.

18 Pannekoek 1920, 16.

psychological conditions were largely random and not necessarily connected to the actual distribution of stars. Furthermore, because pattern creation was influenced by the observer's prior investigations of the Milky Way, an effect that could not be lessened by further observations: 'No repetition of the work, no matter how often, can help there; personal style will not be reduced, but will only impress itself stronger and clearer.'[19]

Pannekoek's views on human psychology and the role of prior knowledge in the creation of the Milky Way phenomenon resonated with his particular Marxist philosophy. The foundational principle of Marxist philosophy is that human consciousness is ultimately determined by external material factors. What exactly encompassed these material factors, however, remained a point of contention among Marxists. Pannekoek's interpretation was remarkably broad: for him, everything that was objectively observable was material, including ideas, thoughts, and theories. These were observed through conversations or texts and could have a notable influence on the further development of thoughts and ideas.[20] In the case of the Milky Way, this meant that any knowledge of earlier observations, either through memory or by looking at drawings, would inevitably influence the perceived structure. The resulting image of the Milky Way would then mimic preconceived notions of how it should look. Escaping this influence of earlier knowledge was impossible and so observations of the Milky Way were always altered by purely psychological conditions.[21]

Although Pannekoek did not appear to be too concerned about personal differences due to optical-anatomical or physiological-psychological conditions, differences caused by purely psychological conditions were a problem to him, precisely because they were both substantial and random. In 1897, he discussed various recently published Milky Way drawings and drew attention to the fact that, despite remarkable agreement on certain features, there were also great discrepancies in the structures they depicted. At times it was even hard to recognize that they were meant to represent the same object at all as a result of differences in the way observers recognized and recorded features, and differences in style and method of drawing.[22] Pannekoek was not alone in noticing the discrepancies among Milky Way drawings. A few years earlier, for example, Edward Emerson Barnard, a pioneer in Milky Way photography, argued: 'Eyes differ so much, and astronomers, as a rule, are such poor artists,

19 Pannekoek 1920, 16.
20 Pannekoek 1937, 451.
21 Pannekoek 1920, 16.
22 Pannekoek 1897a, 40-41.

that we may never expect to get anything like a fair delineation of the Milky Way by the human hand alone'.[23] Pannekoek disagreed with this sentiment, however. As we have seen, he considered it valuable to create a representation of the Milky Way based on visual observations. Such a representation, he believed, could be constructed by combining the work of many different independent astronomers in such a way that eliminated personal biases while preserving the inherent advantages of human perception.[24]

This section has revealed some striking interrelations between Pannekoek's scientific research and Marxist epistemology. By considering the latter, we can better understand methodological and epistemic choices he made in the former. It elucidates why Pannekoek believed it was important to capture the Milky Way as it was observed by the human eye, despite the fact that it was an optical illusion, and despite the considerable discrepancies among different observers. Intuitive abstraction was, after all, an inherent virtue of being human, and if the Milky Way aspect proved to be valuable for the investigation of the general structure of the distribution of stars, then it was worthy of scientific research. It also reveals why Pannekoek thought it was impossible to eliminate personal interpretation from visual observations. Since, as he explained in his Marxist writings, ideas and memories are material factors that determine human thought, subsequent observations of the Milky Way would only reinforce this interpretation, as they were unavoidably influenced by earlier impressions. It should be stressed, though, that neither belief was unique to Marxism and that Pannekoek had already begun to develop his ideas on the Milky Way before he had turned to Marxism. What the interrelations indicate, however, is that Pannekoek had a coherent epistemology that connected the practice of science with political and ethical philosophy.

At the same time, we can relate Pannekoek's extensive description of the various anatomical, physiological, and psychological circumstances that create the Milky Way phenomenon to how astronomers reflected on their own role in astronomical observations from the mid-nineteenth century onwards. By this time, due to the increasing precision of astronomical observations, astronomers began to notice that different observers recorded different stellar coordinates when using the trusted eye-and-ear method in transit observations.[25] These

23 Barnard 1890, 312; also quoted in Pannekoek 1897a, 41.

24 Pannekoek 1897a, 42.

25 The ear-and-eye method is a method of measuring the right ascension of a star by following its movement across reticles in the telescope while listening to a ticking clock.

so-called 'constant differences' forced astronomers to acknowledge that even among the most skilled and educated observers, inherent differences could occur. Astronomers started to reflect on themselves as an intricate part of their astronomical instrumentation. They each had their own characteristics and variations that could be measured and had to be corrected for, as in the case of any systematic instrumental error. Crucially, different beliefs on what caused constant differences led to different strategies to eliminate them. When it was believed that the effect was caused by psychological factors, the proposed solution was to minimize it by emphasizing discipline, skill, and education. When the effect was believed to be due to physiological factors, on the other hand, it became an inherent characteristic of the observer that could not be eliminated. It could, however, be standardized and accounted for by introducing mechanical methods and keeping track of who made each measurement. This ultimately led to the measurement of each observer's characteristics in order to calculate their so-called 'personal equation'.[26] According to Pannekoek, both psychology and physiology played a substantial role in creating the appearance of the Milky Way. Accordingly, we will see combinations of both strategies in his research. Psychological conditions could be reduced through proper methods and collaboration. Physiological conditions, on the other hand, could only be eliminated through photography.

## How to Represent the Milky Way

Pannekoek's solution to the problem of providing a visual representation of the Milky Way that everyone could agree upon, was to make use of collaborative effort. By combining various independent drawings and descriptions of the Milky Way, it would be possible to filter out random personal patterns, which were restricted to a single observer, while preserving those features that were present in the work of multiple observers. The resulting image, Pannekoek argued, would then be far more objective than any individual image.

> Here, the importance of many independent works becomes apparent. Their differences give an impression of the objective uncertainties of faint particulars, which far exceeds the limits of subjective certainty. On the other hand, their agreement can secure faint details that each observer individually would be inclined to consider doubtful. In the

26 Hoffmann 2007. For more on the personal equation, see Schaffer 1988; Canales 2001.

> average of various representations, the accidental-subjective, the style of each observer, disappears to a large extent. What is retained, is not an objective image of the Milky Way, but that which one could call the mean-subjective image [*durchschnittlich-subjektive Bild*], the objective image as it is altered by the general physiological-psychological observation conditions. The connection with an objective Milky Way image is then at least significantly easier to find.[27]

The method of combining the observations of multiple observers to create a single composite image was common in late-nineteenth-century astronomy. Similar projects had been undertaken, for example, by William Parsons, the third Lord Rosse, in his drawings of nebulae, and by Arthur Ranyard and William Wesley in their depictions of the solar corona.[28] In both these cases, the final image was extracted by a single astronomer whose task it was to determine the true shape of the astronomical object based in their careful visual inspection of the various observations. Pannekoek, as we will see, took a far more mechanical approach in his pursuit for the mean subjective image; an approach that was closely connected to his ideas of how the Milky Way should be represented in the first place.

A requirement for constructing a collaborative representation of the Milky Way was that there were observations by other astronomers in the first place. In 1897, when he was still a student in Leiden, Pannekoek published a series of articles in popular astronomy journals that encouraged amateur astronomers to record their observations of the Milky Way and outlined a method that they should follow while doing so. Prior to observing, Pannekoek asserted, observers had to take proper precautions. They had to ensure that there was no artificial illumination nearby and that the sky was clear and cloudless, but more importantly, they also had to avoid learning about any previous research: 'For [the Milky Way's] great faintness makes it very easy to see what we expect to see, and preconceived ideas will soon vitiate the results'.[29] This is a clear example of how Pannekoek believed thoughts and ideas could have a real influence on scientific research. It should be noted, however, that this epistemic fear of prior knowledge altering what was seen was quite common among astronomers of his time. Milky Way researcher Otto Boeddicker, for example, wanted to exclude the influence of prior knowledge to the point that he avoided looking at

27 Pannekoek 1920, 16-17.

28 For nebula drawings, see Nasim 2013, 38-65; for the solar corona, see Pang 2002, 96-105.

29 Pannekoek 1897b, 77.

any earlier drawing of the Milky Way, including his own, so that he could 'remain as long as possible in ignorance of [the Milky Way's] appearance *as a whole*'.[30]

To record observations of the Milky Way, Pannekoek proposed a dual method that combined verbal descriptions with visual diagrams. To record particular features, it was important to investigate only small parts of the Milky Way at a time, and describe in detail, the position, boundaries, and interconnections of each Milky Way stream and cloud. Often, it was advantageous not to look at a bright spot directly but slightly next to it, as indirect vision could reveal details that were not seen by direct vision. Recording these details could best be done by written descriptions, as Pannekoek considered these to be much more intelligible and certain than drawings, for which it was never clear whether particular features were actually seen by the observer or the result of an inaccurate rendering by the draughtsman.[31] To record the general distribution of brightness in the Milky Way, Pannekoek recommended the use of isophotes – lines of equal brightness – which could be produced as follows: 'After having examined the region thoroughly, a boundary line is picked out, and its course is followed along the Milky Way, everywhere tracing the places of equal brightness. After having finished such a line, and after having marked its course upon the chart, another is chosen, shaping its course along a track of greater or lesser brightness'.[32] The number of isophotes should be limited to only a few in order to avoid confusion. They also should be supplemented with systematic photometric estimates that had to be made by repeatedly comparing distant sections of the Milky Way to each other. The dual method had the advantage of catering to both astronomers who wanted to track changes in the visual appearance of the Milky Way, where minute details were important, as well as those who wanted to use the Milky Way as a guide for researching the overall structure of the galactic system, for which the general distribution of light was more useful.

Pannekoek abandoned his own research on the appearance of the Milky Way in 1899, when he was hired as observer at the Leiden Observatory. When he picked up the subject again in 1910, he noticed that he had failed to cover the whole of the northern Milky Way in his observations, which he attributed to the fact that he had deliberately avoided looking back at his earlier observations during this research. From 1910 to 1913, he worked

30 Boeddicker 1889, 13; emphasis in the original.
31 Pannekoek 1897b, 78-79.
32 Pannekoek 1897b, 79.

**Figure 11.1 Naturalistic drawing of a section of the Milky Way by Pannekoek**

Source: Anton Pannekoek, *Die nördliche Milchstrasse*, Annalen van de Sterrewacht te Leiden, 11:3 (Haarlem: Joh. Enschedé en Zonen, 1920)

on the missing areas until he finally covered the Northern Milky Way in its entirety. The results of his observations were only published in 1920. Throughout this period, Pannekoek's ideas on how to represent the Milky Way continued to develop. He concluded that the dual method of verbal descriptions and isophotic diagrams was insufficient; they had to be supplemented with naturalistic white-on-black drawings that showed the Milky Way 'as it appeared to [Pannekoek's own] eyes' (Figure 11.1).[33] This inclusion is significant as these naturalistic drawings would have been by far the most difficult and expensive to reproduce, while serving no immediate scientific purpose like the isophotic diagrams and verbal descriptions did. Isophotic diagrams could be used in comparison with statistical star counts in order to probe the three-dimensional structure of the star system, while verbal descriptions could be recorded over a prolonged period of time in order to track minute changes in particular features of the Milky

33 Pannekoek 1920, 11.

Way.[34] Instead, the naturalistic drawings were included because they had aesthetic value. Conveying this aesthetic value was important, according to Pannekoek, because it was what often stimulated interest in astronomy in the first place: 'For modern man [...] the aesthetic element undeniably helps to arouse love for the night sky, all the more because the pleasure that direct observation provides us, [...] is further validated and enriched by knowledge'.[35]

Pannekoek's observations of the northern Milky Way prompted German astronomer Josef Hopmann to observe the southern Milky Way as part of his 1922 solar eclipse expedition to Christmas Island.[36] Hopmann explicitly followed Pannekoek's method in making and recording his observations. He also presented his results in the form of an isophotic diagram, which he later supplemented with numerical values for the surface brightness.[37] Pannekoek, however, was sceptical of Hopmann's results. The latter's photometric values for those areas that overlapped with the northern Milky Way were not consistent with the values that Pannekoek had found. Furthermore, Pannekoek doubted the truthfulness of the incredibly rich and detailed structure displayed in Hopmann's southern Milky Way.[38] When the Dutch Royal Academy of Sciences organized an expedition to Palembang in the Dutch East Indies for the 1925 solar eclipse, Pannekoek saw it as an ideal opportunity to observe the southern Milky Way himself.

Prior to his expedition to the Dutch East Indies, Pannekoek had never been able to follow his own instructions in earnest as he had been well acquainted with the appearance of the northern Milky Way prior to his first recorded observations. Now, with the southern Milky Way, he could truly start with a blank canvas. He soon discovered that there were practical problems to being unfamiliar with the area under investigation. It took him several days to get familiar enough with the stars of the southern hemisphere to be able to observe the southern Milky Way without constantly having to reorient himself. Moreover, he realized that even when looking at a completely unfamiliar sky, there were still ways in which implicit bias altered his observations. Increased knowledge of the importance of absorbing nebulae, for example, made him more inclined to mark dark features as real resolved objects. Nevertheless, he was satisfied with his method as it

34 Pannekoek 1897b, 79-80; for an analysis of how Pannekoek used the appearance of the Milky Way for his research on the statistical distribution of stars, see Tai 2017, 230-240.

35 Pannekoek 1916, 3.

36 Ferrari d'Occhieppo 1977.

37 Hopmann 1923; 1924.

38 Pannekoek 1925.

provided him with a systematic method of handling observational data, which in turn led to a more successful representation of the Milky Way. He also mentioned the valuable contribution of his wife, Anna Pannekoek-Nassau Noordewier, who acted as an observational assistant and penned down the verbal descriptions he dictated.[39] Significantly, one of the main conclusions of his research was that the richness of the southern Milky Way, which Pannekoek had dismissed in the work of Hopmann was indeed accurate. In a letter to Easton, he described how he had been stunned by the beauty of the southern Milky Way, further reinforcing the continued presence of aesthetics in Pannekoek's Milky Way research.[40]

Of course, presenting his own Milky Way observations was only the first step of the process for Pannekoek. His ultimate goal was to produce the collaborative mean subjective image. In 1920, Pannekoek did exactly that for the northern Milky Way, making use of the earlier observations of multiple independent observers, most prominently those by Otto Boeddicker, Cornelis Easton, and J.F. Julius Schmidt.[41] Pannekoek had initially intended to present the mean subjective image in the form of separate reproductions of each individual drawing from which the readers could draw their own conclusions about the structure of the Milky Way by comparison.[42] By 1920, however, Pannekoek had grown more ambitious in his plans for the mean subjective image. His new strategy was to make use of the numerical properties of isophotic drawings. He wanted to mimic the image that would emerge if these drawings had been made on translucent paper, placed on top of each other. He believed he could simulate this effect numerically by measuring isophotic diagrams of the drawings and calculating the arithmetic mean.[43]

For his own observations and those of Easton, isophotic diagrams were already available, but those of Schmidt and Boeddicker had to be specially created from the original drawings.[44] When these were done, however, Pannekoek realized that the brightness estimates in the drawings of Boeddicker and Schmidt were far from systematic, making their absolute values

39 Pannekoek 1928, 6. I have not found any other instance where Anna Pannekoek-Nassau Noordewier assisted Anton Pannekoek in his astronomical research.

40 Pannekoek to Easton, 19 April 1926, CE.

41 The drawings of Boeddicker and Easton had been published in 1892 and 1893 respectively. The drawings of Schmidt were unpublished in 1920. Pannekoek and De Sitter eventually managed to get them published as: Schmidt 1923.

42 This strategy was later used by Fritz Goos, see Goos 1921.

43 Pannekoek to Willem de Sitter, 11 August 1920, WdS 45.1, 80-82.

44 Pannekoek to Willem de Sitter, 21 July 1919, WdS 45.1, 57-59; Pannekoek to Willem de Sitter, 24 January 1920, WdS 45.1, 67-70.

**Figure 11.2 Isophotic diagram of the mean subjective image**

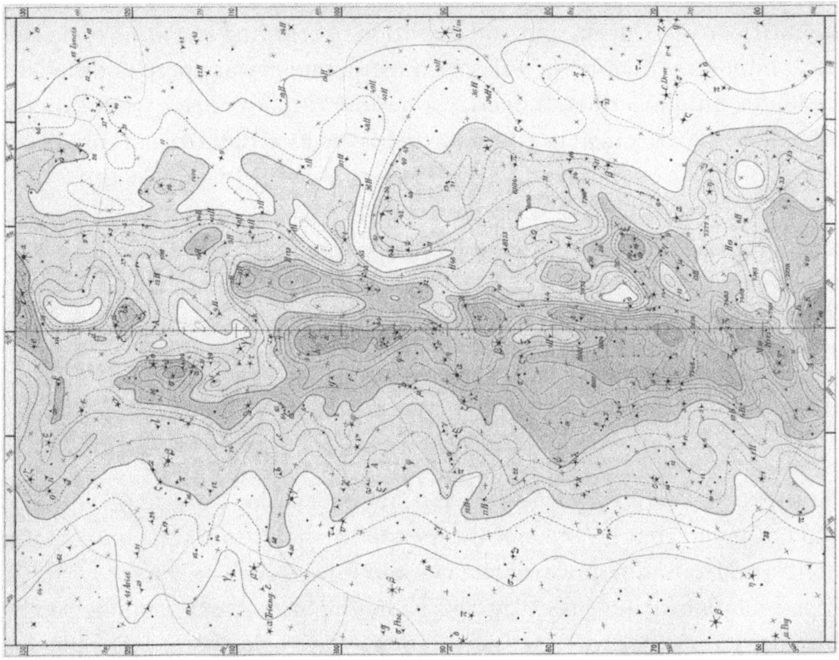

Source: Anton Pannekoek, *Die nördliche Milchstrasse*, Annalen van de Sterrewacht te Leiden 11:3 (Haarlem: Joh. Enschedé en Zonen, 1920)

unreliable. Yet, at the same time, their drawings were often richer and better in their finer structures than those of Easton and Pannekoek. To make the most of the benefits of each drawing, Pannekoek ignored the work of Boeddicker and Schmidt for the general structure of the Milky Way, while attributing greater weight to them in the case of particular feature rich areas – a striking example of how he relied heavily on his own judgement in creating the mean subjective image.[45] Pannekoek was very pleased with the end result, which he believed rose far above that of any one observer in depicting the Milky Way structure, making it ideal for comparison with photographic results.[46] The calculated mean subjective image was presented both in the form of an isophotic diagram (Figure 11.2) and as a numerical table. Additionally, for each section of the Milky Way, verbal descriptions by multiple observers were placed side by side.

45 Pannekoek 1920, 90.

46 Pannekoek to Willem de Sitter, 20 September 1920, WdS 45.1, 83-84.

Pannekoek's strategy for constructing the mean subjective image from existing depictions of the Milky Way elucidates his views on the ethos of scientific investigation and collaboration. According to Pannekoek, the most important quality for Milky Way astronomers was not their excellent vision or innate genius. Indeed, such individual qualities were exactly what Pannekoek sought to eliminate in his creation of the mean subjective image. Instead, he implored astronomers to show self-restraint and follow the proper method in describing the Milky Way. Doing so would make their contribution to the combined image that much more valuable. And ultimately this combined image, the mean subjective image, was much more trustworthy than any individual observer could ever hope to produce.

## Photography as an Observational Tool

Pannekoek's extensive work on visual observations of the Milky Way did not mean that he was not interested in photography.[47] Quite to the contrary: from 1919 onwards, he worked for decades on a photographic representation of the Milky Way. This photographic research was noteworthy because Pannekoek was not interested in wide-angle photography like his contemporaries. Instead, he used extrafocal photography, which meant that the photographic plate was intentionally placed outside the focal plane. Furthermore, the presentation of this research was remarkably similar to that of his visual observations. It came in the form of isophotic diagrams and naturalistic drawings, and not, as one might expect, in the form of photographic reproductions. Analysing Pannekoek's photographic method of representing the Milky Way provides crucial insight into the application of astrophotography in the early twentieth century and the impact it had on the daily practice of astronomy.

When photography was first introduced in astronomy, it was primarily the domain of amateur astronomers, who had the freedom to experiment with photographic techniques, while professionals remained mainly focused on precision measurements using large visual refractors.[48] Even in the depiction of visually striking objects, like nebulae, planetary surfaces, or the solar corona, professional astronomers generally preferred drawings based on visual observations over photography. These were considered more

47 For an overview of Pannekoek's ideas on the role of photography in the historical development of astronomy, see Jennifer Tucker, 'Popularizing the Cosmos', in this volume, 173-195.

48 Lankford 1981; 1984.

trustworthy because the human eye was considered better at capturing large-scale structures and evaluating large differences in brightness.[49] Photography did have one major advantage over visual observations, however: photographic plates could be taken in large numbers and then be stored for later use.[50] This, in turn, enabled a division of labour among astronomical institutions. Since observatories with photographic instruments managed to produce far more photographic plates than they could possibly reduce, they could send photographic plates to institutions lacking photographic equipment. It even became possible to found astronomical institutes that lacked any kind of observatory, like the Astronomical Laboratory of Jacobus C. Kapteyn in Groningen. The success of collaborative projects like the *Cape Photographic Durchmusterung*, which was based on photographic plates taken by David Gill at the Cape Observatory and measured by Kapteyn in Groningen, helped to convince professional astronomers of the advantages of photography. By the early twentieth century, professional astronomers had started to embrace photography as new techniques and methods were developed that could work around its limitations. Meanwhile, drawing and visual observation increasingly became the domain of amateurs.[51] The case of the Milky Way, however, illustrates that the epistemic concerns surrounding photography persisted well into the twentieth century.

The Milky Way provided an interesting challenge for astronomers wanting to study it photographically, because telescopes – which were required to focus light onto the photographic plate – generally resolved the Milky Way into the many tiny individual stars that formed it. In the late nineteenth century, Barnard found that he was able to capture unresolved Milky Way clouds on the photographic plate using a wide-angle lens. Around the same time, German astronomer Max Wolf used a similar lens to obtain photographs of Milky Way clouds and other extended bodies in the night sky. Pannekoek considered these photographs a 'revelation' because they had provided definitive evidence that the Milky Way was formed by the combined light of countless stars too faint to see with the naked eye.[52] At the same time, the image these early photographic recordings revealed of the Milky Way was fundamentally different from what could be seen with the naked eye; it was much more detailed and irregular in structure.

49 For nebulae, see: Nasim 2013; for Mars, see: Lane 2011; Tucker 2005; for the solar corona, see: Pang 2002; Becker 2000; 2013.

50 This is particularly evident in the case of the Carte du Ciel, which is explicitly conceived as a photographic atlas of the stars that can serve as an archive for future astronomers. See Daston 2017.

51 Lankford 1984.

52 Pannekoek 1951, 409-411.

To some astronomers, this indicated that visual observations should no longer be trusted. Barnard, in particular, believed in the inherent value of photography: '[N]o matter how erroneous the various theories concerning the constitution of the Milky Way, the photographs are supposed to tell their own story, from which the student can judge for himself how well the theories fit into the actual appearance of this wonderful zone of stars'.[53] As we have seen, Pannekoek continued to value visual observations, but he was also enticed by the possibilities of Milky Way photography.

Pannekoek started his efforts to create a photographic representation of the Milky Way in 1919 while he was still refining his ideas on the mean subjective image, and many similarities exist between the two methods. The goal of both was to represent the large-scale distribution of galactic light. Wide-angle photography, as employed by Wolf and Barnard, was unsuited for this purpose because it emphasized minute structure over the general distribution of light. Pannekoek's alternative was extrafocal photography. The method of extrafocal photography was mainly developed by Karl Schwarzschild for photographic photometry of individual stars. As plates were taken out of focus, the light of stars was spread over a larger area, which allowed more accurate photometric measurements.[54] Pannekoek realized that this technique could be used to effectively produce the theoretical Milky Way – the Milky Way altered only by optical-anatomical conditions – as it would cause the light of the countless faint stars composing the Milky Way to overlap on the photographic plate.[55] While the mean subjective image could only eliminate the purely psychological conditions, extrafocal photography promised to also eliminate personal physiological-psychological conditions.

Since Pannekoek lacked his own observatory, he had to rely on the assistance of other astronomers for the implementation of his extrafocal photographic project, leading to its own set of logistical problems. For the northern Milky Way, the extrafocal plates were taken by Max Wolf in Heidelberg. The first batch of these plates, which arrived in 1920, turned out to be unsuited because they were not taken sufficiently out of focus.[56] Subsequent attempts were more successful, but even then, individual photographic plates were often found to have flaws and had to be replaced.

53 Barnard 1909, 89.

54 For more on Schwarzschild's extrafocal method, see Habison 2000.

55 Pannekoek 1923, 19.

56 Pannekoek to Max Wolf, 20 December 1920, MW.

All this meant that coverage of the northern Milky Way was not completed until 1928.[57] For the southern part of the sky, it took even longer. In 1926, Pannekoek instructed Joan Voûte of the Bosscha Observatory in Lembang on how to take the extrafocal plates (see Figure 11.3). Because the main telescope of the observatory was also used for other purposes, it took three years before Pannekoek received the plates.[58] Again, many of the photographic plates were found to have flaws and had to be retaken in 1933 and in the winter of 1938-1939. An added complication was that the southern-most part of the sky was not sufficiently visible from Lembang. For that part, Pannekoek had to turn to Harlow Shapley, director of the Harvard College Observatory, who agreed to have the plates taken at the Boyden Station in Mazelspoort, South Africa. These plates could only be taken in 1942 and by this time, they could not be shipped to the Netherlands until 1945 as a result of World War II. After they arrived, two of the Boyden-plates had to be rejected and retaken in 1946, finally completing the entire Milky Way.[59]

Getting a hold of the photographic plates was only the first step of the process, however. The plates first had to be systematically measured using a microphotometer.[60] These measurements then had to be corrected for both general systematic errors that resulted from the extrafocal method, as well as plate-specific systematic errors, which had to be determined empirically for each plate. To be able to combine the measurements and get a meaningful scale for the surface brightness, a reduction curve had to be derived separately for each individual plate.[61] For most of the Milky Way, multiple plates overlapped, and the average value was calculated. All these measurements and calculations were conducted by Pannekoek's long-time calculator David Koelbloed.[62] Pannekoek himself drew the isophotic diagrams, for which he used an episcope that projected the photographic plates onto paper. The isophotes were then drawn by tracing the features that the episcope had projected (see Figure 11.4).

57 Pannekoek 1933, 1-4; see also the Pannekoek-Wolf correspondence in MW.

58 These plates included exposures of the Large and the Small Magellanic Clouds, which were reduced by Gijsbert van Herk, then a student of the Astronomical Institute in Amsterdam, and published as van Herk 1930.

59 Pannekoek and Koelbloed 1949, 1-3.

60 A microphotometer is an instrument for measuring photographic plates that allowed both the coordinates and the blackening of the plate to be accurately determined.

61 The reduction curve is a formula that gives the relation between the incident light intensity of an object and the blackening it causes on the photographic plate.

62 Pannekoek 1933, 6-35; Pannekoek and Koelbloed 1949, 5-26.

**Figure 11.3 Extrafocal photographic plate of a portion of the southern Milky Way, taken at the Bosscha Observatory in Lembang by Joan Voûte**

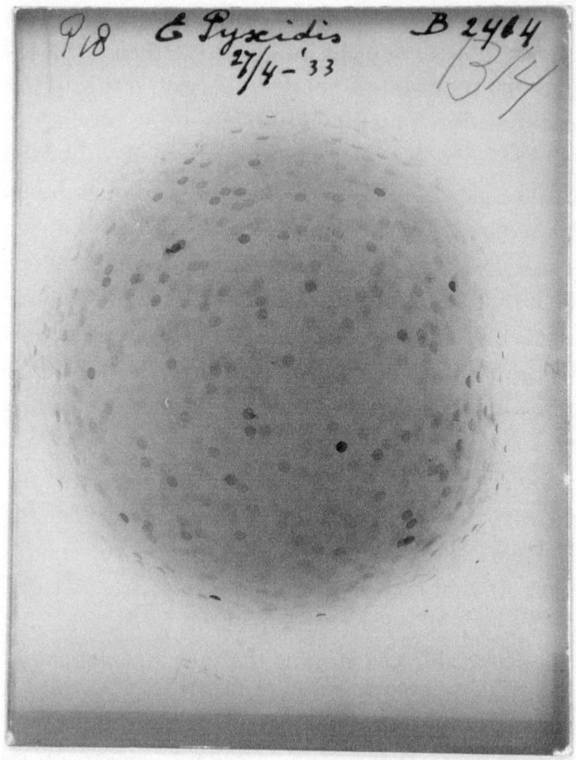


Source: Archive of the Anton Pannekoek Institute for Astronomy, University of Amsterdam

Throughout the entire measurement process, experience and expert judgement played a vital role. Pannekoek made this clear in a letter to Shapley that was sent only two days after the liberation of the Netherlands in World War II. In this letter, he requested that the remaining plates be sent as soon as safely possible, explaining that he had to finish the work himself 'during the years that will be allowed to me' as he was the only one with the skill and expertise needed to draw the isophotic diagrams. Similarly, he argued that only Koelbloed was capable of conducting the required measurements and calculations for this project.[63] Pannekoek's emphasis on the importance of his own hand in drawing the isophotic lines underlines a crucial aspect of his method of photographic photometry: it was never

63 Anton Pannekoek to Harlow Shapley, 7 May 1945, HCO.

**Figure 11.4 Small part of one of Pannekoek's working sheets for photographic photometry**

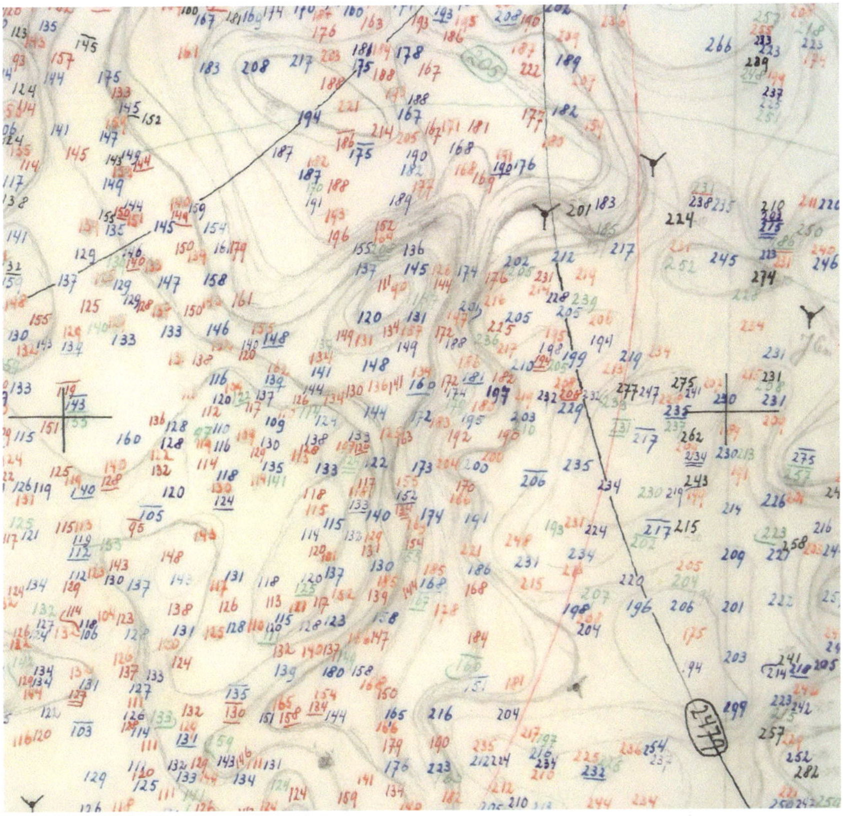

On these working sheets, Pannekoek traced isophotic lines based on multiple photographic plates. The lines are supplemented with numerical measurements of the same plates. The different colours represent information from different photographic plates.

Source: Archive of the Anton Pannekoek Institute for Astronomy, University of Amsterdam

meant to be objective in the sense that nature would represent itself. Not only should mechanical instruments mimic the human eye, expert judgement also remained crucially important.

In the presentation of the photographic research on the southern Milky Way, Pannekoek included naturalistic drawings of the Milky Way based on photographic photometry (Figure 11.5).[64] This inclusion reinforces what we have noticed throughout Pannekoek's photographic method: photographic

64 Pannekoek 1933.

**Figure 11.5 Naturalistic drawing of the southern Milky Way based on measurements of photographic plates**

Source: Anton Pannekoek and David Koelbloed, *Photographic Photometry of the Southern Milky Way*, Publications of the Astronomical Institute of the University of Amsterdam 9 (Amsterdam: Stadsdrukkerij, 1949)

plates were not intended to replace drawings as a way of depicting the Milky Way. Instead, they were meant to take over the role visual observations had played in Pannekoek's construction of the mean subjective image. Pannekoek's visual and photographic programmes displayed a clear continuity as they shared the principle aim of representing the Milky Way as it was seen by human eyes. This continuity from visual observations to photography was certainly not unique to Pannekoek, it can be seen in many astronomical subjects where photography made its entry.[65]

It is important to note that photographic plates were never meant to supplant visual observations completely. Pannekoek worked on both projects simultaneously throughout the 1920s and their results were intended to be complimentary. This was made when he discussed the differences between the two methods. While visual observation was better at revealing the general structure of the Milky Way, individual minor features were more clearly visible using the extrafocal photographic method. As such, the results of the extrafocal method occupied the space between visual observations and focal photography:

> We might describe the picture [produced by extrafocal photography] as the aspect the Milky Way would present to eyes that were far more sensitive to faint glares of light than ours and at the same time able to distinguish smaller details. A comparison with the focal photographs of Barnard and Ross shows a smoothing of all sharp detail, thus gaining a true representation of the surface intensity which is lacking there.[66]

Comparing the visual observations with photographic exposures had an additional practical benefit. Because photographic plates were more sensitive to blue light than the human eye, the difference in surface brightness found through both methods made it possible to determine the colour index of Milky Way clouds. For the Scutum cloud, for example, this colour index was found to be 0.43, similar to an F-type star. Evidently, the Scutum cloud had a similar constitution to the surroundings of the sun.[67] Being able to draw such conclusions illustrated the importance of providing both visual and photographic observations of the Milky Way light.

## Conclusions

Despite the fact that Pannekoek acknowledged the artificial nature of the Milky Way phenomenon as an optical illusion created by the nature

65 See, e.g., Nasim 2011 for the case of depicting nebulae.

66 Pannekoek and Koelbloed 1949, 28.

67 Pannekoek 1923, 23-24.

of human physiology and psychology, he was convinced that an accurate description of the Milky Way was still scientifically relevant. It showed how the eyes and the mind processed the light of many faint stars into a coherent image, which in turn could be used for further scientific research. As he explained in his Marxist philosophy, usefulness, not truth, was his main criterium for scientific knowledge. The Milky Way image may have been a human construct, but then so were all scientific laws.

Because the Milky Way was intangible, many different representational methods were needed to capture all of its features. Pannekoek's depictions of the Milky Way ranged from naturalistic drawings and verbal descriptions to isophotic diagrams and numerical tables of surface brightness. This variation also reflected the various ways in which the Milky Way image could be useful. Verbal descriptions could be used to track changes in minor features of the Milky Way over time, while isophotic diagrams and numerical tables could be used for comparison with statistical research on the distribution of stars. Finally, naturalistic drawings were meant to display the aesthetic value of the Milky Way. The latter was important because aesthetics often proved to be an important incentive to pursue scientific research, as was demonstrated by Pannekoek's own career in astronomy.

Notably, photography was not one of the methods of depiction. Drawing and photography are often presented as distinct and competing methods of representation, but as Pannekoek's research makes clear, this was not always the case. This is worth emphasizing since mechanically produced photographic images were often used by advocates of mechanical objectivity to argue that one should let nature represent itself without human intervention.[68] According to Pannekoek, however, photography was inherently incapable of representing the Milky Way without human intervention. Before photography could produce scientific results, measurement and expert judgement was required from the astronomer. The drawings that resulted from this critical engagement with photography were not the result of nature unveiling itself, but constructed images highlighting the structure of the system. Photography, in this case, replaced visual observation, but not drawing.

Both Pannekoek's visual method as well as his photographic method of observing the Milky Way were developed to make optimal use of the desirable qualities of human perception. As he explained in his scientific writing as well as in his Marxist philosophy, human perception depended both on how information was received by the senses and on how it was transferred and

68 For more on photography and mechanical objectivity, see Daston and Galison 2007, 161-173; cf. Pang 1997; Tucker 2008; Wilder 2009b.

interpreted by the human brain. Individual psychological conditions were undesirable here, but as in the case of constant differences, their effects could be minimized in visual observations. In the case of the Milky Way, this was achieved through a combination of adhering to proper methodology and combining the work of independent observers. The resulting mean subjective image was capable of presenting the Milky Way as it was seen by the average human eye, unaltered by purely psychological effects. The goal of Milky Way photography, on the other hand, was to also remove physiological effects, much like mechanization had done in the case of the personal equation. By mechanizing observation, the image of the Milky Way would no longer be affected by personal physiological conditions like the strength of the eye's stimulus threshold. Crucially, in both photographic photometry and the mean subjective image, Pannekoek sought to eliminate personal alterations of the Milky Way image while striving to preserve the shared optical-anatomical conditions; these he considered crucial for the way that humans interpreted the Milk Way. In isolation, such a dichotomy can be difficult to understand, but it makes perfect sense in light of his Marxist philosophy of mind. Even if individuals could be led astray, without the interpretive and analytic abilities of the human mind, nothing could be known at all.

## Archives

| | |
|---|---|
| CE | Correspondentie en Manuscripten van Cornelis Easton. Museum Boerhaave, Leiden. |
| HCO | Harvard College Observatory, Records of the Director, Harlow Shapley, 1921-1956, UAV 630.22. Harvard University Archives. |
| MW | Nachlass Max Wolf, Heid. Hs. 3695 E. Universitätsbibliothek Heidelberg. |
| WdS | Leiden Observatory Archives, directorate W. de Sitter. Leiden University Library. |

## Bibliography

Ashbrook, Joseph. 1984. *The Astronomical Scrapbook: Skywatchers, Pioneers, and Seekers in Astronomy*. Ed. by Leif J. Robinson. Cambridge: Cambridge University Press.

Barnard, Edward Emerson. 1890. 'On Some Celestial Photographs Made with a Large Portrait Lens at the Lick Observatory'. *Monthly Notices of the Royal Astronomical Society* 50(5): 310-314.

—. 1909. Review of *Die Milchstrasse*, by Max Wolf. *Astrophysical Journal* 29(1): 89-90.

Becker, Barbara J. 2000. 'Priority, Persuasion, and the Virtue of Perseverence: William Huggins's Efforts to Photograph the Solar Corona without an Eclipse'. *Journal for the History of Astronomy* 31(3): 223-243.

—. 2013. 'Historic Photographs by William and Margaret Lindsay Huggins'. *Journal for the History of Astronomy* 44(4): 481-484.

Bigg, Charlotte. 2000. 'Photography and Labour History of Astrometry: The Carte Du Ciel'. In *The Role of Visual Representations in Astronomy: History and Research Practice*, ed. by Klaus Hentschel and Axel D. Wittmann. Thun: Harri Deutsch, 90-106.

Boeddicker, Otto. 1889. 'Note to Accompany a Drawing of the Milky Way'. *Monthly Notices of the Royal Astronomical Society* 50(1): 12-15.

Canales, Jimena. 2001. 'Exit the Frog, Enter the Human: Physiology and Experimental Psychology in Nineteenth-Century Astronomy'. *British Journal for the History of Science* 34(2): 173-197.

—. 2002. 'Photogenic Venus: The "Cinematographic Turn" and Its Alternatives in Nineteenth-Century France'. *Isis* 93(4): 585-613.

Daston, Lorraine. 2008. 'On Scientific Observation'. *Isis* 99(1): 97-110.

—. 2017. 'The Immortal Archive: Nineteenth-Century Science Imagines the Future'. In *Sciences in the Archives: Pasts, Presents, Futures*, ed. by Lorraine Daston. Chicago, IL: University of Chicago Press, 159-182.

Daston, Lorraine, and Peter Galison. 2007. *Objectivity*. New York: Zone Books.

Ferrari d'Occhieppo, Konradin. 1977. 'Josef Hopmann, Nachruf'. In *Almanach für das Jahr 1976: 126. Jahrgang*. Vienna: Österreichischen Akademie der Wissenschaften, 518-535.

Galison, Peter. 2004. 'Image of Self'. In *Things That Talk: Object Lessions from Art and Science*, ed. by Lorraine Daston. New York: Zone Books, 257-296.

Galison, Peter, and Lorraine Daston. 2008. 'Scientific Coordination as Ethos and Epistemology'. In *Instruments in Art and Science: On the Architectonics of Cultural Boundaries in the 17th Century*, ed. by Helmar Schramm, Ludger Schwarte, and Jan Lazardzig. Berlin: Walter de Gruyter, 296-333.

Goos, Fritz. 1921. *Die Milchstrasse*. Hamburg: Henri Grand.

Habison, Peter. 2000. 'Schwarzschild's Investigations of "Out-of-Focus Photometry" between 1897 and 1899 at Kuffner Observatory in Vienna'. In *The Role of Visual Representations in Astronomy: History and Research Practice*, ed. by Klaus Hentschel and Axel D. Wittmann. Thun: Harri Deutsch, 107-117.

Herk, Gijsbert van. 1930. 'Photographic Photometry of the Magellanic Clouds'. *Bulletin of the Astronomical Institutes of the Netherlands* 6(209): 61-64.

Hoel, Aud Sissel. 2016. 'Measuring the Heavens: Charles S. Peirce and Astronomical Photography'. *History of Photography* 40(1): 49-66.

Hoffmann, Christoph. 2007. 'Constant Differences: Friedrich Wilhelm Bessel, the Concept of the Observer in Early Nineteenth-Century Practical Astronomy and the History of the Personal Equation'. *British Journal for the History of Science* 40(3): 333-365.

Hopmann, Josef. 1923. 'Eine neue Milchstraßenkarte'. *Astronomische Nachrichten* 219(12): 189-200.

—. 1924. 'Auswertung der Isophotenkarte der Milchstraße'. *Astronomische Nachrichten* 222(6): 81-94.

King, Henry C. 1958. 'The London Planetarium'. *The Observatory* 78(903): 69-72.

Lane, K. Maria D. 2011. *Geographies of Mars: Seeing and Knowing the Red Planet.* Chicago, IL: University of Chicago Press.

Lankford, John. 1981. 'Amateurs and Astrophysics: A Neglected Aspect in the Development of a Scientific Speciality'. *Social Studies of Science* 11(3): 275-303.

—. 1984. 'The Impact of Photography on Astronomy'. In *The General History of Astronomy: Volume 4, Astrophysics and Twentieth-Century Astronomy to 1950: Part A*, ed. by Owen Gingerich. Cambridge: Cambridge University Press, 16-39.

—. 1987. 'Photography and the 19th-Century Transits of Venus'. *Technology and Culture* 28(3): 648-657.

Lundmark, Knut. 1957. 'A Swedish Chart of the Photographic Milky Way'. *Sky and Telescope* 16(5): Center.

Nasim, Omar W. 2011. 'The "Landmark" and "Groundwork" of Stars: John Herschel, Photography and the Drawing of Nebulae'. *Studies in History and Philosophy of Science* 42(1): 67-84.

—. 2013. *Observing by Hand: Sketching the Nebulae in the Nineteenth Century.* Chicago, IL: University of Chicago Press.

—. 2018. 'James Nasmyth on the Moon: Or, on Becoming a Lunar Being without the Lunacy'. In *Selene's Two Faces: From 17th Century Drawings to Spacecraft Imaging*, ed. by Carmen Pérez González. Leiden: Brill, 147-187.

Pang, Alex Soojung-Kim. 1997. '"Stars Should Henceforth Register Themselves": Astrophotography at the Early Lick Observatory'. *British Journal for the History of Science* 30(2): 177-202.

—. 2002. *Empire and the Sun: Victorian Solar Eclipse Expeditions.* Stanford, CA: Stanford University Press.

Pannekoek, Anton. 1897a. 'On the Existing Drawings of the Milky Way and the Necessity of Further Researches'. *Journal of the British Astronomical Association* 8(1): 39-42.

—. 1897b. 'On the Best Method of Observing the Milky Way'. *Journal of the British Astronomical Association* 8(2): 77-80.

—. 1906. 'The Position and Significance of J. Dietzgen's Philosophical Works'. In *The Positive Outcome of Philosophy*, by Joseph Dietzgen, ed. by Eugene Dietzgen and Joseph Jr. Dietzgen, trans. by Ernest Untermann. Chicago, IL: Charles H. Kerr & Co., 7-37.

—. 1916. *De wonderbouw der wereld. De grondslagen van ons sterrekundig wereldbeeld populair uiteengezet*. Amsterdam: S.L. van Looy.

—. 1917. 'Twee natuuronderzoekers in maatschappelijk-geestelijke strijd'. *De Nieuwe Tijd* 22: 300-314, 375-392.

—. 1920. *Die Nördliche Milchstrasse*. Annalen van de Sterrewacht te Leiden, 11:3. Haarlem: Joh. Enschedé en Zonen.

—. 1923. 'Photographic Photometry of the Milky Way and the Colour of the Scutum Cloud'. *Bulletin of the Astronomical Institutes of the Netherlands* 2(44): 19-24.

—. 1925. 'Some Remarks on the Relative Intensities of the Two Sides of the Milky Way'. *Bulletin of the Astronomical Institutes of the Netherlands* 3(86): 44-46.

—. 1928. *Die Südliche Milchstrasse*. Annalen van de Bosscha-Sterrenwacht, 2:1. Amsterdam: De Bussy.

—. 1932. 'Das Wesen des Naturgesetzes'. *Erkenntnis* 3(1): 389-400.

—. 1933. *Photographische Photometrie der Nördlichen Milchstrasse*. Publications of the Astronomical Institute of the University of Amsterdam 3. Amsterdam: Stadsdrukkerij.

—. 1937. 'Society and Mind in Marxian Philosophy'. *Science & Society* 1(4): 445-453.

—. 1951. *De groei van ons wereldbeeld*. Amsterdam: Wereldbibliotheek.

Pannekoek, Anton, and David Koelbloed. 1949. *Photographic Photometry of the Southern Milky Way*. Publications of the Astronomical Institute of the University of Amsterdam 9. Amsterdam: Stadsdrukkerij.

Ratcliff, Jessica. 2008. *The Transit of Venus Enterprise in Victorian Britain*. London: Pickering & Chatto.

Rothermel, Holly. 1993. 'Images of the Sun: Warren De La Rue, George Biddell Airy and Celestial Photography'. *British Journal for the History of Science* 26(2): 137-169.

Schaffer, Simon. 1988. 'Astronomers Mark Time: Discipline and the Personal Equation'. *Science in Context* 2(1): 115-145.

Schickore, Jutta. 2007. *The Microscope and the Eye: A History of Reflections, 1740-1870*. Chicago, IL: University of Chicago Press.

Schmidt, J.F. Julius. 1923. *Dessins de la Voie lactée faites à Athènes par J.Fr. Julius Schmidt dans les années 1864-1876*. Annalen van de Sterrewacht te Leiden, 14:2. Haarlem: Joh. Enschedé en Zonen.

SMBA. 2016. *Jeronimo Voss – Inverted Night Sky*. SMBA Newsletter 146.

Tai, Chaokang. 2017. 'Left Radicalism and the Milky Way: Connecting the Scientific and Socialist Virtues of Anton Pannekoek'. *Historical Studies in the Natural Sciences* 47(2): 200-254.

Tai, Chaokang, and Jeroen van Dongen. 2016. 'Anton Pannekoek's Epistemic Virtues in Astronomy and Socialism: Personae and the Practice of Science'. *BMGN – Low Countries Historical Review* 131(4): 55-70.
Tucker, Jennifer. 2005. *Nature Exposed: Photography as Eyewitness in Victorian Science*. Baltimore, MD: Johns Hopkins University Press.
—. 2008. 'Objectivity, Collective Sight, and Scientific Personae'. Review of *Objectivity*, by Lorraine Daston and Peter Galison. *Victorian Studies* 50(4): 648-657.
Wilder, Kelley E. 2009a. *Photography and Science*. London: Reaktion.
—. 2009b. 'Photography and the Art of Science'. *Visual Studies* 24(2): 163-168.

## About the Author

**Chaokang Tai** is a PhD researcher in history of science at the University of Amsterdam. He is a member of the Vossius Center for History of Humanities and Sciences, University of Amsterdam, and the Descartes Centre for the History and Philosophy of the Sciences and the Humanities, Utrecht University. His current research focuses on the astronomical research of Anton Pannekoek in historical and societal context.

# 12 The Labour of Handwork in Astronomy

## Between Drawing and Photography in Anton Pannekoek

*Omar W. Nasim*

**Abstract**

This chapter discusses the crucial role of handwork in historical practices of astronomical representation, focusing particularly on Anton Pannekoek's Milky Way drawings. Using a range of cases, it explores how the acts of seeing, knowing, and drawing interacted to benefit observers, especially as a form of scientific labour. This functions as background for understanding the role of drawing and photography in Pannekoek's graphical work. This paper activates the notion of handwork in relation to labour to make it historically relevant for astronomy's representational practices, but also to connect these to broader political and epistemological trends. It will be shown that Pannekoek's emphasis on manual labour acted as a bridge between photography and drawings, and more generally, as an important cross-over point between Pannekoek-the-socialist and Pannekoek-the-astronomer.

**Keywords:** Anton Pannekoek, socialism, labour, astronomy, drawing, photography, practice

Anton Pannekoek (1873-1960) was a socialist thinker and an astronomer, making major contributions in both domains. However tantalizing this combination might be for a historian of science, it remains to be seen how exactly these two sides of Pannekoek's life might have productively interacted. One of the challenges is that Pannekoek seems to have kept both sides separate and even wrote two different memoirs, one for Pannekoek-the-astronomer and the other for Pannekoek-the-socialist. A reason, perhaps, why he chose

Tai, Chaokang, Bart van der Steen, and Jeroen van Dongen (eds), *Anton Pannekoek: Ways of Viewing Science and Society*. Taylor & Francis Group, 2019
DOI: 10.5117/9789462984349_CH12

to duplicate himself in this way, was that his socialism occasionally hindered his astronomical career; while his status as a professional astronomer was used against him in political and ideological disputes. And it does not help that these different sides of his life and work continue to be separated even in subsequent, scholarly works on Pannekoek. All this makes it difficult to see how these sides can be recombined so as to be fruitful to the history of science – difficult, perhaps, but not impossible.

One possibility for connecting the two halves of Pannekoek's oeuvre has recently been proposed by Chaokang Tai. He attempts to connect a particular kind of radical Marxist philosophy of mind to the epistemic virtues underpinning Pannekoek's astronomical research.[1] Though Tai's thoughtful approach yields important results, and has inspired what follows, I propose another alternative, one which focuses not so much on a theory of mind as on hands and tools. I argue that there is a strong, operational presence of the hand in both Pannekoek's astronomical work and in his socialist theories about human development. After all, the significance of the human hand for Pannekoek is immediately gleaned from both his socialist theories and in his representational preferences in astronomy, especially when it came to depicting the Milky Way – a research pursuit that engaged him for most of his life. I claim that what reconnects the two sides of Pannekoek is the astronomical labour connected to *handwork*. The labour and business- or factory-like character of nineteenth-century astronomy, particularly in large, national observatories, has been examined in previous studies.[2] But the idea of handwork, as instanced in the case of Pannekoek, permits us to extend the useful notion of astronomical labour further into the representational practices of astronomy, like drawing and photography. Each medium will be approached as different but related forms of labour that are linked to the production of knowledge. From this perspective, both drawing *and* photography reveal their productive character as handwork *and* technology. In fact, this approach to Pannekoek will elucidate how labour and handwork related to astronomical practice are implicated in photography; and, on the other hand, how paper and pencil, ink and pen, used in handmade drawings, are technologies in their own right. And all this, thanks to handwork, which acts as a bridge between different media but also between Pannekoek-the-socialist and Pannekoek-the-astronomer.

The first section of this paper summarizes the salient features of Pannekoek's socialist theory of human origins and development. Among the features

1 Tai 2017; see also Tai and van Dongen 2016.

2 Schaffer 1988; Ashworth 1994; Smith 1991.

that will be important for us are the ways in which the labour of hands and tools functioned, socialist thinkers argued, to give rise to the human being, history, science, and technology. But another salient feature is how handwork was thought to overcome ideology and metaphysics, and the dualisms that these are based upon. This section sets the scene for the role of the hand in Pannekoek's socialist framework and for his astronomical practice. In the second section, I aim to motivate what it means for there to be handwork in astronomy. The same section will provide a few cases from nineteenth-century astronomy for understanding the labour involved in the handwork of drawing difficult celestial objects. This is important to do, because, as we shall see in section three, Pannekoek's own astronomical practices are replete with hand-drawings of all kinds, despite the availability of photography, leaving us with the question: why did he not employ photography instead? Section four outlines the broader context for understanding photography's relationship to astronomical labour. It is in the concluding section of this paper that I offer remarks about how handwork acts as a bridge between photography and drawings, a fact that will only go to reinforce how handwork, more generally, acts as an important cross-over point between Pannekoek-the-socialist and Pannekoek-the-astronomer. With my focus on the hand, I do not want to claim that its role is the one and only key to a reunified picture of Pannekoek's oeuvre. Rather, what I offer is a modest proposal about a common feature to both sides of Pannekoek's life and work; which also happens to have implications for how we view astronomical labour especially in relation to representational practices like drawing and photography.

## I

Before we come to see how astronomy and socialism might be connected in Pannekoek's unique career, we need to see how he understood the connections between Darwinism and Marxism. Doing so, will lead us directly to the significance of the hand in Pannekoek's worldview. As general theories, Darwinism and Marxism apply to two different domains, but they can nonetheless complement one another in important ways. As a card-carrying socialist, however, Pannekoek was quick to notice that Darwinism had previously been used to justify a particular view of society that opposed socialism, especially in the guise of Social Darwinism. In order to understand, therefore, how these theories relate, Pannekoek begins by addressing the purported misuse of the theory of evolution by social darwinists like Ernst Haeckel and Herbert Spencer, who take the theory of evolution as a justification of the status-quo authority of the petit bourgeoisie and capitalism. The great mistake all 'bourgeois Darwinists' make, according to Pannekoek, is to incorrectly take a theory that is appropriate and applicable to one domain (the animal kingdom) and uncritically apply it to another (the social world of the human being). Pannekoek believes that this is a category mistake, because human beings have unique characteristics that set them apart from the rest of the animal world. By recognizing these peculiarities, we can also begin to appreciate the importance of maintaining two separate domains: one for Darwinism and the other for Marxism.[3] And in distinguishing the different domains of application, we can, it should be noted, accept Darwinism without having to give up Marxism. The two, writes Pannekoek, 'supplement each other, in the sense that, according to Darwinian theory of evolution, the animal world develops up to the stage of man, and from then on, that is, after the animal has risen to man, the Marxian theory of evolution applies'.[4] Each of these theories explain features of two distinct domains – the animal or human worlds – while remaining continuous and complementary.

But what exactly differentiates the human world from the animal? In answering this age-old question, Pannekoek puts forward the usual suspects: language, society, and abstract thought. However, the crucial driver of human development is uniquely adopted hands, conducive to the use of tools. It is by means of his hands, for example, that 'primitive man, at his lowest stage', distinguishes some objects as tools that assure his survival. As tools begin to vary in complexity and application, 'primitive man' begins to discern different kinds of objects, no longer treating the world as 'a single unit', as do animals.[5] Being so important to his survival, these tools are designated by sounds and thus named in some primordial language, and in this way, they are shared and their memory passed on. We have already moved from the hand to tools to language. Consciousness appears when the new being distinguishes not just between tools and objects, but also different sorts of intentions manifested in different functions, permitting the development of tools that are ever more task-oriented and refined. But at the same time, the development of more refined tools makes thought itself more nuanced. There is therefore a progressive 'circuit' between material and mental development in the human being – it is this circuit that contributes to human progress, even up to our own day, according to Pannekoek.[6]

3 Pannekoek 1912.
4 Pannekoek 1912, 33.
5 Pannekoek 1912, 49.
6 Pannekoek 1912, 46.

Central to the human hand and its associated tools is the idea of labour, which lies at the heart of this story of progress. It is 'by his labour', writes Pannekoek, '[that] the primitive ape-like man has risen to real manhood'.[7] It should to be noted that Pannekoek was not alone in arguing for the key role played by labour in the very origins of the human being. We find something very similar in others, like Ludwig Noiré and Frederik Engels.[8] In fact, this is just one of the many ways in which Pannekoek echoes Engels, who, in an unfinished work, *The Part Played by Labor in the Transition from Ape to Man* (1876), argued that 'labour created man himself';[9] while labour itself emerged with the making of tools. It is in this way that for Engels the hand is central to any story of the development of the human, for 'the hand is not only the organ of labour, *it is also the product of labour*'.[10] Labour, for both Engels and Pannekoek, is what propels history forward.

Pannekoek, furthermore, takes up the theme of the hand as the organ of labour in order to argue that while Darwinism applies to animal organs, it does not apply in the same way to the 'artificial organs of men', which has its own unique laws of development and progress (i.e., Marxism). Darwinism and Marxism can thus be further contrasted when one considers the temporal domains of each: In the case of the former, the continuous evolution in the animal world is 'infinitely slow, as dictated by biological laws'. In the case of Marxism, on the other hand, human tools 'can be transformed quickly, and technique makes such rapid strides that, in comparison with the development of animal organs, it must be called marvelous [sic]'. The artificial organs of the human being, i.e. tools, are 'free from the chain of biologic laws'.[11] And just as in the case of the animal, organic world, progress – however rapid – in human tools is actually the result of a struggle that leads to the 'ever greater perfection of tools' such that '[t]hose races whose technical aids are better developed, can drive out or subdue those whose artificial aids are not developed. The European race dominates because its external aids are better'.[12] Pannekoek observes that although each theory applies differently to their respective domains, Marxism and Darwinism actually share 'the same principle [which] underlies both theories'; namely, the survival of

7 Pannekoek 1912, 50.

8 The hand has also been of interest in more recent times to, see: Napier 1970; 1980; Wilson 1998; McGinn 2015. The last, though an essay in 'philosophical anthropology', meant to be a 'hymn' to the hand, does not cite, let alone engage with the ideas of Engels, Noiré, nor Pannekoek.

9 Engels [1876] 1950, 7.

10 Engels [1876] 1950, 9; emphasis in the original.

11 Pannekoek 1912, 51.

12 Pannekoek 1912, 53.

the fittest – it is, in fact, the warrant for technological determinism but also imperial expansionism.[13] But, unlike capitalism (the bedbug of Social Darwinism), which according to Pannekoek creates a world that resembles the 'rapacious' animal world, socialism aims rather to externalize these struggles out towards Nature and away from 'our own kind'.[14]

The hand therefore represents more than just an organic limb of the human body for Pannekoek. It distinguishes humans from other animals, and when it is coupled, in its labour, with artificial tools (and later technologies and machines) it is the driver of progress and history. But this handwork can also connect mind and matter, thought and being, and thereby overcome ideologies arising from such dualisms. This is especially true for our everyday existence in the world as practical beings. But it is also true for the sciences, despite their over-emphasis on the intellectual at the expense of the hand. The last point was already made by Engels, when he observed that the more human society advanced, the more did these advances appear to be the products of the human mind rather than the 'more modest productions of the working hand'. This lack of appreciation of the hand's role in the history of the sciences, has consequences for Engels. In particular, when the mundane role of the hand is overshadowed by the glories of the mind we are led to idealism, an ideology that dominates 'even the most materialistic natural sciences of the Darwinian school', which is also the main reason that '[the Darwinians] do not recognize the part that has been played by labour'.[15] By bringing handwork into the foreground, therefore, we bring back into balance handwork and headwork, according to Engels, so that we might avoid the idealism lurking in our dualist accounts of even the most materialist theories of science.

Pannekoek, too, echoes the same sentiments about the intrusion of ideology into the sciences and their histories. So, for example, he takes to task the early nineteenth century, Scottish anatomist, moralist, and surgeon Charles Bell, and his widely read Bridgewater treatise, *The Hand* (1837). According to Pannekoek, one of the things that Bell fails to appreciate – in a book on the hand, no less – is that touch, by means of the fingers, is not just a passive but also an active 'energy'. The reason that Bell does not see this, claims Pannekoek, is because 'the practical life of manual labour is outside of [Bell's] orbit and his interest'.[16] And again, though Bell sung 'a hymn of praise to the

13 Pannekoek 1912, 54.
14 Pannekoek 1912, 56, 58; given Pannekoek's position, the last remains ambiguous.
15 Engels [1876] 1950, 16.
16 Pannekoek [1944] 1953, 14 fn.

human hand', it has remained hollow, because Bell, according to Pannekoek, specialized in 'mental and scientific effort, [such that, if] practical work with tools and the manual labour of the millions producing goods had not been entirely outside his orbit, and if consequently the hand's destination to hold and direct tools had been clear to him, how much deeper a note of world power his hymn of praise would have acquired'.[17] As with Engels on the Darwinists, the irony is that even when a book-length treatise on the hand is imbued with an idealist ideology (e.g. in the case of Bell it is a naturalized religion), it too can underestimate the significance of labour – and even handwork. The implication is clear: we can only ennoble and enrich our sciences, and their histories, with a proper, hands-on appreciation of the labour deeply implicated in handwork. As we shall see below, Pannekoek's astronomical practices exemplified just this kind appreciation.

Pannekoek regarded the sciences themselves as forms of labour, that when understood to include handwork, might eschew lurking ideologies by overcoming, at least in practice, the dualisms of mind and matter, thought and being – dualisms that find their origin in the separations made between physical labour and mental labour, a separation so detrimental to the egalitarian spirit of socialism.[18] Consider the following passage:

> [Scientists] deal with nature in their practical activities by acting upon her and making her part of their existence: Through his labour man does not oppose nature as an external or alien world. On the contrary, by the toil of his hands he transforms the external world [...] The object of his thinking is that which he himself produces by his physical and mental activities and which he controls through his brain.[19]

I take Pannekoek to be shifting our attention from treating scientists as passive receivers to those who actively toil with their hands so as to make nature a 'part of their existence'. The intimate nature of this labour combats dualisms, precisely because it is rooted in the concreteness of handwork. The

17 Pannekoek [1944] 1953, 17. Recent studies on Sir Charles Bell have shown that as a surgeon he had first-hand experiences of the injuries sustained by factory workers, especially to their hands and fingers; see Capuano 2015, especially chapter 2. But Bell was also very much hands-on when it came to teaching and presenting anatomy by way of models and drawings that he himself made with his own hands; see Berkowitz 2015, especially chapter 2.

18 It is interesting to note that the hand continues to be seen, even to our own day, as a way to overcome dualisms of many different sorts, see for example Radman 2013. Also see the incomplete but influential efforts made by Merleau-Ponty (1968).

19 Pannekoek 1942, 7.

level of labour-intensity involved and the resulting intimacy and concreteness of experience implicated in this account is certainly reminiscent of one of Pannekoek's major philosophical influences: the socialist German tanner and proletarian philosopher, Joseph Dietzgen (1828-1888).

In Dietzgen's *Das Wesen der menschlichen Kopfarbeit* (1869) – a work that, in later editions, included an elaborate introduction by Pannekoek – we encounter a materialist theory of mind grounded in the concreteness of sensuous experience. But unlike the empiricism of the philosophers, this experience is understood as practical, process- and action-oriented; in a word, we might describe it as rooted in handwork. Dietzgen, for example, regards any meaningful, universal statement as having to be grounded in individual objects such that 'we must *handle* definite and concrete objects or phenomena'.[20] Even more strikingly, Dietzgen regards any abstract 'science of understanding' that deals with 'all objects' as also being so grounded: 'But all objects which this science may wish to analyse theoretically must first be handled practically. According to their special natures, they must either be *handled in various ways, or carefully inspected, or scrutinized* by intent listening, in short, they must be *thoroughly experienced in some way*'.[21] A proper understanding of science, therefore, does not disconnect headwork (*Kopfarbeit*) from handwork – Pannekoek would have not just agreed but as we shall soon see this standpoint was a productive source of astronomical knowledge.[22]

## II

At first blush it would seem that astronomy poses a fatal challenge to handwork in the sciences. Certainly, each science will have its own manner of handling objects, as implied by Dietzgen above. It is easy to see how smelling or tasting might be involved in the handwork associated with chemistry; physics might incorporate listening or touching; while the handwork associated with geology might include rubbing, boring, or crushing. However, unlike chemistry, physics, geology, and the many other sciences besides, astronomy does not have the luxury of having its objects near-to-hand. Its objects are so distant that there is no obvious way we can poke or prod them, twist or turn them, let alone taste, smell, or rub them. Astronomers seem to

20 Dietzgen 1906, 141; emphasis added.
21 Dietzgen 1906, 72; emphasis added.
22 See also: Sohn-Rethel 1989.

be forever restricted to the mere optical appearances of astronomical objects (from behind computer screens or optical lenses, mirrors, or atmospheres), separated in space and time from an observer, who stands passive and ready to receive. But whether as a socialist or as an astronomer, this is exactly the view of external objects that Pannekoek rejects.

So how are astronomical objects handled so as to be implicated in handwork? The short answer is that the astronomer's handwork is part-and-parcel of routine acts of recording and representing objects. Since the most ancient times and up to the present, astronomers actively engaged their eyes, minds, and hands. Babylonian astronomers, for instance, literally *chiselled* astronomical records into stone (i.e. cuneiform); while today astronomers use digital technologies to twist and turn three- or four-dimensional objects on their computer screens; in fact, they can even listen to sounds emanating from the universe in order to identify and discover the nature of cosmic objects.[23] When we understand the hand and its tools as a single unit – as Pannekoek suggests we do – we see in these examples cases of astronomical handwork and handling. But in the long history of astronomy, one of the most interesting forms of handwork is handmade drawings, sketches, or even paintings of astronomical objects.[24] At least in Pannekoek's own extensive research into the Milky Way, apart from the tables, numerical, and verbal descriptions, one of his primary forms of handwork and handling was drawings and sketches. But before we come to Pannekoek's astronomical handwork, allow me to briefly outline a few examples of how astronomical drawings were made and used as research tools in the nineteenth century, particularly in the case of the nebulae and star clusters, objects thought to have been as elusive as the Milky Way. Doing so will not only cast light on Pannekoek's own labours but will also help us to see how labour so grounded – in the hands and its tools – contributes to astronomical knowledge.

Published drawings of astronomical objects are both records and representations. Once, however, they are printed and the ink has dried on them, it is easy to forget that these were produced over long periods of time, over many nights and days of difficult work. We can therefore see published astronomical drawings, found in many major publications, as either finished products whose surfaces are beheld and treated as mere illustrations; or as the results of long and often toilsome observational procedures that are the highly polished products of many layers of scribbling and sketching, groping and exploring, despair and exhilaration – that is, we can either treat them

23 Kessler 2012; Vertesi 2015; Hadhazy 2014.

24 Even for recent astronomical work image-making is crucial, see Roy 2017

as products or as processes of production.[25] In my own work on nineteenth-century drawings of the nebulae, I have opted to explore astronomical images from the perspective of processes or picture-making rather than as pictures.[26] It is from the perspective of process that I believe we can best, for our present purposes, explore the astronomer's labour of handwork.

When we turn to the laborious processes of picture-making, we find diverse preliminary, preparatory, tentative sketches made by astronomers, which reside as records in their private observational notebooks that were used to build up a final image-product for publication. These provisional images acted as mutable tools to see better or more with, to expand what is possible in the object or to limit how the object is seen, or simply to direct us to look in particular places in our next set of observations. These paper-tools were especially effective for extremely faint and notoriously difficult observational objects like the nebulae, star clusters, or the Milky Way. Image-making techniques like drawing with paper and pencil were used as observational tools, and it is for this reason that I have elsewhere called such a practice, 'observing by hand'.

Such observational tools as stylus and paper, moreover, could be refined and adjusted not just by the pressure applied to the pencil, chalk or ink, or the type and texture of paper selected, but also by means of switching between colours, shades, and positive or negative images of the object. Though there are many examples of this practice found in the observing books of nebular observers of the nineteenth-century, one of the best comes from the labours of George P. Bond (1825-1865) at the Harvard Observatory. Using multiple media, Bond dedicated nearly six, very intensive years drawing the nebula in Orion (M42) (Figure 12.1). In 1858, he began by plotting out 262 stars on paper, which provided the support for the entry of the faint nebulosity. The latter was entered after the area surrounding the nebula was divided into four charts on dark ground so that the nebulosity could be traced in, using white chalk and watercolours. Once these parts were entered, the four distinct charts were then recombined and checked as a whole against the object as seen through the telescope, only to be corrected accordingly. In the third year of this procedure, he continued to draw the brighter parts of the nebulae using white chalk on dark paper, so that he could adjust for the different intensities of light in relation to the darkness. But at the same time, almost as a check, Bond drew the same areas in the negative, where

25 For the classic treatment of astronomical images as mere illustrations to be beheld by perception, see Sheehan 1988.

26 For a much more detailed account of what follows, see Nasim 2013.

**Figure 12.1 Published drawing of the nebula in Orion as drawn by George P. Bond**

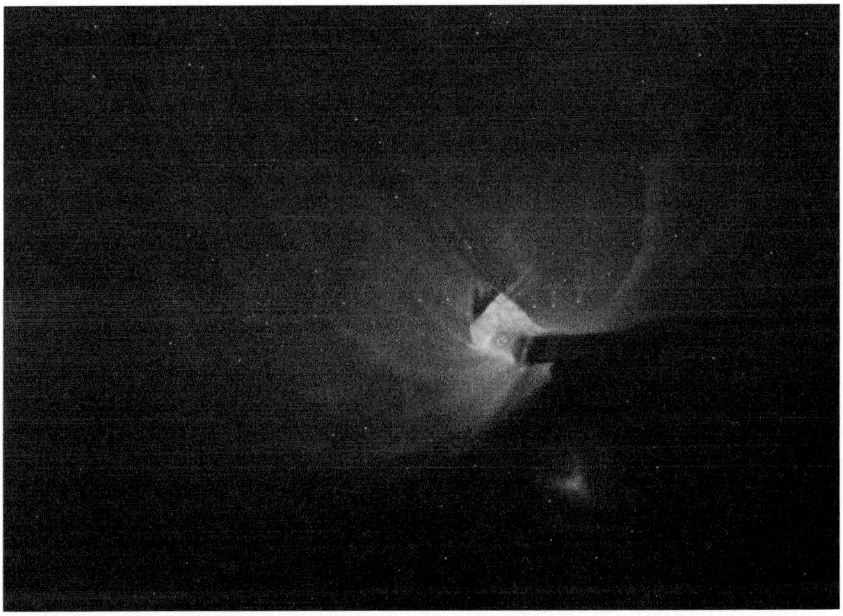

Source: William C. Bond, George Phillips Bond, and Joseph Winlock, *Results of Observations*, Annals of the Astronomical Observatory of Harvard College 8 (Cambridge, MA: John Wilson and Son, 1876), courtesy of Linda Hall Library of Science, Engineering & Technology

the light parts were drawn in dark media against a white background. It was by means of this back-and-forth between media that Bond discovered the supposed spiral character of M42.[27]

Besides the novel features of the nebula that Bond revealed to himself (and eventually to others as well), what stands out is his dedication and handwork in producing an image of it over a period of six years, despite his failing health. Since 1858, Bond had suffered from tuberculosis and it seems that his 'monomania obsession' to draw the nebula by hand only hastened his untimely death. A month before he died in 1865, at the age of 39, Bond wrote to his assistant:

> My disease makes progress, and leaves me little hope of putting the materials of my work on Orion – to which I had devoted so much labour – into condition such that another could prepare them for the press. In truth, I am becoming resigned to the idea that most of it is destined to oblivion.[28]

27 Bond 1861.

28 Letter from Bond to Asaph Hall, January 7, 1865; quoted in Sheehan and Conselice 2016, 97.

It was two years after his death that the full account and the final resulting drawing of M42 was engraved in steel and published.[29] The labour that it cost Bond was widely noted, and as late as 1882, Edward S. Holden, director of the Washburn Observatory (and soon to be director of the Lick Observatory) used Bond's published image of M42 as a frontispiece to a whole book dedicated to hundreds of drawings of this nebula, declaring it to be 'the most satisfactory representation of any celestial object which has yet been produced'.[30]

The years that Bond spent, ceaselessly experimenting with media in order to make out the subtleties of the notoriously intricate nebula in Orion, was time and labour essential in coming to know the unfamiliar. Bond's example shows how the hand and eye can work together through multiple materials and media. But there are also examples of how the hand and eye can be further supplemented in observing and drawing astronomical objects by particular conceptions of the mind. At Yale College, in the summer of 1839, E.P. Mason and H. Smith built themselves a telescope for the express purpose of observing and drawing nebulae. Essential to the observational procedure used to compose the drawings, were chains of triangles and isolines laid down on paper – one was used to survey and plot the stars trigonometrically onto paper, and the other to aid in the entry of various gradations of light and dark involved in the nebulae. Each of these artificial aids were referred to as 'conceptions' that worked together to build up, over time and many observations, a unified picture of a nebula.[31] Indeed, the configurations evidenced in Mason's practice between the hand, eye, and mind echo one of Pannekoek's fundamental claims: 'Since the tool stands between man and outside objects, thought must arise between the impression and the performance [...] This material circuit causes the mental circuit; the thoughts leading to a certain act are the results of the tools necessary for the performance of the act.'[32]

But it is the example of another nineteenth-century observer of the nebulae who, independently of Mason, took a similarly conceptual approach to another level and who's practice nicely exemplifies Pannekoek's claim, quoted above. This was none other than John F.W. Herschel, who, while situated with his family and his twenty-foot reflecting telescope at the Cape of Good Hope, spent four years (1834-1838) drawing and

29 Bond 1867.

30 Holden 1882, 82.

31 Mason 1841. For further detail, see Nasim 2013, 126-137.

32 Pannekoek 1912, 46.

**Figure 12.2 Printed descriptive map of η Argus, published in John Herschel's *Cape Results***

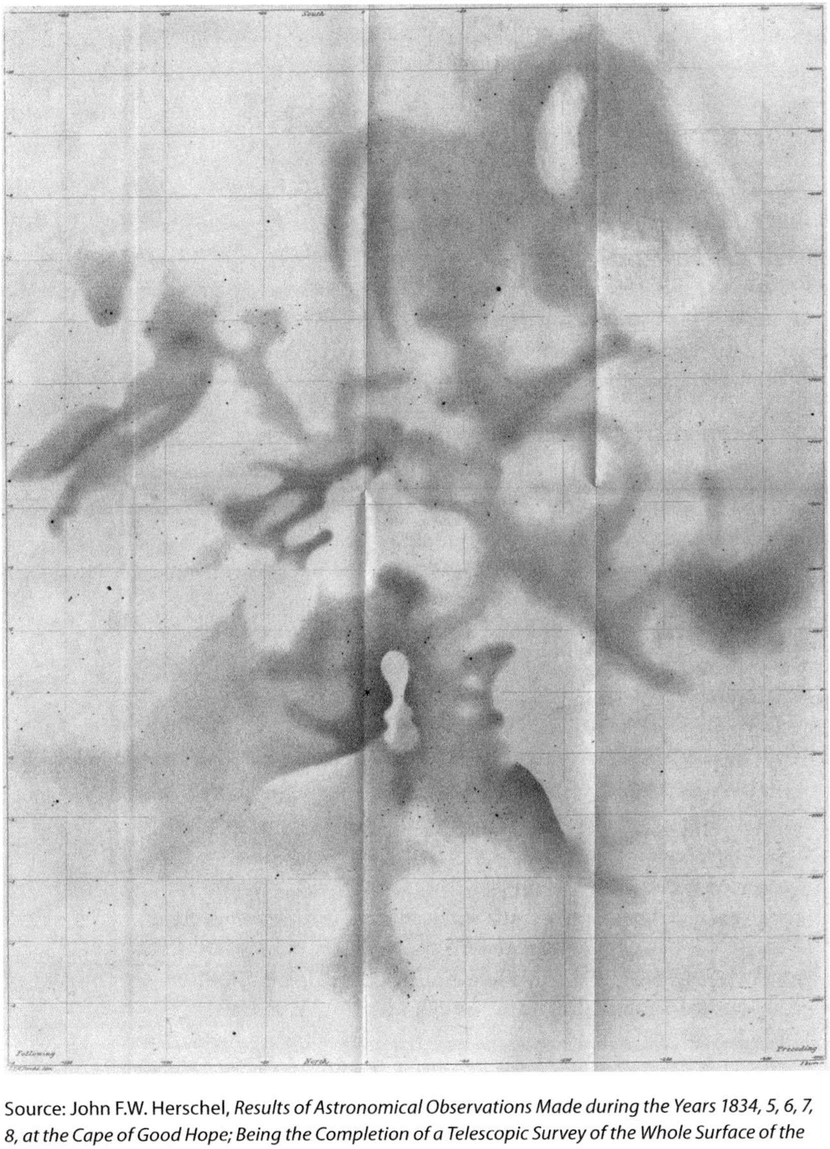

Source: John F.W. Herschel, *Results of Astronomical Observations Made during the Years 1834, 5, 6, 7, 8, at the Cape of Good Hope; Being the Completion of a Telescopic Survey of the Whole Surface of the Visible Heavens, Commenced in 1825* (London: Smith, Elder & Co., 1847)

cataloguing the nebulae and clusters of the southern hemisphere. Even after he returned home to England, he continued to catalogue, calculate, and draw for at least another six years, when in 1847 he finally published the results that included a number of splendid prints of the nebulae, clusters, and even the Milky Way.[33] From just the printed image-surfaces of these published pictures one cannot surmise all that went into their production (Figure 12.2). Turning to Herschel's backstage work, however, one is struck by the use of pencilled in dots and lines that form chains of triangles and grids. These triangles and grids are the 'working skeletons' that were employed to triangulate the approximate, relative positions of the stars and nebulous material. Each one contains layers of handwork. In some cases, as many as 23 successive 'working skeletons', each on a separate piece of paper, were used in the production of just one final image of a nebula. The skeletons provide consistency, regularity, measure, and scale to his pictures of deep-sky objects that are notoriously hard to see and measure. But above all, they provide a way to coordinate what the eye sees and what the hand enters onto paper, allowing Herschel to scrutinize each and every single part of a nebula in a systematic and highly attentive manner. I have also argued elsewhere that these conceptual tools on paper corresponded to Herschel's own understanding of the specific mental processes employed in the construction of external objects. But for our purposes, the handwork implicated in this observational procedure should not be underestimated. When writing about all the work involved in picturing just one nebula, Herschel shifts to a personal, intimate style in order to explain his labour:

> The accurate representation of this nebula with its included stars has proved a work of very great difficulty and labour [...] To say that I have spent several months in the delineation of the nebula, the micrometrical measurement of the co-ordinates of the skeleton stars, the filling in, mapping down, and reading off the skeletons when prepared, the subsequent reduction and digestion into a catalogue, of the stars so determined, and the execution, final revision, and correction of the drawing and engraving, would I am sure, be no exaggeration. Frequently, while working at the telescope on these skeletons, a sensation of despair would arise of ever being able to transfer to paper, with even tolerable correctness, their endless details.[34]

The level of intensity and intimacy of handwork, exhibited in Herschel's observational procedures, was essential, however, to the production of accurate published images that could then go on to be used by scientists. In

33 Herschel 1847.
34 Herschel 1847, 37.

addition to this, the disciplined handwork behind each of Herschel's drawings was necessary in order to safeguard the results from the detrimental influences of the intrusion of the self.[35]

The handwork required in the production of these visual representations did not always rely on one individual observer but could extend to include the hands and tools of many. Employing a very similar set of working skeletons as Herschel, William Parsons (the third Earl of Rosse (1800-1867)) and his assistants, which included his son Lawrence Parsons (Lord Oxmantown, 1840-1908), Robert S. Ball (the future Astronomer Royal for Ireland, 1840-1913), and the artist Samuel Hunter, all contributed to the production of a celebrated drawing of the nebula in Orion (M42), published in 1868 (Figure 12.3). The groundwork for the elaborate drawing was laid down using micrometrical measurements of 150 stars in and around the nebula, taken by Otto Struve, the director of the Pulkovo Observatory, near St Petersburg. Back in Ireland, Lord Rosse's team of observers used this data to create the paper-ground for the construction of a grid that would be used to then plot in other stars not included in Struve's catalogue. With the paper so prepared, the observers entered, over many nights, the minutia and complexity of this nebulous object in a way that was governed by the stars and the lines of the grid. Rosse's 'Skeleton Map' of M42 coordinated *seven years of work*, which included multiple entries from a number of hands, and different telescopes. Besides the detailed drawing of the nebula, which was engraved and printed for the publication, a topographical map of the nebula was also included, one that only showed the outlines of the nebula and identified by labels all the main stars and regions to be found therein. The painstaking labour of this handwork should not be underestimated. After having made 74 observations between 1860-1864 for the purposes of drawing the nebula in excruciating detail, Samuel Hunter, who did the large bulk of the work, reportedly became so ill that he had to leave the Earl's employ.[36] But in introducing the publication that included the drawing and map of M42, it is precisely this exhausting level of handwork that Lawrence Parson highlights, especially as a reason to regard the new figure of the nebula as accurate, efficacious, and true.[37]

35 See Nasim 2013, 137-167.

36 Besides the work for M42, Hunter was also observing the moon, collating the notebooks, and continuing with the observations for the survey of the nebulae in general. See Nasim 2013, chapter 1.

37 Oxmantown 1868.

**Figure 12.3 Drawing of the nebula in Orion by Samuel Hunter based on collaborative efforts of Robert Ball, William Parsons, and Lawrence Parsons**

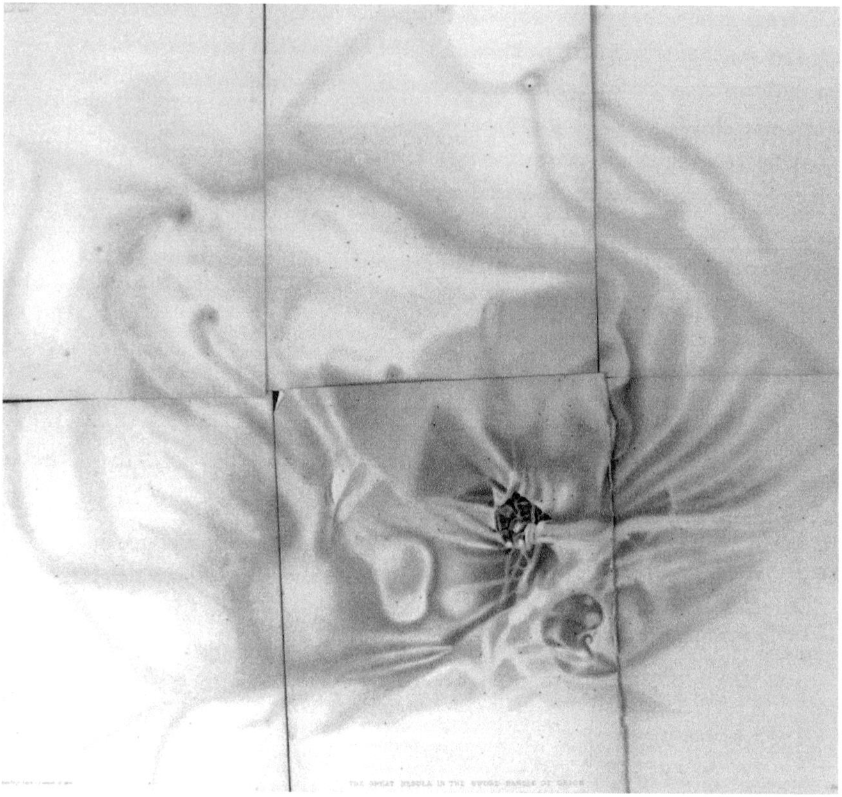

Source: The Rosse Papers, Birr Scientific and Heritage Foundations, courtesy of the Earl of Rosse

We can provide other examples of how handwork was implicated in astronomical knowledge. But doing so would only belabour the point that I hope has now been sufficiently made, that the act of drawing was a productive way for astronomers to come to know distant objects in ways that were intimate, intense, and laborious. That through the interactions of the eye, mind, and hand, observers made astronomical objects a part of their personal existence. That these systematic interactions were so demanding, only meant that they led not just to despair or illness but to visual representations that were so disciplined so as to show what was really there – or so it was thought. Indeed, that they expended so much energy and life into these drawings only goes to show the immense value that they placed on the products and the processes. And all this thanks to the sheer amount of time spent using one's hands in relation to reams of paper and notebooks, telescope, and styli (whether pencil, ink and pen, or brush), over and over again. According to Pannekoek, then, this is not just how human beings first emerged and came to dominate nature. It is also how they, even in their more advanced sciences, come to know the world:

> A new and powerful influence emanates from the handling of tools to the organs of perception and consciousness, and thereby to mental life. It supplies a new experience of the exterior world. The delicate sense of touch vested in the fingers comes into action when gripping and guiding the tool which is used to operate on the outside world by some such as beating, pressing, rubbing, and boring [and we might add drawing] [...] The exterior world reacts, as its resistance.[38]

I take the examples provided above, as instantiating these claims. Let us now turn to Pannekoek's own representational practices to find something similar.

## III

Over the span of his long astronomical career, Pannekoek made many sorts of drawings of the Milky Way, an object that occupied his hands for nearly sixty years. His first systematic attempts at observing the Milky Way date back as early as the years 1889 to 1890.[39] During this period, the young Pannekoek already preferred to depict the Milky Way using a series of isophotic lines – lines representing numerically the same level of brightness – in order to describe the different gradations of light on paper blackened so that the stars and lines could appear white (by using white lead). This media was selected primarily because it made it easier to draw in the dark, once the naked eyes were night-adopted and directly perceiving the Milky Way. Afterwards, Pannekoek advises that these charts be 'copied at home' upon another set of charts made on white paper and constructed on the basis of Albert Marth's catalogue of the main stars in the neighbourhood of the Milky Way.[40] From all the information gained, a number of drawings of three different regions of the Milky Way were made between 1890-1892. It was based on these experiences that Pannekoek published a note in 1897 recommending astronomers to use 'new charts for inserting the Milky

38 Pannekoek [1944] 1953, 61.
39 I take this timeline chiefly from the introduction to Pannekoek 1920.
40 Pannekoek 1898, 527.

Way'. These were essentially the same pre-prepared charts that he had previously used for his own observations and drawings, and he suggested that copies, as lithographs, would be made available to other observers around the world, who were also drawing the Milky Way. Quite early in his astronomical career, therefore, Pannekoek opted to pursue a visual strategy that was defined by handwork.

Pannekoek's isophotic maps of the Milky Way do not show the phenomenon as it is seen in the sky, however. For this, much more naturalistic and detailed drawings had to be made. The observations for these were begun in 1893 but had come to an end in 1899, due to Pannekoek's routine duties at the Leiden observatory, which were, he writes, 'too exhausting for me'. Between 1910-1913, Pannekoek resumed his observations and found that he could compose 'a total picture of the Milky Way'.[41] The total picture was divided into three different regions of the Milky Way. What served as the basis for these unified compositions were not just years of notes, charts, stars, drawings – both naturalistic and schematic – but also 128 select points whose specific levels of brightness were estimated by the eye and given a numerical value that could be used as standard for the determination of light intensities throughout the rest of the Milky Way. It was finally in 1920 that Pannekoek published the results of the above observations. In it, Pannekoek presented naturalistic drawings of three regions of the Milky Way, all printed in the positive (Figure 12.4). The same regions were also figured in three detailed isophotal maps (Figure 12.5). But the most notable figures were what he called 'the mean subjective images', which schematically combined his own drawings with those of at least three other major observers of the Milky Way (Figure 12.6). All in all, these different visual approaches, or as Pannekoek refers to them, 'lines of research', were intended to complement one another: He explained:

> When on a star chart we draw the lines of equal brightness by shading the region between them with increasing deepness, then blending these shades into one another at their boundary lines, we have a picture with good distribution of brightness, but showing only the more general details, though it gives the general appearance very well. Upon this background, we can draw the minutest peculiarities, taken from the results of the studies [...] we have a picture that contains all that the observer has been able to see, and which still shows the brightness of the different parts in a very exact proportion.[42]

41 Pannekoek 1920, 2.

42 Pannekoek 1898, 528.

**Figure 12.4 Naturalistic drawing of a region of the northern Milky Way by Pannekoek**

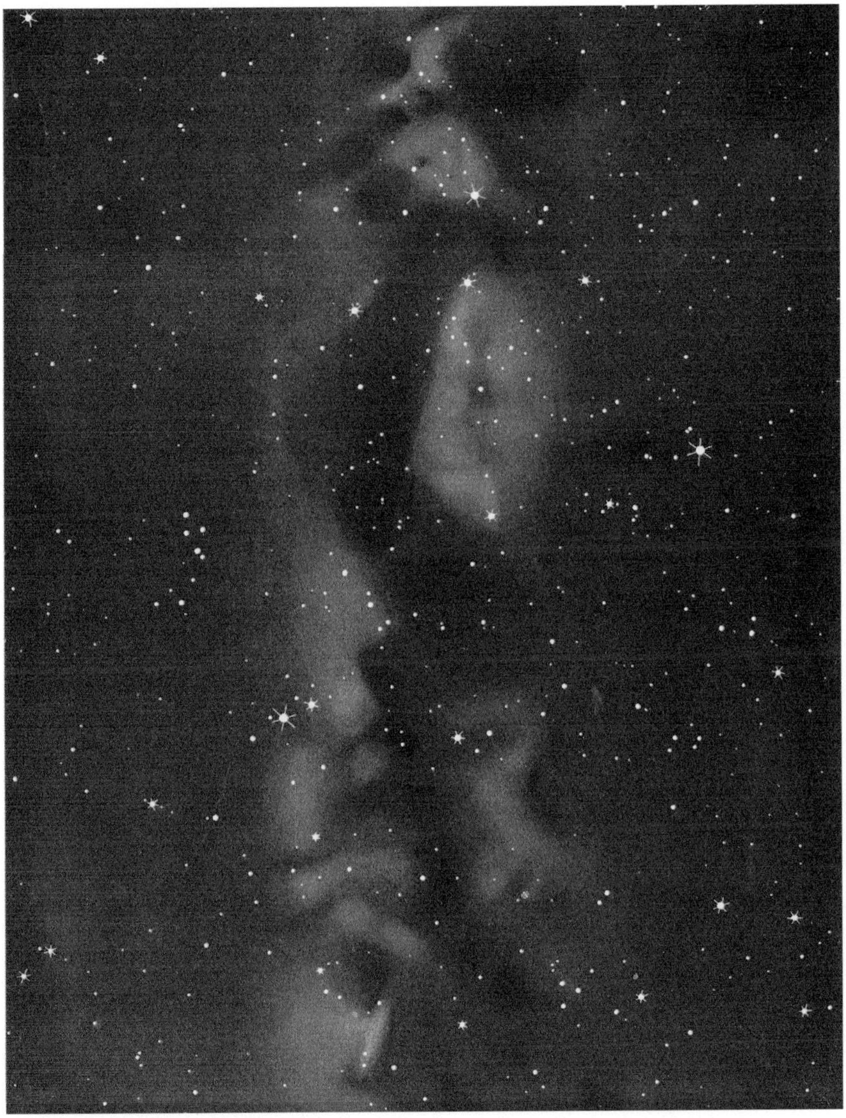

Source: Anton Pannekoek, *Die nördliche Milchstrasse*, Annalen van de Sterrewacht te Leiden 11:3 (Haarlem: Joh. Enschedé en Zonen, 1920)

**Figure 12.5 Isophotal map by Pannekoek of the same region of the northern Milky Way based on his own observations**

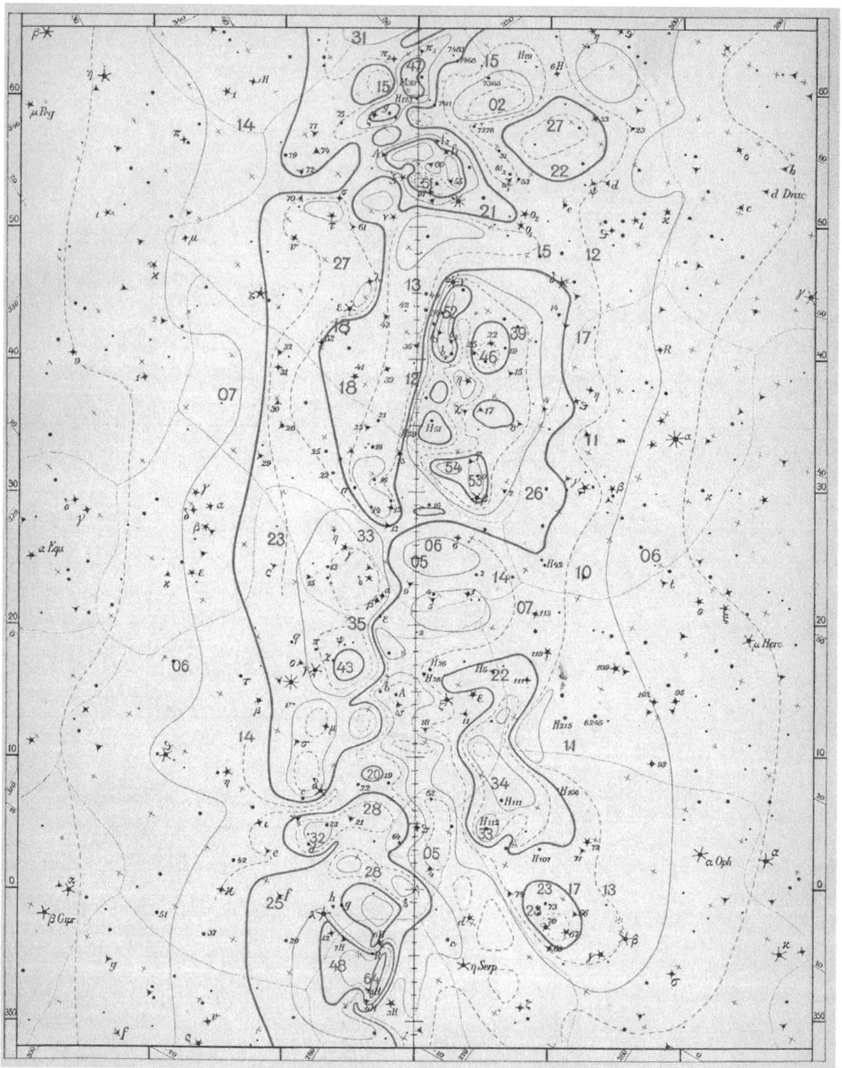


Source: Anton Pannekoek, *Die nördliche Milchstrasse*, Annalen van de Sterrewacht te Leiden 11:3 (Haarlem: Joh. Enschedé en Zonen, 1920)

**Figure 12.6 The mean subjective image of the northern Milky Way, which combined the drawings of Pannekoek with those of others**

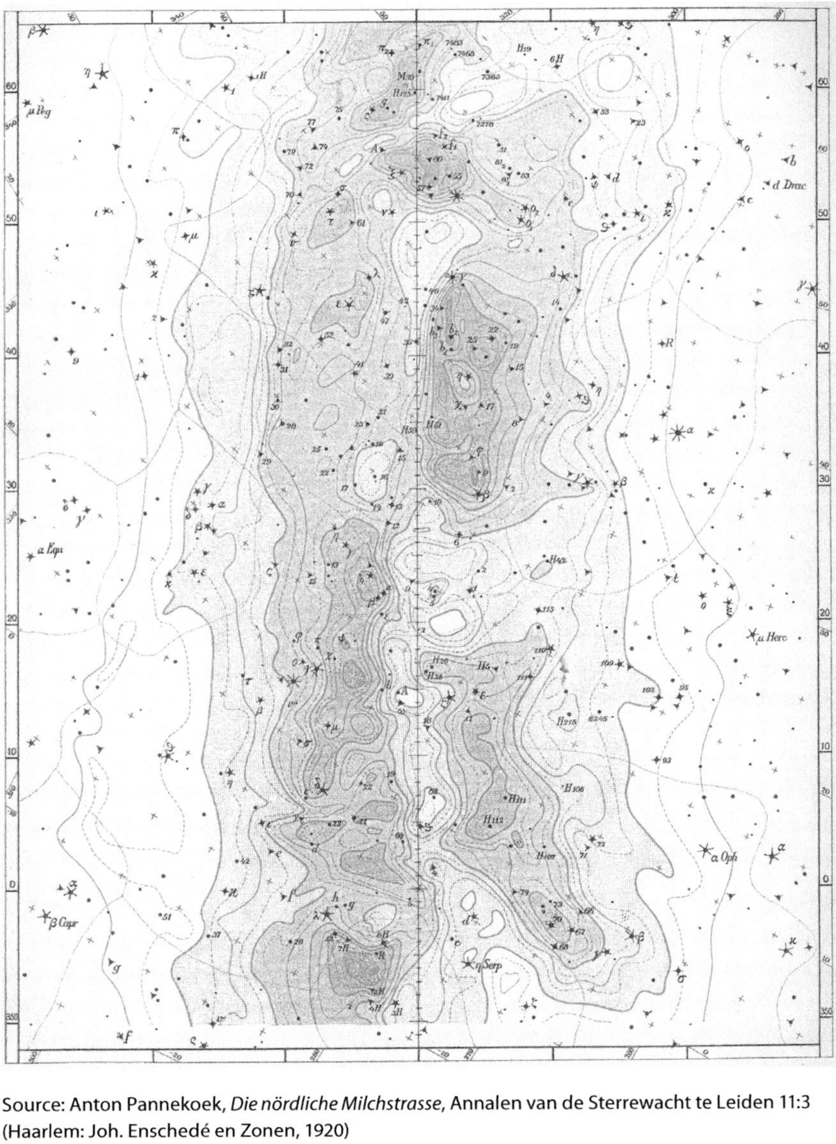

Source: Anton Pannekoek, *Die nördliche Milchstrasse*, Annalen van de Sterrewacht te Leiden 11:3 (Haarlem: Joh. Enschedé en Zonen, 1920)

**Figure 12.7 Naturalistic drawing of a region of the southern Milky Way, published in negative**

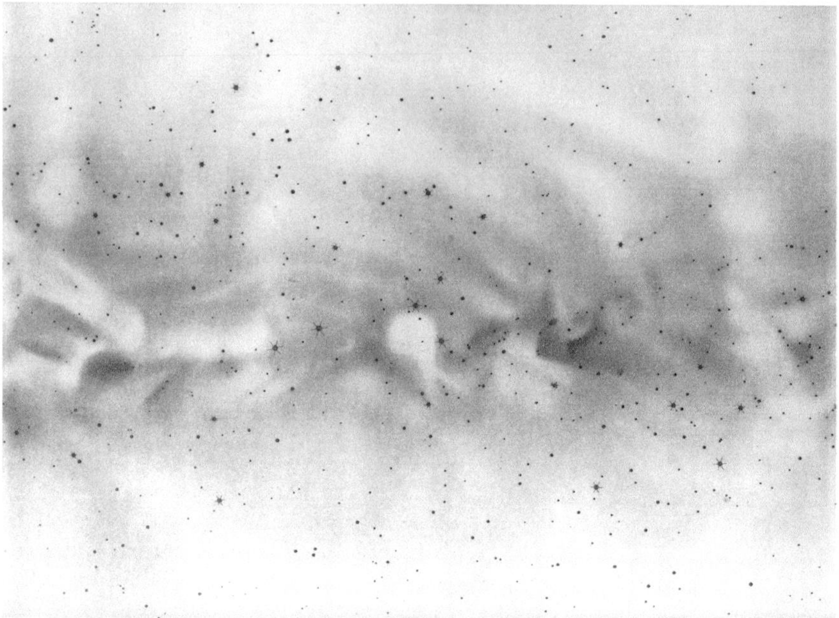

Source: Anton Pannekoek, *Die südliche Milchstrasse*, Annalen van de Bosscha Sterrenwacht 2:1 (Amsterdam: De Bussy, 1928), plate 2

The active alternation between different media and techniques of representing – naturalistic and schematic – is certainly indicative of drawings used not just for presentation but as tools for observing. Again, we have not just pictures but means to observe more attentively and differently with. The artefacts of Pannekoek's handwork also speak to the intensity of labour, time and handicraft implicated in his observational procedures. Given the significance of handwork for his procedures, it is little wonder, then, that the next time Pannekoek published drawings of the Milky Way, as seen from the southern hemisphere, he remained unsatisfied with them, because they were produced in a relatively short period of time.[43] These were observations made within just six months, between 1925-1926 at the Bosscha observatory in Lembang, Indonesia. Thanks to bad weather and a short stay, Pannekoek could not properly engage in the same amount of labour-intensive scrutiny of the Milky Way as he

43 See Pannekoek 1928.

**Figure 12.8 The corresponding isophotic map of the same region of the southern Milky Way**

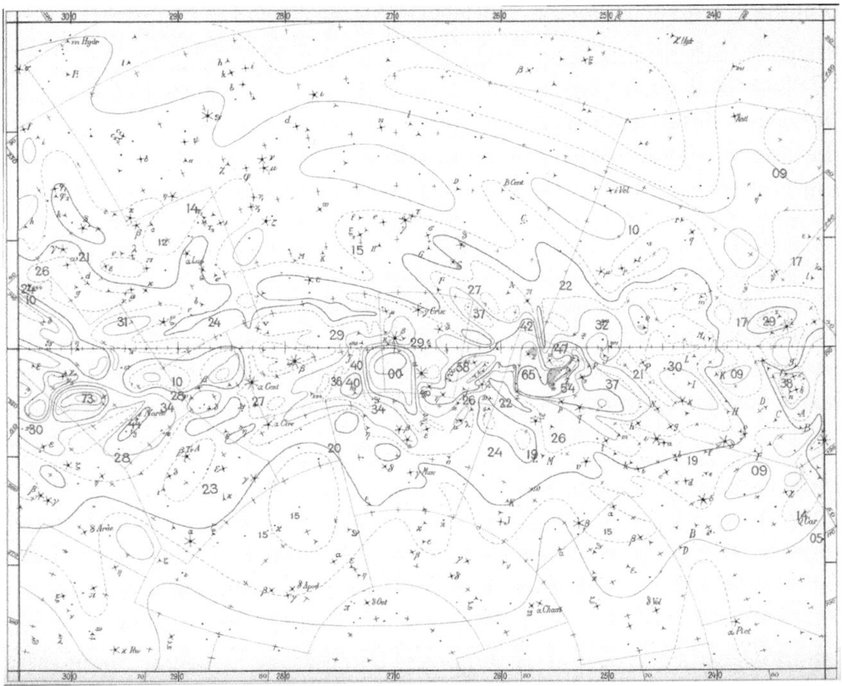

Source: Anton Pannekoek, *Die südliche Milchstrasse*, Annalen van de Bosscha Sterrenwacht 2:1 (Amsterdam: De Bussy, 1928), plate 5

had done previously. Though three naturalistic figures (all printed, this time, in the negative; Figure 12.7) and three corresponding isophotic maps (Figure 12.8) were published for three regions of the southern Milky Way, these, for Pannekoek, remained incomplete precisely because they lacked a sufficient amount of labour. One might fairly ask at this juncture: Why did Pannekoek not just use photography? After all, Pannekoek already had some experience working with photography; and that by this time there were already a handful of successful and exemplary photographs taken of the Milky Way. In order to approach an answer to this question, we need to step back a bit and reconsider the use of photography in the sciences from a wider angle, one that will put us in a position to see dualism lurking around the corner; dualisms that Pannekoek wished, due to his socialist orientation, to avoid.

## IV

It might appear that we have moved very far from socialism. But let us recall that the human hand and its tools were not just central to Pannekoek's view of science but also to his socialism. Indeed, the hand and its tools have been emblematic of early socialist theories, especially as they arose in reaction to the rise of automatic 'iron fingers' powered by steam engines. The common rhetoric in defence of these machines should be familiar enough: they save time and human labour; and they supplant handwork so that more time can be shifted to more privileged kinds of work, like headwork. Intriguingly but unsurprisingly, a similar rhetoric can be found in the rise of astrophotography. So, for example, in contrast to the tireless eyes of photography, the astronomer and popularizer Robert Proctor sardonically refers to those who made drawings of the nebulae as 'our *laborious telescopists* [who] wait and watch until at least the true shapes of these mysterious mist masses had been determined. But with long looking comes only more confused vision'.[44] Or take the case of Edward S. Holden, who describes photography as the 'servant' of astronomy, which does not tire in its faithful and 'automatic register'. He continues: 'Another important advantage of the new methods [of photography] is that they do not require highly skilled observers [...] The skill of the astronomer is reserved for real difficulties, and the merely laborious work can be done in duplicate, if necessary, by younger men'.[45] By 'real difficulties' the famously heavy-handed director of Lick Observatory was presumably referring to the headwork of the astronomer, rather than the deskilled labour of the hand.

Consider once again the case of George P. Bond, who, before starting on the laborious project of drawing the nebula in Orion, was among the first to employ photography to capture the stars. In contrast to what we have seen above, Bond applauds the relative ease of photography, especially as a labour-saving mechanism: 'On a fine night', he writes, 'the amount of work which can be accomplished, with entire exemption from the trouble, vexation and fatigue that seldom fail to attend upon ordinary observations, is astonishing. The plates, once secured, can be laid by for future study by daylight and at leisure. The record is there, with no room for doubt or mistake as to its fidelity'.[46] But despite these high praises, Bond was soon to quit his experiments with photography, due to a number of challenges, including

44 Proctor 1883, 447; emphasis added.
45 Holden 1886, 467, 468.
46 Bond 1890, 301.

the prohibitive costs involved in the venture. What does continue, however, are the tropes of leisure and fidelity, closely associated with photography well into the twentieth century.

Perhaps the most striking are the remarks of Sir Robert S. Ball, who, as a young man, was engaged with Lord Rosse's grand project of drawing the nebula in Orion. Many years after his work with the Earl, the Astronomer Royal for Ireland recollected that it was

> with infinite patience, Lord Rosse devoted years to making a drawing of the Great Nebula [in Orion]. Those were not the days of astronomical photography [...] It is an exquisite piece of work. It was repeatedly compared with the actual object in the heavens, and corrected or altered until accuracy was attained. In some respects we may say it is unique. Never before was so much pains bestowed on the drawing of a celestial object, and never again will equal pains be devoted to the same purpose. In an hour or two the photographic plate will now record much more than the most accomplished astronomer can observe, even though his repeated observations cover a period of several years.[47]

The same rhetoric can be found in the twentieth century, where, even in Pannekoek's writings we find claims such as: '[T]he first photographs of the sun and the moon showed, in short exposures, an abundance of detail that would have demanded hours and months of observing in direct drawing and mapping'.[48] These fine photographic details, finally, could be examined at one's leisure during the day and in the comfort of one's own office. Astrophotography radically challenged the labour practices of the 'old astronomy' and introduced brand-new labour relations.[49] More importantly, however, photography's purportedly hands-free labour, and time-reducing character, seem to directly mitigate against our notion of handwork in astronomy – a notion that operates on the basis of the intensity of long and trying labour of the astronomer's hands and tools. And just as in the case of ideologically governed rhetoric surrounding the steam engine, here too we have an ideologically driven rhetoric, one wherein astrophotography is more conducive to headwork than handwork. In fact, things have been turned on their head: while the painstaking labour of the previous generation's handwork (in drawing nebulae, for instance)

47 Ball 1915, 69.

48 Pannekoek 1961, 336; see also 373.

49 Clerke 1888; see also Nasim 2016.

was seen as a guarantor of representational reliability; now, at the end of the nineteenth century, the same painstaking handwork was seen as mitigating against the reliability of what was produced, especially in light of photography. Photography, therefore, seems to have been pitted against the handwork manifest in drawings – we have, in other words, another series of hazardous dualisms. Given this situation, it is of some significance to notice Pannekoek's own reaction to these dualisms within his own astronomical practice: true to form they are made to merge by means of not less but more handwork.

## V

Allow me to conclude by addressing this challenge, for doing so will help to further reinforce the overall significance of handwork for Pannekoek's socialism and astronomy. There are a number of related ways one can react: One can simply observe, to begin with, how the rhetoric of photography conflicts with the vast number of manuals dedicated to photography, which explicitly included the labour of handwork into the very processes of photography. Photography, in other words, is not wanting in hands and handling.[50] Another is to simply point out the fact that Pannekoek maintains an important space for the visual in contrast to photographic work in observational astronomy. After contrasting, for instance, the photographic from visual atlases of the moon, Pannekoek argues, as late as 1951, that given certain limitations of lunar photography, 'visual work should not be abandoned'.[51] And in the case of the Milky Way, especially given photography's failure to adequately show the overall brightness and milkiness of the Milky Way, the images produced by visual means remains absolutely crucial.

This leads me to the final point. Although Pannekoek seems to divide the visual and photographic, his characteristic aversion to dualism and his commitment to handwork bring these two kinds of images into a complementary relationship within his own practice. One of the most distinctive things about Pannekoek's two most important publications on the photographic photometry of the Milky Way is just how many hand-drawn, schematic

50 This is so also for astrophotography, see, for instance, Scheiner 1897; Abney 1893. For more on hands and photography in astronomy, see Nasim 2018. For more on labour and photography generally, see Edwards 2006.

51 Pannekoek 1961, 374.

maps there are as opposed to photographs – and this in a book based on photographic methods. In the first publication on the northern Milky Way (1933), there are just eight photographs of stars printed on two plates (Figure 12.9), compared to 35 separate plates for schematic translations (by hand) of the photographs used (Figure 12.10), and another eight entire plates showing the isophotal maps derived (again by hand) from these photographs (Figure 12.11).[52] The second major publication is dedicated to extending the same methods to the southern Milky Way (1949).[53] It contains four photographs of the stars on one plate, two plates with naturalistic drawings, and fifteen plates of isophotal maps. The quantitive and qualitative information visualized by means of multiple schematic maps, all made by hand, are derived from photographs showing the stars of the Milky Way as no human eye can see them (Figure 12.12). From these extrafocal photographs of the stars, Pannekoek maps out a number of quantities of light intensity, magnitudes, and gradations by hand and eye, all in gruelling detail. And even though all the photographic plates used for these publications – spanning a period of nearly thirty years – come from a variety of cameras and optics, telescopes and astronomers, climates and geographies, they are harmonized by means of a battery of reductions and projections in order that the information be depicted in a series of coherent hand-drawn maps. But it is not just schematic maps that are derived; Pannekoek goes as far as to actually translate the series of extrafocal photographs of stars into naturalistic, shaded drawings of the Milky Way (Figure 12.13). Again, we have a variety of media made to supplement and complement one another. Most importantly, however, is the fact that even in the case of photography, Pannekoek grounds the work back into laborious handwork.

Even behind Pannekoek's photographical labours, therefore, there is a preference for handwork, reflecting a phenomenology that hearkens back to Dietzgen's epistemically productive and intimate handling of scientific objects, which afford a level of scrutiny otherwise not available by other, more automatic means. It is through patient and painstaking handwork that the objects of science are brought into the very fibres of the scientist's existence. It is for this reason that even photographs are anchored into handwork; or as Pannekoek puts it in a pithy but revealing statement: 'Photography, with all its documentary value [...] misses [however] the direct contact with the happenings of every moment'.[54] In

52 Pannekoek 1933.

53 Pannekoek and Koelbloed 1949.

54 Pannekoek 1961, 416.

**Figure 12.9 Published reproductions of the northern Milky Way photographs**

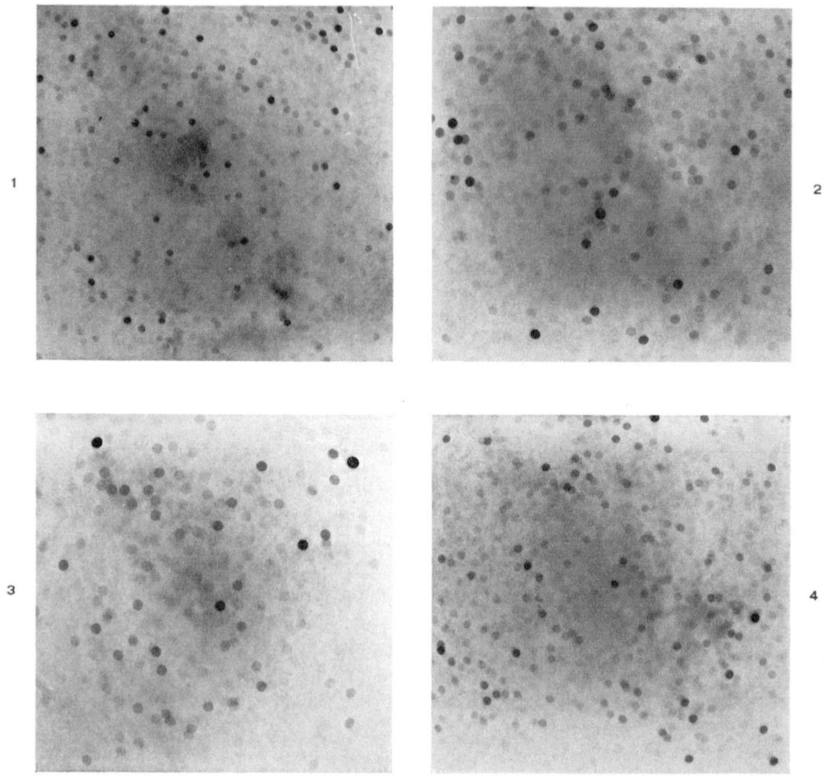


1. *C* 579 $\delta$ Scuti
3. *C* 521 $\beta$ Cygni

2. *C* 836 $\gamma$ Aquilae
4. *C* 533 $\pi^2$ Cygni

Source: Anton Pannekoek, *Photographische Photometrie der nördlichen Milchstrasse*, Publications of the Astronomical Institute of the University of Amsterdam 3 (Amsterdam: Stadsdrukkerij, 1933)

terms of handwork, that is to say, there is a surplus of experiences born of laborious and time-consuming handling by means of the hands and its tools – a surplus that photography seems to lack when employed by itself. Whether it is Marxism or Darwinism, thought or being, drawing or photography, these dualisms can, according to both Pannekoek-the-astronomer and Pannekoek-the-socialist, be superseded and, indeed, be made complimentary by human hands and tools. Handwork is crucial to both sides of Pannekoek's oeuvre.

**Figure 12.10 A plate with schematic translations of one of the photographs used**

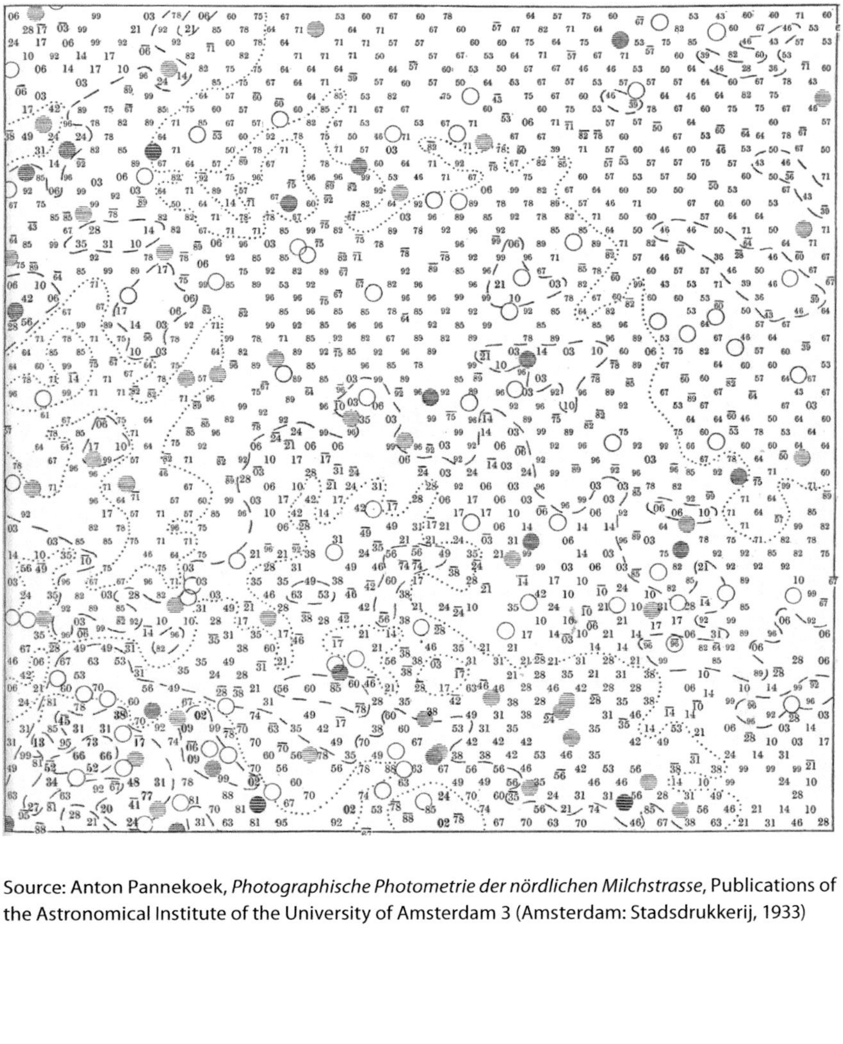

Source: Anton Pannekoek, *Photographische Photometrie der nördlichen Milchstrasse*, Publications of the Astronomical Institute of the University of Amsterdam 3 (Amsterdam: Stadsdrukkerij, 1933)

**Figure 12.11 Isophotic map that is derived from the schematic translations of the Milky Way photographs**

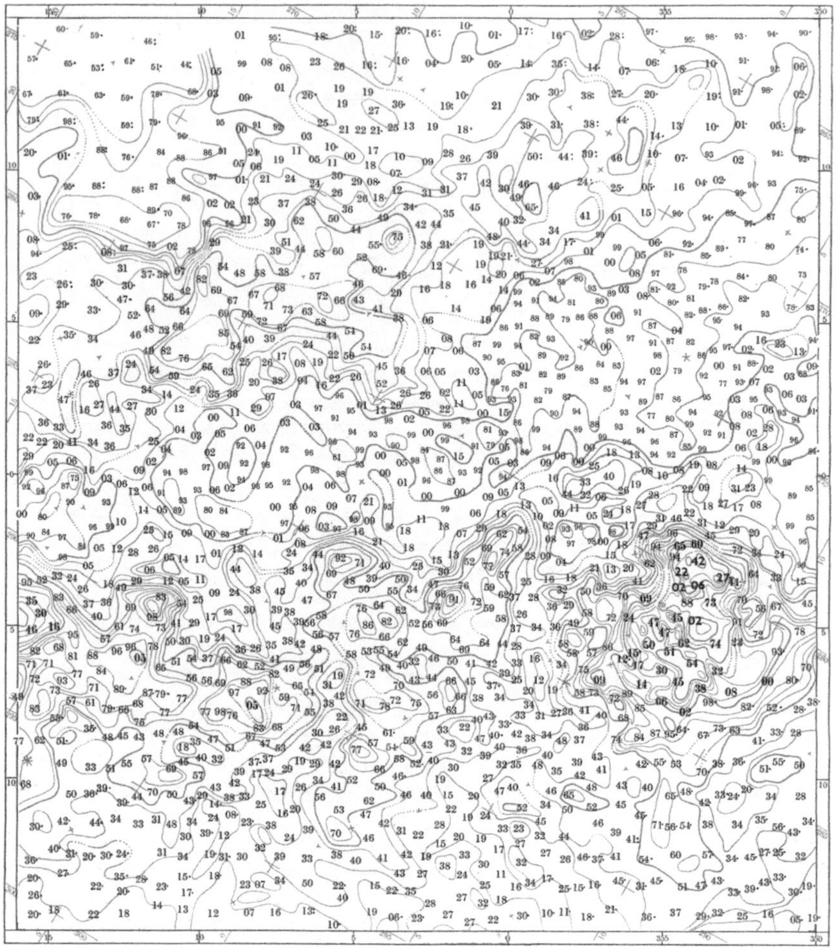

Source: Anton Pannekoek, *Photographische Photometrie der nördlichen Milchstrasse*, Publications of the Astronomical Institute of the University of Amsterdam 3 (Amsterdam: Stadsdrukkerij, 1933)

**Figure 12.12 Isophotic map derived from photographs of the southern Milky Way**

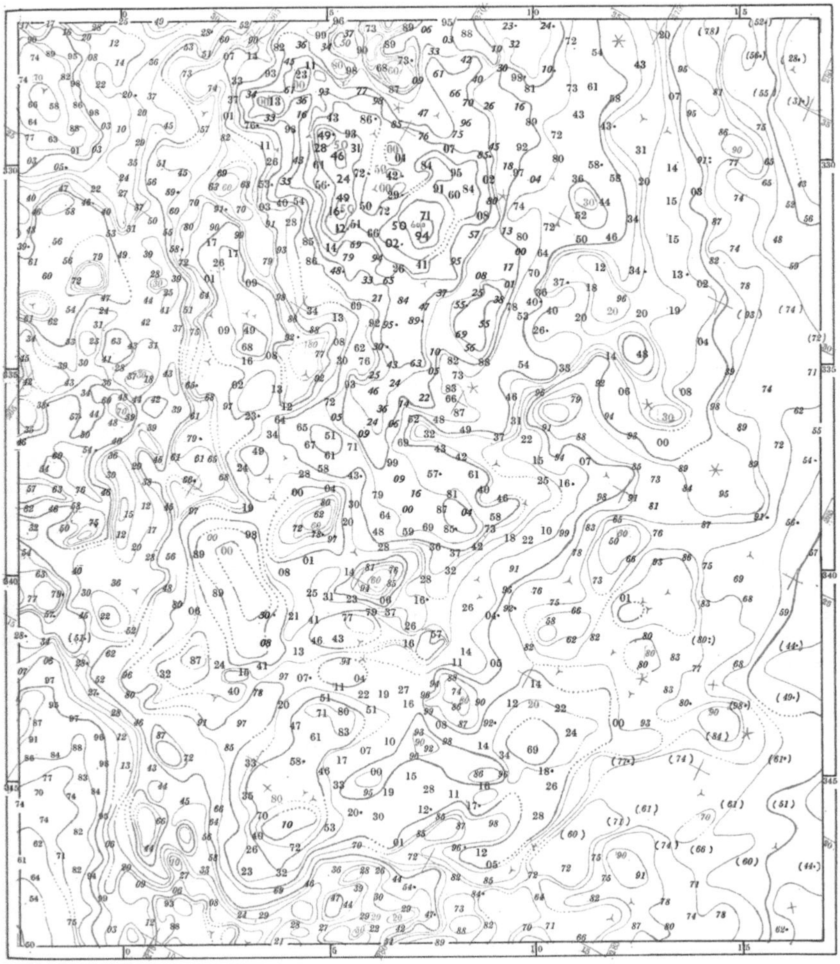


Source: Anton Pannekoek and David Koelbloed, *Photographic Photometry of the Southern Milky Way*, Publications of the Astronomical Institute of the University of Amsterdam 9 (Amsterdam: Stadsdrukkerij, 1949)

**Figure 12.13 Naturalistic drawing of the southern Milky Way derived from photographs of the southern Milky Way**

Source: Anton Pannekoek and David Koelbloed, *Photographic Photometry of the Southern Milky Way*, Publications of the Astronomical Institute of the University of Amsterdam 9 (Amsterdam: Stadsdrukkerij, 1949)

## Bibliography

Abney, William de Wiveleslie. 1893. *A Treatise on Photography*. 9th edition. London: Longmans, Green & Co.

Ashworth, William J. 1994. 'The Calculating Eye: Baily, Herschel, Babbage and the Business of Astronomy'. *British Journal for the History of Science* 27(4): 409-441.

Ball, Robert. 1915. *Reminiscences and Letters of Sir Robert Ball*. Ed. by W. Valentine Ball. Boston: Little, Brown & Co.

Berkowitz, Carin. 2015. *Charles Bell and the Anatomy of Reform*. Chicago, IL: University of Chicago Press.

Bond, George P. 1861. 'On the Spiral Character of the Great Nebula in Orion'. *Monthly Notices of the Royal Astronomical Society* 21(7): 203-207.

—. 1867. *Observations upon the Great Nebula of Orion*. Ed. by Truman H. Safford. Annals of the Astronomical Observatory of Harvard College 5. Cambridge, MA: Riverside Press.

—. 1890. 'The Future of Stellar Photography: Extract from a Letter Written in 1857 to Wm. Mitchell by G.P. Bond'. *Publications of the Astronomical Society of the Pacific* 2(11): 300-302.

Capuano, Peter J. 2015. *Changing Hands: Industry, Evolution, and the Reconfiguration of the Victorian Body*. Ann Arbor, MI: University of Michigan Press.

Clerke, Agnes Mary. 1888. 'Astronomical Photography'. *The Sidereal Messenger* 7(46-47): 138-153, 181-191.

Dietzgen, Joseph. 1906. *The Positive Outcome of Philosophy: The Nature of Human Brain Work; Letters on Logic*. Chicago, IL: Charles H. Kerr & Co.

Edwards, Steve. 2006. *The Making of English Photography: Allegories*. University Park, PA: Pennsylvania State University Press.

Engels, Friedrich. (1876) 1950. *The Part Played by Labor in the Transition from Ape to Man*. New York: International Publishers.

Hadhazy, Adam. 2014. 'Heavenly Sounds: Hearing Astronomical Data Can Lead to Scientific Insights'. Scientific American. https://www.scientificamerican.com/article/heavenly-sounds-hearing-astronomical-data-can-lead-to-scientific-insights/ (accessed 23 January 2017).

Herschel, John F.W. 1847. *Results of Astronomical Observations Made during the Years 1834, 5, 6, 7, 8, at the Cape of Good Hope; Being the Completion of a Telescopic Survey of the Whole Surface of the Visible Heavens, Commenced in 1825*. London: Smith, Elder & Co.

Holden, Edward S. 1882. *Monograph of the Central Parts of the Nebula of Orion*. Washington, DC: Government Printing Office.

—. 1886. 'Photography the Servant of Astronomy'. *Overland Monthly* 8(47): 459-470.

Kessler, Elizabeth A. 2012. *Picturing the Cosmos: Hubble Space Telescope Images and the Astronomical Sublime*. Minneapolis, MN: University of Minnesota Press.

Mason, E.P. 1841. 'Observations on Nebulae with a Fourteen Feet Reflector, Made by H.L. Smith and E.P. Mason, during the Year 1839'. *Transactions of the American Philosophical Society* 7: 165-213.

McGinn, Colin. 2015. *Prehension: The Hand and the Emergence of Humanity.* Cambridge, MA: MIT Press.

Merleau-Ponty, Maurice. 1968. *The Visible and the Invisible: Followed by Working Notes.* Evanston, IL: Northwestern University Press.

Napier, John. 1970. *The Roots of Mankind.* Washington, DC: Smithsonian Institution Press.

—. 1980. *Hands.* Princeton, NJ: Princeton University Press.

Nasim, Omar W. 2013. *Observing by Hand: Sketching the Nebulae in the Nineteenth Century.* Chicago, IL: University of Chicago Press.

—. 2016. 'Astrofotografie und John Herschels "Skelette"'. In *Zeigen und/oder Beweisen? Die Fotografie als Kulturtechnik und Medium des Wissens.*, ed. by Herta Wolf. Berlin: De Gruyter, 157-178.

—. 2018. 'James Nasmyth on the the Moon: Or on Becoming a Lunar Being without the Lunacy'. In *Selene's Two Faces: From 17th Century Drawings to Spacecraft Imaging*, ed. by Carmen Pérez González. Leiden: Brill, 147-187.

Oxmantown, Lord [Lawrence Parsons]. 1868. 'An Account of the Observations on the Great Nebula in Orion, Made at Birr Castle, with the 3-Feet and 6-Feet Telescopes, between 1848 and 1867. with a Drawing of the Nebula'. *Philosophical Transactions of the Royal Society of London* 158: 57-73.

Pannekoek, Anton. 1898. 'III. On the Best Method of Observing the Milky Way'. *Popular Astronomy* 5(50): 524-528.

—. 1912. *Marxism and Darwinism.* Trans. by Nathan Weiser. Chicago, IL: Charles H. Kerr & Co.

—. 1920. *Die Nördliche Milchstrasse.* Annalen van de Sterrewacht te Leiden, 11:3. Haarlem: Joh. Enschedé en Zonen.

—. 1928. *Die Südliche Milchstrasse.* Annalen van de Bosscha-Sterrenwacht, 2:1. Amsterdam: De Bussy.

—. 1933. *Photographische Photometrie der Nördlichen Milchstrasse.* Publications of the Astronomical Institute of the University of Amsterdam 3. Amsterdam: Stadsdrukkerij.

—. 1942. '[J. Harper, Pseud.] Materialism and Historical Materialism'. *New Essays* 6(2): 9-25.

—. (1944) 1953. *Anthropogenesis: A Study of the Origin of Man.* Amsterdam: North-Holland.

—. 1961. *A History of Astronomy.* London: George Allen and Unwin.

Pannekoek, Anton, and David Koelbloed. 1949. *Photographic Photometry of the Southern Milky Way.* Publications of the Astronomical Institute of the University of Amsterdam 9. Amsterdam: Stadsdrukkerij.

Proctor, Richard A. 1883. 'The Photographic Eyes of Science'. *Longman's Magazine* 1(4): 439-462.

Radman, Zdravko. 2013. 'Beforehand'. In *The Hand, an Organ of the Mind: What the Manual Tells the Mental*, ed. by Zdravko Radman. Cambridge, MA: MIT Press, xix-xxii.

Roy, Jean-René. 2017. Unveiling Galaxies: The Role of Images in Astronomical Discovery. Cambridge: Cambridge University Press

Schaffer, Simon. 1988. 'Astronomers Mark Time: Discipline and the Personal Equation'. *Science in Context* 2(1): 115-145.

Scheiner, Julius. 1897. *Die Photographie der Gestirne*. Leipzig: Wilhelm Engelmann.

Sheehan, William. 1988. Planets & Perception: Telescopic Views and Interpretations, 1609-1909. Tucson: The University of Arizona Press.

Sheehan, William, and Christopher J. Conselice. 2016. *Galactic Encounters: Our Majestic and Evolving Star-System, from the Big Bang to Time's End*. New York: Springer.

Smith, Robert W. 1991. 'A National Observatory Transformed: Greenwich in the Nineteenth Century'. *Journal for the History of Astronomy* 22(1): 5-20.

Sohn-Rethel, Alfred. 1989. *Geistige und körperliche Arbeit: zur Epistemologie der abendländischen Geschichte*. Weinheim: VCH Acta Humaniora.

Tai, Chaokang. 2017. 'Left Radicalism and the Milky Way: Connecting the Scientific and Socialist Virtues of Anton Pannekoek'. *Historical Studies in the Natural Sciences* 47(2): 200-254.

Tai, Chaokang, and Jeroen van Dongen. 2016. 'Anton Pannekoek's Epistemic Virtues in Astronomy and Socialism: Personae and the Practice of Science'. *BMGN – Low Countries Historical Review* 131(4): 55-70.

Vertesi, Janet. 2015. *Seeing like a Rover: How Robots, Teams, and Images Craft Knowledge of Mars*. Chicago, IL: University of Chicago Press.

Wilson, Frank R. 1998. *The Hand: How Its Use Shapes the Brain, Language, and Human Culture*. New York: Pantheon.

## About the Author

**Omar W. Nasim** is Professor of the History of Science at the University of Regensburg. He is the author of *Observing by Hand: Sketching the Nebulae in the Nineteenth Century* (2013), which won the History of Science Society's Pfizer Award in 2016. Nasim is currently completing a book on the cultural history of observing chairs in astronomy and another book on the history of astrophotography. Nasim directs the Master's programme in the History of Science at the Institute of Philosophy at the University of Regensburg.

# 13 Cross-Fading the Milky Way

*Johan Hartle in conversation with Jeronimo Voss about universalism and a realist economy of time**

**Abstract**

Inspired by the work of Anton Pannekoek Johan Hartle and Jeronimo Voss discuss the relationship between astronomy and communism in the work of Jeronimo Voss. In this light, astronomy provides several figures of thought also for understanding the inner necessities of artistic production. Thus, the universe appears as a code for discussing the possibility of universalism, which informs both the histories of communist politics and the paradigm of artistic realism. Also, the futurist horizon, the projection of worlds to come is classically identified with the open cosmos, reflecting upon the possibility and failure of radical progress. In such ways, so the text stresses, the stars can become a source of political and artistic imagination.

**Keywords:** communism, universalism, futurism, progress, temporality, realism

## Inverted Night Sky

***Johan Hartle:*** The title of your exhibition 'Inverted Night Sky' at the Stedelijk Museum Bureau Amsterdam (SMBA, 15 May – 26 June 2016) is inspired by Anton Pannekoek's drawings of the Milky Way from the 1920s, currently archived at the Anton Pannekoek Institute for Astronomy in Amsterdam. Can you explain how your artistic work brought you to the institute in the first place and how these drawings eventually inspired your project?

* This conversation is based on a public interview by Johan Hartle of artist Jeronimo Voss that took place at the Royal Netherlands Academy of Arts and Sciences, Amsterdam, on 9 June 2016.

Tai, Chaokang, Bart van der Steen, and Jeroen van Dongen (eds), *Anton Pannekoek: Ways of Viewing Science and Society*. Taylor & Francis Group, 2019
DOI: 10.5117/9789462984349_CH13

***Jeronimo Voss:*** I was researching the history of the Zeiss Planetarium and its projection system when I visited the archives of the Institute in Amsterdam for the first time in 2013. There I met astronomer Edward van den Heuvel and science historian Chaokang Tai. Van den Heuvel explained to me that these drawings, made by the astronomer and radical socialist Anton Pannekoek, are the most precise depictions of the Milky Way as it appears to the naked eye in a clear and dark night sky. This was the reason why the Zeiss Planetarium translated these drawings into its projection system in order to simulate the Milky Way as it would be visible to the naked eye.

The Zeiss planetarium was invented at a time when the view of the stars had become more and more diffused in metropolitan areas, due to crossfading urban electrification. Today, the Milky Way is even less visible compared to the 1920s. We estimated that the number of viewers that saw Pannekoek's mapping of the Milky Way in the Zeiss projection systems must run into the hundreds of millions. His radical politics never achieved any comparable audience, among others due to his exclusion from the international communist movement in 1921.[1] I was already working with full dome projections for a while, and I wondered how I could translate Pannekoek's drawings of the Milky Way into an experimental dome installation. To me, the fact that the night sky is becoming more and more diffused to the naked eye is nothing to be nostalgic about. I rather see it as an opportunity to develop a more realist view.

## A Realist Universe

***JH:*** It is striking that three perspectives converge in your project: art, astronomy, and radical politics. One aspect that unites these three fields is that they all strive for a certain kind of universality. One of the biggest and probably most abysmal questions in artistic practices is the question of *binding norms* and where to find them. Socialism very explicitly deals with universal equality, as difficult as this is to conceptualize and achieve.

1 In 1920, Vladimir I. Lenin, chairman of the Communist Party in Russia, published *'Left-Wing' Communism: An Infantile Disorder* in which he discredited Anton Pannekoek and other council communists of the time, such as Sylvia Pankhurst, Henriette Roland Holst, Herman Gorter, and Amadeo Bordiga. They were subsequently excluded from the Communist International.

**Figure 13.1 Charcoal Drawing of the Milky Way by Anton Pannekoek**

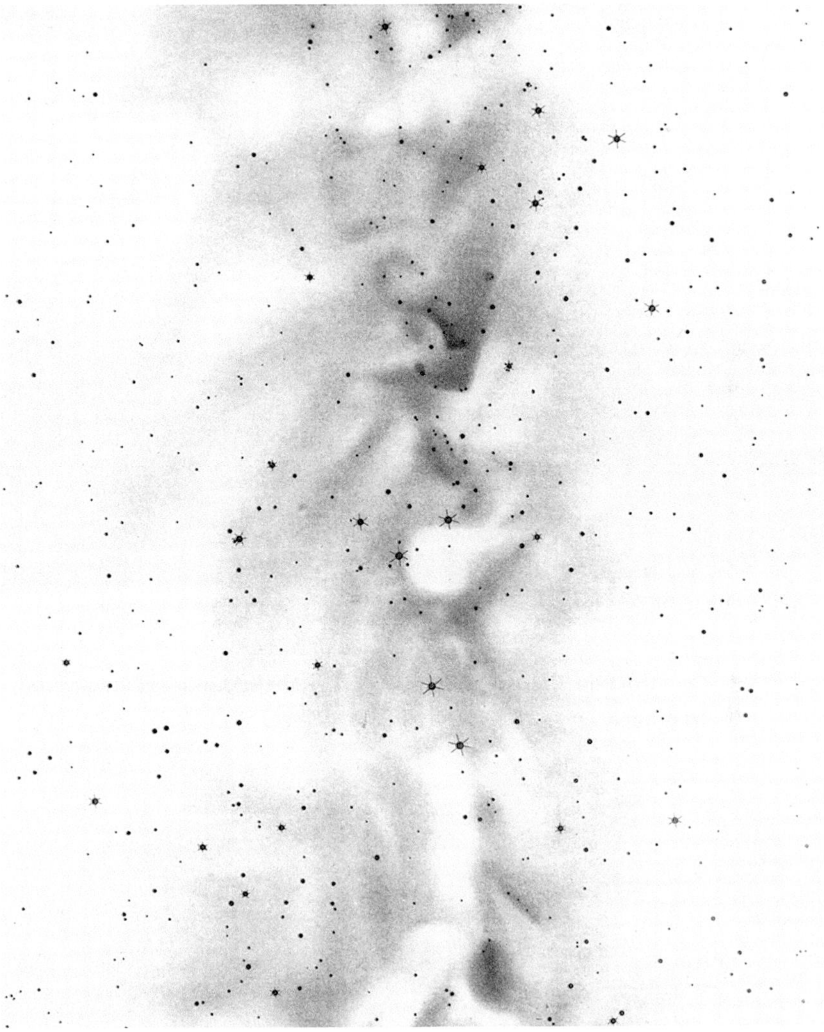

Source: Anton Pannekoek, *Die südliche Milchstrasse*, Annalen van de Bosscha Sterrenwacht 2:1 (Amsterdam: De Bussy, 1928)

Constructing and interpreting the universe through astronomical means is, in some way, the model case for any kind of universalism. But this is only one possible way of constructing a link between the three perspectives. How do you see this relationship? What conceptual links matter for your artistic practice?

**Figure 13.2 Construction of the Zeiss Planetarium in Jena, 1924**

Source: Archive of the Zeiss Planetarium in Jena

**Figure 13.3 Construction Plans of a Milky Way Projector in the Zeiss Planetarium from 1964**

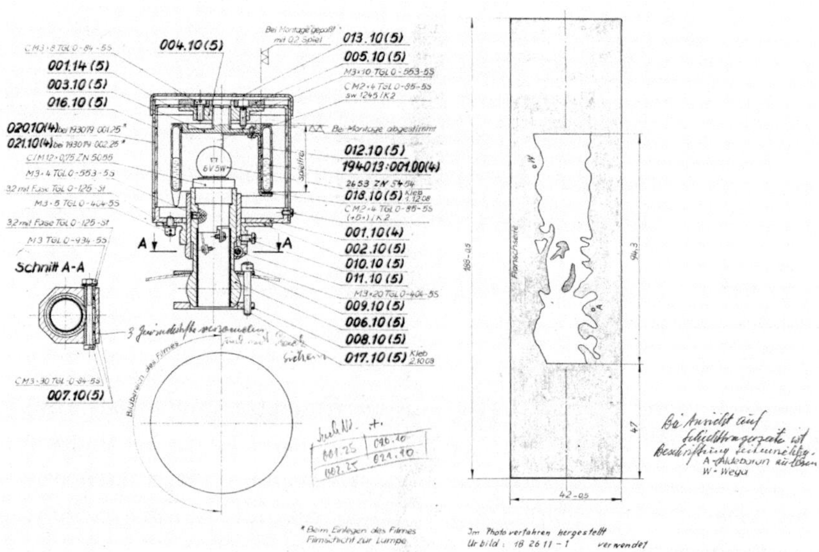


Source: Archive of the Zeiss Planetarium in Jena

**Figure 13.4 Milky Way projection film of the Zeiss Planetarium, 1927**

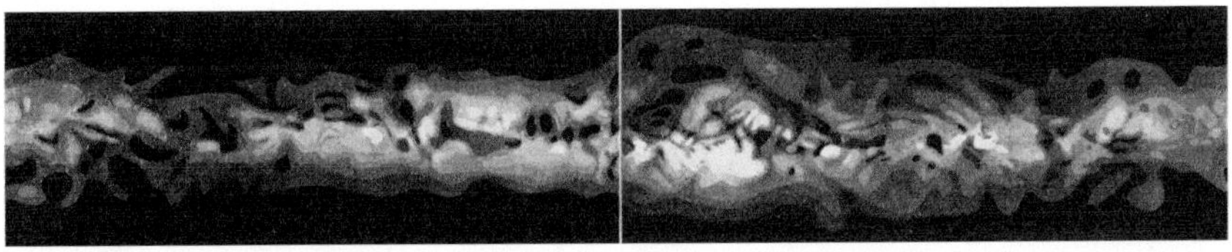

Source: Archive of the Zeiss Planetarium in Jena

*JV:* I think it's about Realism – and with that I don't mean the common understanding of the traditional naturalist documentarism that sticks to the given. I see Realism, inside and outside the field of art, as a radical expansion of one's own standpoint or perspective. These different fields you mentioned might be separated for practical or moral reasons. But from a realist point of view there is an infinity of overlappings. So I agree, this implies a universalist perspective that goes beyond and connects the traditionally separated domains of visual art, documentarism, science, politics, and everyday life. Thus, during my research in Amsterdam, I was not only interested in Pannekoek's statistical Milky Way astronomy but also in the socialist debates he was involved in.

*JH:* The link between astronomy and radical politics has already inspired you in a series of projects, for example 'The Sun was Captured, yet no Victory' from 2011, or 'Eternity through the Stars' in the Planetarium of the Orangerie for documenta 13. Both deal with models of social transformation that are reflected in a form of cosmology. How exactly do you address the question of universalism in these works?

*JV:* Eternity through the Stars is based on *L'Éternité par les Astres: Hypothèse Astronomique* published in 1872, after the defeat of the Paris Commune, by one of its protagonists, Louis Auguste Blanqui. The book lays out a worldview in which every possible decision or event actually exists in an infinite number of variations within the material reality of physical space. In this view, history is not a product of progress, of an independent flow of time moving from the past through the present toward the future. Instead it appears as a result of decisions. All I added was a small but obvious supplement to Blanqui's hypothesis by tying it closer to its historical context of the Paris Commune of 1871. My interpretation claims that if all kinds of life have already been realized on various far-away Earth-like planets, we can also imagine an infinity of versions where the Commune didn't fail.

**Figure 13.5 Eternity through the Stars (2012) by Jeronimo Voss**

Source: Anders Suneberg/documenta 13

Blanqui's hypothesis suggests a very activist worldview, in which every second offers an abundance of options, thus everything could fundamentally change at any given moment, depending on collective decisions. This may also explain why its author spent most of his life in prison for participating in various insurrections and coups d'états. Nevertheless, it is a view that goes beyond fatalist or absolute conceptions of time. His cosmic model introduces an infinite space of possibilities. In many ways, it is the opposite of belief in progress as an independent linear development.

Within the Realism Working Group, we had discussed this topic of progress before, in the context of a collaborative project dealing with the futurist opera *Победа над солнцем* ('Victory over the Sun', premiered in St Petersburg in 1913). The original opera revolved around a futurist revolution in which an airplane takes down the sun from the sky in order to spark a new society. The opera

**Figure 13.6 The Sun was Captured, yet no Victory (2011) by Jessica Sehrt and Jeronimo Voss**

Source: Markus Tretter/KUB Kunsthaus Bregenz

anticipated a certain spirit of communist futurism that was linked to a strong belief in industrial progress. From today's point of view, we could say that the sun has in fact been captured – with which I mean that technological progress has been tremendous. But this didn't lead to the expected 'victory'. This led us to approach the opera's central theme differently. We programmed an artificial sun, trapped in a circle of stage lights – a silent colour loop constantly repeating the same dramaturgy of pulsating light.[2]

*JH:* Cosmic space is often referred to, directly or indirectly, as a means of addressing utopian possibilities, or even concrete political alternatives. In the works of Russian avant-gardists, futurists and even Bolsheviks such as Alexander Bogdanov – who wrote the Bolshevik utopian sci-fi novel *Red Star* – space exploration and the conquest of the galaxy is equated with technological and social progress.[3] These progressivist models of utopia have inspired much of the Soviet fascination for space exploration, but is only one pole of what we could call astro-utopianism.

2 www.anfanggutallesgut.net (accessed 8 November 2016).

3 See: Bogdanov 1984.

On the other side of the spectrum, we see authors such as Walter Benjamin who instead equate space with the idea of intangibility, of letting go. Using the term allegory or constellation, Benjamin argues that we build constellations with words, images, and concepts to describe things that ultimately remain ungraspable. This notion is often discussed in close relationship with cosmology: just as we characterize the locations of the stars in terms of constellations that remain fundamentally arbitrary and external to the stars themselves, so do allegories reveal themselves as provisional and incomplete representations.[4]

Both Walter Benjamin and Theodor Adorno have claimed that this way of thinking contains a utopian promise because it leaves the things that are discussed at a respectful distance, as if they were stars: dignified but ungraspable. This view, however, strongly contradicts the optimistic and progress-oriented perspective on the stars that we find in the Russian avant-garde.

*JV:* I agree, but for me this means that 'Victory over the Sun' could be read as a dystopian scenario, too. Our intention was to look at the opera as the collective experiment that it was, but to put it in a different, present-day, setting. The technological futurism that formed the crescendo of the opera was realized in a state-socialist regime, which led to the 'electrification of the whole country' (Lenin) – without soviet power. Technological progress in Russia was a progress of national capital.

Today, we have smart phones, efficiently managing our extended work schedules – but does it make sense to speak of this as progress? Especially since the global economy seems to be stuck in a drift of continuous slowdown – 'There is, alas, no progress!' This phrase from Blanqui's astronomical hypothesis is truer today than it was in his own time. I believe that it may be more useful to appropriate, rearrange, and adjust the tools we already have at hand instead of waiting for any kind of continuous progress.

*JH:* Your notion connects a number of intellectual traditions. Walter Benjamin was similarly sceptical of the dogmatic belief in progress that was central to much of classical Marxism. He also commented critically on 'imperialist' dreams of cosmic conquest, evoking J.J. Grandville's caricature of a capitalist industrialist walking from planet to planet across iron bridges. One might venture the interpretation that for Benjamin, cosmic space needed to be defended as a realm of dreams rather than a raw material for economic expansion.[5] If I understand correctly, you suggest that one should see Soviet

4 See: Benjamin 1998, 159 ff.

5 As Benjamin writes in One-Way Street: Benjamin 1979, 104. See also Benjamin 1999, 65.

**Figure 13.7 Illustration by J.J. Grandville from *Un autre monde*, 1844**

Source: J.J. Grandville, *Un autre monde* (Paris, H. Fournier, 1844)

constructivist optimism as a broken optimism which lives through dreams rather than direct expectations.

This would mean two things: First, that there is no radical break between the progressivist idea of conquering space and the emphasis on intangibility, because both were meant allegorically. Both are based on a radical divide between what can (and should) be revealed and signified and what will (and must) remain intangible and undisclosed.

The second point throws an interesting light on your own work and on that of Pannekoek. Your work uses hard-edged geometrical forms, projections of archival materials ('transparent archives'), and, of course, advanced media technologies (complex projections, digitally produced soundscapes). These technological means seem to suggest an optimistic view on technological progress or even a rationalist take on your subject matters. At the same time much of what you are dealing with – including Pannekoek and his record of the Milky Way – seems very anachronistic, as if you were uncovering traces from an almost forgotten past. The historical struggles for socialism and communism have faded away almost as much as the Milky Way has for the naked eye. By its use of technology, your work seems to be reaching out for concrete forms of knowledge and resources while at the same time being slightly melancholic, restoring lost causes and forgotten histories. Your way of approaching the stars and the history of radical politics would then be as much constructivist and quasi-scientific as it would be allegorical.

*JV:* I am neither interested in lost causes nor do I feel the need to approach them with melancholia. The historical material I deal with is urgent and necessary for me to navigate the present and to at least imagine a future. It is about the search for clarifications and overview, which, I believe, goes along with a certain kind of conceptual and formal coherence. For me, this navigation starts with dismantling the absolute conception of time that not only structures the wage workday but also aims at dictating everything we know about life. It declares even the most obvious potentials to be over, lost in the past and forever out of reach. Surely, we might have good reasons to mourn these few years around the 1920s, that in certain metropolitan areas seemed like a golden age of emancipation on various levels of everyday life. But these progressive times collapsed and were followed by a world economic crisis with its escalation into the rise of European fascism. Today, we're in the midst of an even more substantial global crisis – and fatalism will certainly not prepare us for what we will have to deal with in the future. So my research does not focus on some kind of forgotten history or lost cause – this material is available and very present, and it can be used and rearranged for whatever purpose.

Today, critical theory seems to offer little more than the prediction that the future of class society will be more violence, competition, and destruction – the global retro-fascism of the present states pretty much the same, with the difference that it optimistically embraces this future, while most of the left is still stuck in its own parliamentary traditions. What is missing is a concrete alternative, a realist picture of how a classless global commune could actually work, so as to replace the current conditions. Otherwise we're left with the prospects of a bourgeois futurism building some kind of techno-habitat, probably not on Mars but on Earth, in order to protect itself from the environment it destroyed and the populations it considers superfluous.

## Self-Regulation

*JH:* One of the reasons why the stars have often played a role in socialist ideals was a certain scientific understanding among socialists of how society works. According to this view, laws of nature are projected onto the dynamics of human activity and social organization.

For Charles Fourier, a pioneering author of nineteenth-century socialism, the cosmic order served as a model for social order. Fourier saw similar mechanisms at work in the cosmic order, the natural world and in human society. He was greatly admired by the Surrealists, by Walter Benjamin and by the Situationists.

This idea of a profound isomorphism between cosmic order, nature, human society, and even the individual subject was highly utopian: it promised the liberation from repressive laws of organization, a liberation that would be driven by natural forces (inclinations, affects, and passions). It held the promise of a society based on a self-regulating system of material forces.

Both the idea of isomorphism and of self-regulation are not as outdated as they might seem. The Frankfurt School philosophers Alexander Kluge and Oskar Negt argue very similarly in their recently reworked study *History & Obstinacy*. Not only do they show great interest in cosmological and astronomical metaphors, they are also intent on reconstructing the general course of history as a system of 'self-regulations'. In their work, human history becomes a history of physical, bodily, mental, and social forces and the inner strife between them, leading towards new forms of organization. The argument heavily relies on the assumption of an isomorphism between the physical, the social and the psychological life. If such a fundamental isomorphism between the stars and society is assumed, the free movement of stellar, social, or individual bodies becomes an immensely utopian image.[6]

Furthermore, the views of Fourier and Negt/Kluge suggest a certain aesthetic strategy too. The self-regulation of social and psychological forces informs the ways in which they write. It leads to the question of how to aesthetically arrange material, how to work with it and how to present it. The idea of self-regulation implies its own directive.

Another link between socialism and interest in the cosmos can be seen in the life and work of Pannekoek. The science historian Chaokang Tai analyses Pannekoek's method of reconstructing the Milky Way. For Pannekoek, the Milky Way was mainly a visual construction of what in fact is an immense collection of stars. In reconstructing the human image of the Milky Way, Pannekoek sought to combine the viewpoints of different people. He asked multiple amateur astronomers to keep a detailed account of their individual night-sky observations. These subjective observations were then averaged and inter-compared in order to represent the real aspect of the Milky Way as true as possible.[7] Pannekoek called it the 'mean subjective image' of the

6 Fourier 1971, 101-110; Negt and Kluge [1981] 2001, 55.

7 Tai 2017.

Milky Way. This bottom-up approach seems to correspond to Pannekoek's political ideas of workers' self-organization in councils.

In this sense you can speak of a certain political ethos, even of a method of construction, one could even say of *representation* – in light of the double meaning of the term in aesthetic and political contexts. I wonder how this inductive method of construction and implicit politics is to be seen in light of your own ideas on realism.

*JV:* Chaokang Tai emphasizes that Pannekoek's concept of historical materialism was closely related to the inductive, bottom-up research method that he utilized in his astronomical work – one should add that Pannekoek was an astronomer before he became a socialist.[8] This is a relevant connection, but it is also the reason why I had no interest in doing the same as Pannekoek did in astronomy. I did not want to translate his empirical method of drawing the Milky Way into my dome projection. In my opinion, a realist view takes into account far more perspectives or, if you like, self-regulated forces of motion, than any inductive, bottom-up research method could possibly do.[9] Pannekoek's visual strategy was successful in creating a naturalist night-sky portrait of the faint visual appearance of the Milky Way – my portrait of the night sky starts from there and then expands towards ticking pulsars, fisheye lens video recordings of kitchens, hallways, workplaces, and conflicting concepts of time.

## Economies of Time

*JV:* When I read Pannekoek's socialist writings, and that of his comrades in the Group of International Communists (GIC Holland),[10] I was most interested in the ways in which they discussed time as a fundament of economic systems. This idea was discussed in publications such as the GIC's *Fundamental Principles of Communist Production and Distribution* (1931), where they state that 'the social revolution is nothing more than the introduction of the labour-hour as the unit measure regulating and controlling

8 Mattick 1962.

9 Anton Pannekoek's celebration of the Philosophy of Joseph Dietzgen hints to how closely he was affiliated with the empiricism Dietzgen called 'inductive socialism' – which he defined as: 'The fundamental proposition of inductive socialism may be thus formulated: there is no eternal principle or an *a priori* idea of the divine, just and free; there is no revelation or a chosen people, but there are material factors which govern human society'. Dietzgen [1873] 1917, 85.

10 Pannekoek joined the GIC Holland in 1927.

**Figure 13.8 Aspects of the Milky Way (2016) by Jeronimo Voss**

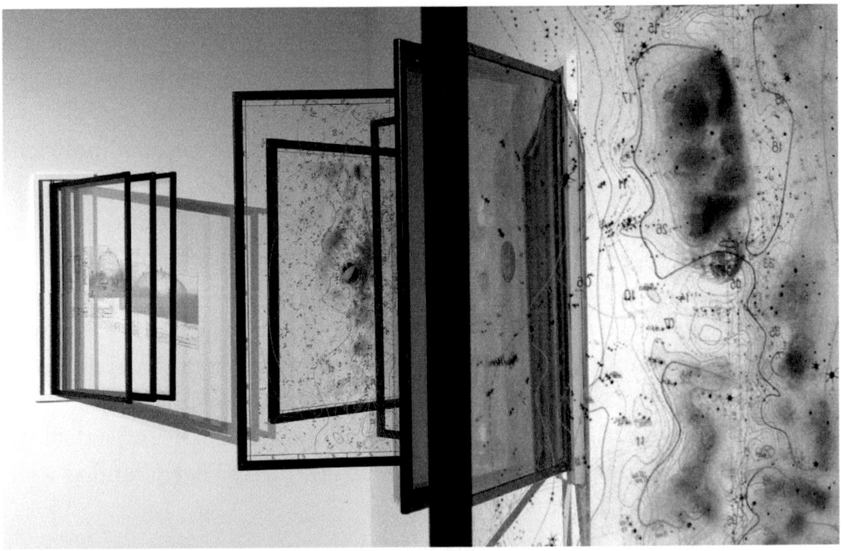

Source: Gert Jan van Rooij/Stedelijk Museum Bureau Amsterdam (SMBA)

the whole of economic life'.[11] In Pannekoek's words the collective planning process 'averages the hours of labour needed and directs the attention to the ways open for progress'.[12] He thus referred to a form of bookkeeping comparable to the way he constructed the Milky Way's 'mean subjective image'.[13] Applying this form of measuring – of budgeting, accounting, and bookkeeping of time – to the economic sphere tends to reduce the focus to measurable units of production. But can all socially necessary activities be quantified and inter-compared in this way? In the GIC's *Fundamental Principles*, activities such as housework merely appear as a side note. I don't believe that this view is realistic, neither today nor back in the 1920s or 1930s. Thus, my aim was to expand the visual Milky Way to the Anton Pannekoek Institute's infrastructure and its broader environment.

I also collected material from the institute's current Milky Way research: light pulses of rotating neutron stars, translated into sound. Today, the astronomy institute engages in the measuring of pulsars from various different areas of the Milky Way in order to determine long gravitational waves, ripples in space-time. These pulsars are very stable clocks rotating and emitting electromagnetic waves – we can see them as light or translate

11 GIC Holland [1931] 1990.

12 Pannekoek 1950, 61.

13 Pannekoek 1920.

them into sound. The sound was adjusted and rearranged by artist colleagues and collaborators, Jessica Sehrt[14] and Martin Stiehl,[15] with whom I work together in the Realism Working Group. They twisted the noise into the subjective beat of a pulse that changes depending on mood and action.

*JH:* The measure of time is of course central to the critique of political economy. Socially necessary labour time determines value, value is basically congealed time, value in process is capital. Which means, in Marx's political analysis, that the crystallization of time into quantifiable temporal units is at the very basis of capitalism. Finding alternative regimes of time is therefore one of the keys to developing alternative models for social organization. One should also think about the famous formulations from the *Grundrisse*, where Marx writes that society is fundamentally based on its conception and distribution of time.[16] And although the *Grundrisse* had not been available in the early 1930s, that has been a central issue for the GIC in Holland as well.

**Figure 13.9 Inverted Night Sky (2016) by Jeronimo Voss**

Source: Gert Jan van Rooij/Stedelijk Museum Bureau Amsterdam (SMBA)

14 http://www.jessicasehrt.com (accessed 24 January 2017).
15 http://institut.gebrauchsgrafik.org (accessed 24 January 2017).
16 Marx 1993.

**Figure 13.10 Various stills from Inverted Night Sky (2016) by Jeronimo Voss**

Source: Jeronimo Voss

*JV:* We have also discussed this with Paul Mattick Jr in a workshop of the Realism Working Group. The GIC's *Fundamental Principles* can't be applied to the current conditions. Today's productivity has increased drastically compared to the times of the 1930s when the book was published. It doesn't make sense to still measure the time you have invested in production in order to get your share of the social surplus product. Today, we live under conditions of chronic overproduction. So why should anyone receive from society only according to the individual units of lifetime that were spent beforehand? A few years ago, in his essay on *The Nature of Time* physicist Julian Barbour stated: 'Unlike the Emperor dressed in nothing, time is nothing dressed in clothes'.[17] Time is thus something we attire and shape in order to handle the changing world around us. There is no reason to blindly obey it.

A realist economy of time could instead serve as an overview of potential scenarios of production and reproduction. Or, as Pannekoek himself has put it: 'All these interconnections of mutually adapted operations may be represented in a well-ordered scheme, a mental image of the actual process'.[18] And if such an image or scheme starts with empirical data, like Pannekoek's construction of the visible Milky Way, it certainly shouldn't stop there.

## Closures

*JH:* Planetarium architecture closes the universe by erecting a half-round closed-off dome structure. This is a paradoxical move, given the fact that the planetarium aims to represent the infinite universe. With your installation 'Inverted Night Sky' (2016), you refer to this architectural tradition and you use this idea of formal closure to discuss astronomical dimensions. I wonder how you deal with this tension.

*JV:* You can feel very small inside the traditional Zeiss planetarium dome, overwhelmed by the naturalism of the stars. I prefer to use the full dome projection tilted at a 45°-60° angle, so that you as the visitor can decide for yourself to what extent you wish to enter the projection. You can also

17 Barbour 2009, 2.

18 Pannekoek 1950, 22. On the same page Pannekoek continues: 'Just as a map or a graph fixes and shows in a plain, to everyone intelligible picture the connections of a complicated totality, so here the state of the total enterprise, at every moment, in all its developments must be rendered visible by adequate representations.'

**Figure 13.11 Communal Villa (2015) by Dogma and Realism Working Group**

Source: Jeronimo Voss

choose to maintain a physical distance to it. 'Inverted Night Sky' is about the inversion of those fatal natural laws that astronomy works with. At the same time, it ponders on the emancipation from the economic order of time that we are currently confronted with. 'Annoying tasks should be accomplished – if at all – with least expenditure of time' is one of the lines rotating in the dome. The organization of temporal orders, and the temporal construction of everyday life – to me, this is what the editing of video time lines is about.

JH: What are your plans for the future?

JV: Currently I am researching the expanding field of time management software, together with artist and programmer Radamés Ajna. Before, in the Realism Working Group, we had been speculating on an economy of time organized through a syndicate of communal villas.[19] As a montage of coding language, software carries the potential to provide flexible tools customized for concrete user demands. What in fact is customized in the current IT landscape are the behaviours of its users according to respective business and consumer software solutions. We are looking for a different approach and methodology to design software applications that actually support our own collective decision-making.

## Bibliography

Aureli, Pier Vittorio, Robert Burghardt, Florian Schmidt, Jessica Sehrt, Martin Stiehl, Martino Tattara, and Jeronimo Voss. 2016. *Dogma + Realism Working Group – Communal Villa: Production and Reproduction in Artists' Housing: Wohnungsfrage, Haus Der Kulturen Der Welt*. Ed. by Christian Hiller, Jesko Fezer, Nikolaus Hirsch, Wilfried Kühn, and Hila Peleg. Leipzig: Spector Books.

Barbour, Julian. 2009. 'The Nature of Time'. *ArXiv:0903.3489* [*Gr-Qc*].

Benjamin, Walter. 1979. *One-Way Street, and Other Writings*. London: New Left Books.

—. 1998. *The Origin of German Tragic Drama*. Trans. by John Osborne. London: Verso.

—. 1999. *The Arcades Project*. Cambridge, MA: Harvard University Press.

Bogdanov, A. 1984. *Red Star: The First Bolshevik Utopia*. Ed. by Loren R. Graham and Richard Stites. Trans. by Charles Rougle. Bloomington, IN: Indiana University Press.

Dietzgen, Joseph. (1873) 1917. 'Scientific Socialism'. In *Some of the Philosophical Essays on Socialism and Science, Religion, Ethics, Critique-of-Reason and the World-at-Large*, ed. by Eugene Dietzgen and Joseph Jr. Dietzgen, trans. by M. Beer and Th. Rothstein. Chicago, IL: Charles H. Kerr & Co., 79-89.

Fourier, Charles. 1971. *The Utopian Writing of Charles Fourier: Selected Texts on Work, Love, and Passionate Attraction*. Ed. by Jonathan Beecher and Richard Bienvenu. Boston, MA: Beacon Press.

GIC Holland. (1931) 1990. *Fundamental Principles of Communist Production and Distribution*. Ed. & trans. by Mike Baker. London: Movement for Workers' Councils.

Marx, Karl. 1993. *Grundrisse*. London: Penguin.

Mattick, Paul. 1962. 'Anton Pannekoek'. *New Politics* 1(winter): 107-114.

Negt, Oskar, and Alexander Kluge. (1981) 2001. *Geschichte und Eigensinn*. Frankfurt am Main: Zweitausendeins.

19 See Aureli et al. 2016.

Pannekoek, Anton. 1920. *Die Nördliche Milchstrasse*. Annalen van de Sterrewacht te Leiden, 11:3. Haarlem: Joh. Enschedé en Zonen.

—. 1950. *Workers' Councils*. Melbourne: Southern Advocate for Workers' Councils.

Tai, Chaokang. 2017. 'Left Radicalism and the Milky Way: Connecting the Scientific and Socialist Virtues of Anton Pannekoek'. *Historical Studies in the Natural Sciences* 47(2): 200-254.

## About the Authors

**Johan Frederik Hartle** is Professor of Art Studies (with a focus on political aesthetics) at the Karlsruhe University of Arts and Design (Hfg). He holds teaching positions at the University of Amsterdam (UvA) and the China Academy of Art (CAA) in Hangzhou. His field of research includes contemporary approaches to the aesthetico-political and institutional approaches to contemporary art. His most recent book publications are *The Spell of Capital. Reification and Spectacle* (2017) and *Aesthetic Marx* (2017; both edited together with S. Gandesha).

**Jeronimo Voss** lives and works in Frankfurt/Main, where he studied visual art at the HfBK Städelschule as well as the Free Class. His artistic production could be described as a modelling of time and parallel worlds. Beyond the ruling promises of progress his installations crossfade the past, present, and future into subjective as well as cosmopolitical perspectives. Voss' recent projects were exhibited in various group and solo shows at international institutions such as Kunstverein Bellevue-Saar in Wiesbaden (2019), Galeria Boavista Lisbon (2018), Fact Liverpool (2017), Stedelijk Museum Bureau Amsterdam (2016), House of the World's Cultures (HKW) in Berlin (2015), Fondazione Trussardi in Milan (2014), MMK Museum für Moderne Kunst Frankfurt am Maim (2013), and dOCUMENTA (13) in Kassel (2012). Voss currently teaches Installation Art and Time-Based Media at the Art Institute HGK FHNW in Basel.

# 14 A Galaxy of Appearances

## Anton Pannekoek and the Planetary Cinema of Jeronimo Voss

*Alena J. Williams*

**Abstract**

Since 1923, the Carl Zeiss optical manufacturing company in Jena had begun producing machines for its newly developed projection planetariums worldwide. Both phantasmagoric illusion and pedagogical tool, the projection planetarium was a hybrid object with affinities to Dutch astronomer Anton Pannekoek's methodological approach towards the visual representation of the Milky Way Galaxy. His approach to preparing this representation of the Milky Way emphasized the subjectivity of perception, and the means by which our visual understanding of the galaxy are governed by a range of influences and contingencies. This chapter examines the remobilization of historical concepts of phantasmagoria in contemporary artist Jeronimo Voss's work in relation to Pannekoek's pioneering studies on the representation of the night sky.

**Keywords:** planetarium, representation, socialism, phantasmagoria, visualization

In the research of Dutch astronomer and socialist Anton Pannekoek (1873-1960), the Milky Way was a bewitching paradox of appearances and disappearances. Within that inky darkness, endless particulates and matter obscure the galaxy's own image. Under ideal conditions for observation, the shape of the Milky Way shoots up from the horizon, cutting obliquely into the sky, and although these elements may render the general contours of the Milky Way visible, there is so much that is not visible. Pannekoek reached towards these vagaries and lacunae in order to grasp the Milky Way's vast immensity. Throughout a large part of his

Tai, Chaokang, Bart van der Steen, and Jeroen van Dongen (eds), *Anton Pannekoek: Ways of Viewing Science and Society*. Taylor & Francis Group, 2019
DOI: 10.5117/9789462984349_CH14

career as a researcher and professor of astronomy, he collected drawings from colleagues and sources dating as far back as Antiquity. He believed that the only true manner to represent this faint, voluminous body was to collect impressions from multiple sources rather than from a single, subjective observer. In fact, so intense was Pannekoek's preoccupation with the accumulation of particularities from these different viewpoints that he believed they might cohere into a general picture, a so-called 'mean subjective image' of the galaxy.[1] Central to this project was his emphasis on the subjectivity of perception, and the fact that our visual understanding of the galaxy is governed by a range of subjective factors. However, Pannekoek's interest in the subjective nature of appearances, while unconventional within astronomy and the natural sciences, was predicated upon ideas that have been fundamental to the analysis of images within the visual arts. This essay examines the relationship between Pannekoek's ideas and the recent work of contemporary artist Jeronimo Voss.

Since completing his studies, Voss has been interested in subverting the prevailing hegemonies of time and progress, by entangling revolutionary history with the phantasmagorical art of projection. In 2012, for the international exhibition documenta 13 in Kassel, he created *Eternity Through the Stars*, an art installation and planetarium projection work that re-examined the speculative text *L'éternité par les astres: hypothèse astronomique* (1872) written by French revolutionary socialist Louis-Auguste Blanqui (1805-1881).[2] Reflecting upon the political implications of Blanqui's hypothesis of infinite worlds and eventualities, Voss exhibited documentary images and artifacts in Kassel's Orangerie, a reconstructed castle in the Karlsaue state park. On oversized transparencies hung in groups vertically

An earlier version of this essay appeared in English and German as Williams 2017.

1 In 1920, Anton Pannekoek said of his method of combining and comparing individual (and thus subjective) drawings of the Milky Way: 'Here emerges the importance of many independent researches. Their differences offer a representation of objective uncertainties, which exceed the borders of subjective certainty. [...] Generally speaking, the accidental-subjective, the manner of each observer, greatly disappears. What remains is not an objective image of the Milky Way, but something one could call the mean subjective image – the objective image as it is transformed through the conditions of general physiological-psychological observation' (Pannekoek 1920, 16, my translation). For an excellent overview of Anton Pannekoek's research on the representation of the Milky Way, see Tai 2017.

2 In *L'éternité par les astres: hypothèse astronomique*, Blanqui – an activist of the French revolts of 1830, 1848, and an influential organizer of the Commune – reached towards the natural philosophy of astronomy in order to mobilize a critique of the contemporary social order. See Blanqui [1872] 2013, 69.

and perpendicular along the walls of the building's exhibition spaces, viewers encountered reproductions of mechanical projection slides, including *The Solar System, showing the Revolution of all the Planets, with their Satellites round the Sun* (London, 1849); street views of *Paris, Rue de Flandre, March 1871* and planetary nebulae; as well as visual montages, like *Barricade Drawing with Jovian Planet* [Jupiter].[3] Within the planetarium itself, twelve synchronized digital projectors threw images of worlds and stars – along with historical images from the Paris Commune and magic lantern slides from the collections – onto the inner surface of the dome, encircling the viewer. A voiceover, in which a woman explains Blanqui's astronomical hypothesis, focused on Voss's understanding of the text's radical subversive potential: that all possible variations of our own past, present, and future are real material facts located within infinite space – a worldview that conceptualizes history as a product of collective decisions rather than as an independent stream of time. Moreover, reflecting Blanqui's idea of infinite return, these correspondences in the work – shuttling between sites of the Paris Commune, the astronomical cabinet in Kassel, the (meta)physical aspects of the universe, and the contemporary viewer – continually stage media within a complex system of (re)invention.

It also bears witness to the fact that the night sky – with its countless stars and heavenly bodies – has been of interest to scientists, philosophers, and visual artists for centuries. Central to that impulse to represent, and by turns, to know the universe has been the evolution of the planetarium, which Voss redeploys today within differing aesthetic and political contexts. Widely popularized in the 1920s, the modern projection planetarium – developed by the Carl Zeiss optics company in Jena in 1923 – signalled a major shift in the system of representation of the stars and planets, away from a range of astrophysical devices, including celestial globes and orreries developed since Antiquity. Housed within a seated theatrical auditorium with a projection device at its centre, the modern planetarium, though incomparable to its infinite vastness, afforded viewers highly illusionistic views of the night sky thrown up against the interior wall of the structure. Pivotal to the development of the project was the introduction of the self-supporting dome, which had a genealogical connection to a number of large-scale mass entertainments of the nineteenth century, namely Irish artist Robert Barker's large-format paintings and panoramas first developed in 1792. It was upon the planetarium dome's inner surface that the projection of tungsten light simulated the night sky as it passed through glass plates with punched-out

3 Palmieri 2012.

**Figure 14.1 Inverted Night Sky (2016) by Jeronimo Voss**

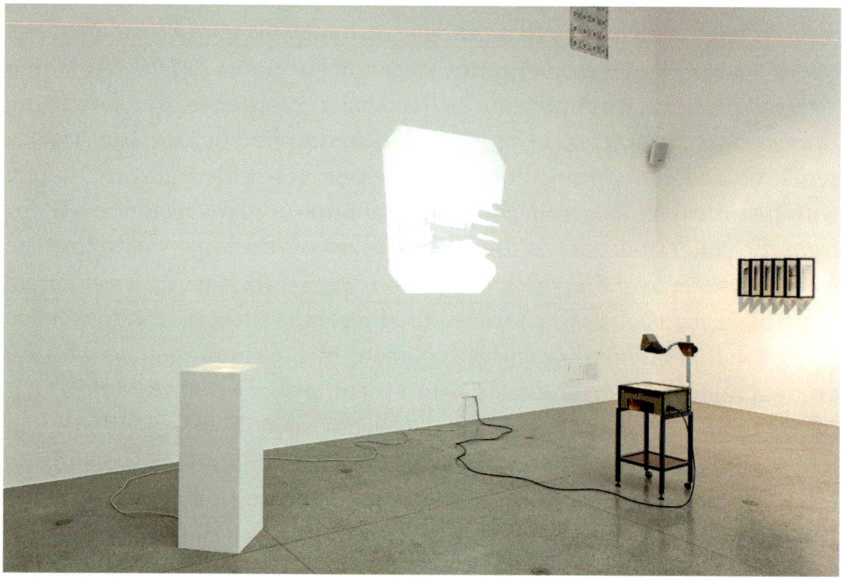

Source: Gert Jan van Rooij/Stedelijk Museum Bureau Amsterdam (SMBA)

copper foils, each representing different star constellations and clusters totaling to nearly 4,500 stars.[4]

Significantly, the Zeiss company integrated Pannekoek's drawings of the contours of the Milky Way Galaxy into the planetarium's projection systems already by 1927.[5] Almost a century later, these charcoal drawings became the wellspring for Voss's *Inverted Night Sky* (2016), a recent multisensory installation the artist created for an exhibition at the Stedelijk Museum Bureau Amsterdam (SMBA). Explicitly engaging with the scientific, social, and political history of the visual Milky Way, the work utilizes three different modes of representation: a dome projection, multiple large-scale transparencies, and sound. The introduction of the dome within the exhibition space migrates the planetarium's *dispositif* into the exhibition context. Voss deploys his planetarium as a portable architectural structure, hanging it at

4 Henry King and Joachim Krausse have comprehensively summarized the complexity of these mechanical and optical systems in their numerous iterations; see King and Milburn 1978 and Krausse 1993. My previous research on the relationship between the planetarium and artistic practice was presented at the conference, *Das Planetarium als Medium kosmologischer Reflexion*, as part of the DFG-Projekt: Zeit – Bild – Raum, Technische Universität Berlin on 26 April 2013. See also Williams 2015.

5 Tai 2017, 201.

a 45-degree angle and suspending it above the floor such that it can both be entered and viewed at a distance. Within it, a mobile spectator encounters contemporary video recordings of interior rooms of the Anton Pannekoek Institute for Astronomy at the University of Amsterdam (see Figure 14.1). Pannekoek's charcoal Milky Way drawings – again seen as an inversion in the digital video – render the stars as sparkling white dots against a dark sky. Directly across the room, Voss also mounted a series of large transparencies onto five large framed panels onto the wall. Printed upon them are reproductions of the scientist's charcoal drawings and different star maps with red outlines and blue areas of contoured colour. These isophotic lines describe the relative brightnesses of the galaxy's areas of light and shadow. Archival documents and documentation of the Zeiss planetarium and its Milky Way projector also appear in these panels. As the viewer leafs through them, the images become superimposed upon one another in real time; one has the sense of assembling a general overview of the stars from a montage of fragments and idiosyncratic particularities. It is a synthetic approach very much in keeping with Pannekoek's methodology of capturing the Milky Way.

In contrast to leading figures in the field who made great efforts to quantify the stars, Pannekoek's astrophysical research emphasizes the value of compiling and synthesizing subjective points of view of the Milky Way from multiple perspectives. Under his proposed system, a nearly untrained observer should, under optimal viewing conditions, render the relative brightness of the sky, and its particularities – specific stars and notable features – by way of naked-eye observations of the Milky Way Galaxy and handmade drawings.[6] At the turn of the twentieth century, the medium of photography offered astronomers a more systematic means of representation; however, Pannekoek found photography difficult in exactly calibrating the relative brightness of stars, and as some scholars have pointed out, the medium was much more useful for the purposes of astronomical classification than capturing astronomical particularities.[7] Although drawings were subjective images, Pannekoek argued that the 'accumulation' of unique observations from a number of independent researchers was the best way to capture the visual semblance of the collection of stars.[8] In 1920, he

6 Pannekoek 1897.

7 In a paper devoted entirely to the subject, Pannekoek (1912) identified many of the difficulties photography poses in the investigation of the structure of the galaxy. See also Omar W. Nasim's discussion of photography versus drawing in Nasim 2013, 229.

8 Pannekoek 1897, 398.

conducted a comparative analysis of existing representations of the Milky Way – including the work of such figures as German astronomer Otto Boeddicker (1853-1937), Dutch astronomer Cornelius Easton (1864-1929), and German astronomer Johann Friedrich Julius Schmidt (1825-1884) – in his publication *Die Nördliche Milchstrasse*. Significantly, the exploitation of variations in sense perception also played an important role in Pannekoek's observational methodology. By Pannekoek's estimation, for example, Boeddicker's drawings, although they were replete with stellar particularities, not only lacked a general sense of the Milky Way's overall distribution of light, but also suffered from poor lithographic reproduction; at the same time, he found the representation of particularities and distribution of light in Easton's and Schmidt's work rich by comparison.[9] Setting these images and verbal descriptions in relation to each other entailed making assessments of each astronomer's findings and modes of representation – a skilled practice that Pannekoek sought to systematize.[10]

As Pannekoek has written, even darkness opens up an enhanced means of viewing and seeing the Milky Way:

> In bright spots it is often advantageous to look at dark lanes, that divide them, and I have always found it best to look not exactly at the point that is being examined but somewhat above or at one side of it. By this indirect vision minute details appear, that are not visible when looked at directly.[11]

Indeed, as scientists later discovered, these dark voids in the sky – much of which results from the absorption, scattering, and polarization of visible light by particles and dust in the interstellar medium – can be tremendous sources of information. In the mid-twentieth century, as scholar Omar Nasim explains, the observations astronomers previously recorded by hand – particularly in drawings of galactic nebulae during the nineteenth century – were eclipsed by the investigation of 'nonvisible wavelengths'.[12]

9 Pannekoek 1920, 12.

10 Tai argues that Pannekoek's scientific persona resonated with what Lorraine Daston and Peter Galison have described as the epistemic virtue of 'trained judgement', which is predicated on a scientist's active and discerning assessment of the evidence at hand. As he writes (2017, 253), Pannekoek 'was part of a growing movement of scientists who increasingly emphasized the need for interpreted structure and systematized data. His ideal astronomer was actively involved in systematizing and analyzing the information provided by instruments or sense perceptions. His task was to recognize characteristic or distinguishing features of particularities and highlight them for other astronomers'. See also Daston and Galison 2007, 309-361.

11 Pannekoek 1897, 525-526.

12 Nasim 2013, 231.

Innumerable electromagnetic waves emanate from the darkness of the Milky Way, revealing the shrouding of light and energy at its centre. Largely impervious to interstellar extinction's distortions and diffusions, the emission of many other types of waves along the electromagnetic spectrum allow new stellar bodies to be 'seen', and even to 'speak'.

Voss's *Inverted Night Sky* exhibition also prospects such lacunae in areas further afield from the study of the distribution of light and stars in the Milky Way. Situated in between the dome projection and the series of panels at either end of the room lies an additional sound work entitled *Relativistic Working Time* (2016). Played back on a clear sound dome overhead, the track unifies the conceptual dynamic between the projection and the transparencies. Made in collaboration with artists Jessica Sehrt and Martin Stiehl – with whom Voss participates in the Realism Working Group, a young collective of activists, artists, and designers based in Frankfurt am Main – the audio reinterprets contemporary recordings of the electromagnetic emissions of selected neutron stars in the Milky Way. Remnants of collapsed massive stars, neutron stars appear to pulse due to the emission of electromagnetic radiation, like radio waves and visible light at their magnetic poles. Significantly, the orientation of the pulsar in relation to the observer (not necessarily confined to Earth but anywhere in the galaxy) determines how it is perceived; were it not for its movements, it would be hardly visible. Mediating between visually and instrumentally observable phenomena, its manipulated beats tick irregularly in critique of English mathematician and physicist Isaac Newton's conception of absolute time. The narrator on the adjacent projection dome's soundtrack paraphrases American sociologist and philosopher Lewis Mumford's 1934 *Technics and Civilization*, as if to echo the audio work's appeal to subjective knowledge and relativistic time: 'Life, instead, has regularities of its own, the beat of the pulse, the breathing of the lungs; these change from hour to hour depending on mood and action.'[13]

While idiosyncratic within astronomy and the natural sciences, Pannekoek's interest in the subjective nature of appearances is also more generally apparent in Voss's multimedia installations, in which historic

13 The original quote can be found in the section entitled 'The Monastery and the Clock' from the publication's first chapter: 'In terms of the human organism itself, mechanical time is even more foreign: while human life has regularities of its own, the beat of the pulse, the breathing of the lungs, these change from hour to hour with mood and action, and in the longer span of days, time is measured not by the calendar but by the events that occupy it'. See Mumford 1934, 15.

optical media confers agency to the viewer. In *Восстание рыбаков* (*Aufstand der Fischer*) (2011, see Figure 14.2), for example, Voss invokes the trope of the 'invisible hand', a metaphor found in the writings of the Scottish political economist Adam Smith, most famously in his 1776 *The Wealth of Nations*.[14] However, in Voss's case, the literal conjuring of a hand no longer represents the free markets, but a history that connects twentieth-century German author Anna Seghers with Erwin Piscator, a political theatre director of Weimar Germany. Piscator's theatre was produced in close rapport with workers – much like that of his contemporary, German playwright Bertolt Brecht. According to Brecht, 'workers judged everything according to the truth of its content; they welcomed every innovation which helped the representation of truth, of the real mechanism of society'.[15] This subjected all theatrical aesthetics to a realist baseline. To meet these new demands, Piscator sought to unmoor the illusionism of theatrical spectacle by way of photographic slides and film projections, which often documented street fights and mass demonstrations in Berlin.

The title of Voss's work refers to Seghers's 1928 novel *Revolt of the Fishermen of St Barbara*, in which fishermen in a rural, seaside town rise up against the market operations that set their wages. Crucially, the story is set without a coherent temporal or geographical specificity,[16] yet in 1934, Piscator completed a film adaption of Seghers's novella as an anti-fascist agitation against National Socialism. In *Aufstand der Fischer*, Voss restages Piscator's dramatic handling of Seghers's text as a scripted narrative, ultimately reflecting on the failure of the film adaptation's own agenda within the popular front against fascism. By way of an operational overhead projection device, Voss suggests that its disembodied female narrator is using the device in real time. As if by magic, her hands cast shadows on the wall as she moves through successive images. In actuality, the entire simulation of movement and presence is rigged up with a concealed digital projector. A similar effect takes place in Voss's *Phantascope* (*Light Archive*) (2013/2014), in which a digital projector surreptitiously imitates the slide projector's analogue techniques with stroboscopic images of historic examples of 1920s Berlin Dada photomontage intercut with present-day newspaper cutouts.

In both works, Voss makes use of the phantasmagoria – one of the most compelling and complex metaphors within Marxist thought for the

14 Smith 1776.
15 Brecht [1967] 2002, 83.
16 Schaub 2015.

**Figure 14.2 Восстание рыбаков (Aufstand der Fischer) by Jeronimo Voss (2011)**

Source: Oliver Ottenschlaeger/Secession Vienna (2013)

mendacious nature of capitalistic effects.[17] Popularized around the time of the French Revolution in the eighteenth century, the phantasmagoria itself was a technical apparatus predicated upon the artifice of theatre. A magic lantern – which could enlarge and project images painted or mounted onto glass slides by way of illumination – was hidden behind a translucent screen and mounted on wheels. Thus, its depicted figures – macabre, including ghosts, skeletons, and the like – were animated and took on lives of their own. For film scholar Tom Gunning, the epistemological implications of the phantasmagoria displays, which hinged upon the viewer's ability to distinguish between reality and illusion, and the uncanny interplay between them, remains one of the central concerns of modernism.[18] In his theatrical work, Piscator sought to empower individuals to free themselves of their passive relationship to the assumed 'benign' and 'benevolent' orchestrations of Smith's capitalist invisible hand by inverting the phantasmagoria's paradoxical simultaneity of illusion and reality. However, Seghers's original work already reflected the breakdown of the revolutionary project – as the narrator points out: 'On the very first page of her book she reveals that the revolt of the fishermen failed'.[19] Unwittingly, Piscator's film *Aufstand der Fischer*, which premiered in Moscow in October 1934, also reflected the nadir of epic theatre and related media arts' tactical efforts – as Voss's narrator concedes: 'By then, the Germans had long – and without any trace of revolt – decided for fascism.'

It is not inconsequential that Smith's first use of this metaphor appeared in one of the economist's early essays of natural philosophy entitled 'The History of Astronomy'.[20] Yet, the tendency found in Smith's work to develop connections between natural philosophy and social economy was neither

17 See 'The Fetishism of Commodities and the Secret Thereof' (section 4 of part 1, chapter 1) in Marx 1906, 83. As noted by scholars Tom Gunning and Margaret Cohen, the English translation inaccurately translates the German word 'phantasmagorische' (phantasmagoric) as 'fantastic'. See also Cohen 1989, 87-107 and Gunning 2005.

18 Referring to the aesthetic practices of modernity, Gunning (2005) writes, 'the [a]vant-[g]arde of the next century and a half [could be seen] as moving between these two poles – a direct and overwhelming address to the senses on the one hand, and the critique of illusion on the other.'

19 Voiceover transcription from Voss 2011.

20 Smith 1967, 30-109. Correlating Smith's notion of natural (economical and social) agency with that of the supernatural in the literary genre of the Gothic novel, scholar Stefan Andriopoulos (1999) argues that Smith's eventual 'inversion from "the invisible hand of Jupiter", disrupting the regular descent of heavy bodies' – as mentioned in his 'History of Astronomy' – to an impersonal 'invisible hand' in *The Wealth of Nations*, 'can be grasped as a naturalization of the supernatural'. Andriopoulos adds (1999, 753-754), 'Although Alec Macfie already drew attention to this first occurence [*sic*] of an "invisible hand" in Smith's writings [...], it has usually been ignored or dismissed as a "curious usage ... of ... limited interest" [...]'.

exclusive to his intellectual persona nor to his economic model of free markets. Socialist figures like Pannekoek and Blanqui, directed inferences to the night sky towards very different philosophical and epistemological ends. During the twentieth century's interwar period, Pannekoek was known, particularly within Germany, for his Marxist writings and his advocacy of council communism. While Blanqui sought to reconcile a theory of astronomical eternity with a revolutionary reality, Pannekoek compartmentalized the scientific and political intellectual threads of his work for much of his life. At periods which alternated with his major publications in the field of astronomy, Pannekoek argued in seminal socialist works, such as *Die taktischen Differenzen in der Arbeiterbewegung* (1909), for mass – as opposed to parliamentary – action, and in *De arbeidersraden* (1946), he meticulously delineated and critiqued a history of exploitative labour set against the backdrop of global geopolitics while promoting the establishment of workers' councils.[21] And yet, despite this division of his energies, 'Pannekoek's conception of the ideal scientist and the ideal Marxist were both rooted in the same epistemic concerns', as historian of science Chaokang Tai argues. To Pannekoek, 'the external world was a continuous and infinitely varied stream of events', of which the human cognition made sense.[22] In *L'éternité par les astres*, Blanqui grasps the methodological problem that gripped Pannekoek throughout this life – that of making sense of appearances and entities which escape or supercede perception. But just as Blanqui excavates political and social meaning from the limitations of vision, Pannekoek eschews statistical fact in favour of a kind of *collective visual speculation*.

Visualization, as we have seen, runs a wide spectrum of practices; and observation requires a sense of knowing and discerning phenomena. In contrast, aesthetic production is a generative process, which issues from the will and agency of the human subject. At the same time, it has also been long presumed (and ultimately encouraged) that works of art reflect an artist's subjectivity and individual point of view.[23] Above all, the sheer immensity of the Milky Way is a determining factor in wide dispersion of astronomers' observations of the night sky. This fact, in addition to the uniqueness and particularities of the subject, is what ultimately drove Pannekoek to his final thesis on its representation: 'The image of the Milky Way Galaxy that we observe is an optical phenomenon, whose materialization interacts with

21 Pannekoek 1909, 1946.

22 Tai 2017, 251.

23 See Daston and Galison's address of this topic in Daston and Galison 2007, 37.

different optical, physiological, and psychological conditions'.[24] However, instead of difference and discrepancy signifying failure, he set them into productive relation, generating new epistemological possibilities.

Jeronimo Voss prospects these illusions in his amplification of the aesthetic significance of technical media, like the phantasmagoria and the planetarium, which introduce new relations between the viewer's body and notions of transparency. In the curatorial statement of Voss's *Inverted Night Sky* exhibition in Amsterdam, the organizers equate Pannekoek's '"bottom-up" approach' towards making observations of the Milky Way Galaxy to 'his view that society ought to be structured by self-organized workers' councils, instead of being governed by state bureaucracy' and identifies this as a principal interest in the artist's project.[25] In a seeming deviation from conventional notions of scientific objectivity, Pannekoek collected and drew from these threads in order to generate an epistemology grounded in collective forms of knowledge production. Voss, too, is in search of new communal approaches towards contemporary society. Initiatives like Project Nika – in which he and his colleagues in the Realism Working Group seek to redevelop and recuperate abandoned commercial space in Frankfurt for a large communal space – signal his deep investments in the reconfiguration of the interrelation of living, engaging, and working.[26] These activities and the forms of interaction mobilized in the work cultivates the productive slippage between representation and a *material* reality that can be tactually manipulated and touched. Rather than merely rehabilitating media from the past, Voss comingles current techniques with (nearly) obsolete optical media in order to actualize a non-linear conception of time, which he infers from his reading of Blanqui's theoretical speculation of infinite worlds. If Pannekoek's work and life questioned the nature of perception and the conditions of collective participation, Voss's undertaking reassesses the socialist past by way of a politicized present.

24 Pannekoek 1920, 14. My translation.

25 SMBA 2016. This reading of his work echoes Tai's summation (2017, 249): '[Pannekoek's] bottom-up conception of the ideal society is reminiscent of the bottom-up method he applied in sidereal astronomy, where individual stars congregated into clusters and the combination of clusters formed the Milky Way Galaxy.'

26 Deutsch 2017. Previous living concepts conceptualized by members of this group were exhibited as part of *Wohnungsfrage* – a group exhibition, publication series, and academy on the contemporary status of housing and its relationship to various artist and housing initiatives and international architectural firms – held at the Haus der Kulturen der Welt in Berlin from 23 October – 14 December 2015. It included a shoptalk and panel discussion with the Realism Working Group and Florian Schmidt, Studio Commissioner of Berlin, held on 30 October 2015, entitled 'Art Studio – Studio Living' on new conceptions of the relationship between artists' living and working spaces.

## Bibliography

Andriopoulos, Stefan. 1999. 'The Invisible Hand: Supernatural Agency in Political Economy and the Gothic Novel'. *ELH* 66(3): 739-758.

Blanqui, Louis-Auguste. 1872. *L'éternité par les astres: hypothèse astronomique*. Paris: Librairie Germer Baillière.

—. 2013. *Eternity by the Stars: An Astronomical Hypothesis*. Trans. by Frank Chouraqui. New York: Contra Mundum Press.

Brecht, Bertolt. (1967) 2002. 'Against Georg Lukács: Popularity and Realism'. Transl. by Stuart Hood. In: Theodor Adorno, Walter Benjamin, Ernst Bloch, Bertolt Brecht, and Georg Lukács. *Aesthetics and Politics: The Key Texts of the Classic Debate within German Marxism*. London: Verso, 68-85.

Cohen, Margaret. 1989. 'Walter Benjamin's Phantasmagoria'. *New German Critique* 48: 87-107.

Daston, Lorraine and Peter Galison. 2007. *Objectivity*. New York: Zone Books.

Deutsch, Oliver. 2017. 'Groß-WG gegen den Leerstand'. *Frankfurter Rundschau*, 26 July 2017, 16.

Gunning, Tom. 2005. 'Illusions Past and Future: The Phantasmagoria and Its Specters'. Programmatic text for 'Refresh! First International Conference on the Histories of Art, Science and Technology', Banff New Media Institute, Banff, AB.

King, Henry C. 1978. *Geared to the Stars: The Evolution of Planetariums, Orreries, and Astronomical Clocks*. In collaboration with John R. Millburn. Bristol: Adam Hilger.

Krausse, Joachim. 1993. 'Das Wunder von Jena. Das Zeiss-Planetarium von Walter Bauersfeld'. *Arch+: Zeitschrift für Architektur und Städtebau* 116: 40-49.

Marx, Karl. 1906. *Capital*, vol. 1. Transl. by Samuel Moore and Edward Aveling. New York: Modern Library.

Palmieri, Rebekka. 2012. 'Eternity through the stars'. In *dOCUMENTA (13): Das Begleitbuch / The Guidebook*. Ostfildern: Hatje Cantz, 226.

Mumford, Lewis. 1934. *Technics and Civilization*. New York: Harcourt, Brace and Co.

Nasim, Omar W. 2013. *Observing by Hand Sketching the Nebulae in the Nineteenth Century*. Chicago, IL: University of Chicago Press.

Pannekoek, Anton. 1897. 'On the Best Method of Observing the Milky Way'. *Popular Astronomy* 5: 524-528.

—. 1909. *Die taktischen Differenzen in der Arbeiterbewegung*. Hamburg: E. Dubber.

—. 1912. 'A Photographic Method of Research into the Structure of the Galaxy'. *Proceedings of the Section of Sciences, Koninklijke Akademie van Wetenschappen te Amsterdam* 14: 579-584.

—. 1920. *Die Nördliche Milchstrasse*. Annalen van de Sterrewacht te Leiden, 11:3. Haarlem: Joh. Enschedé en Zonen.

—. [P. Aartsz, pseud.]. 1946. *De arbeidersraden*. Amsterdam: De Vlam.

Schaub, Christoph. 2015. 'Aesthetics, Masses, Gender: Anna Seghers's Revolt of the Fishermen of St. Barbara'. *New German Critique* 42(124): 163-188.

SMBA. 2016. 'Jeronimo Voss – Inverted Night Sky'. www.smba.nl/en/exhibitions/jeronimo-voss-inverted-night-sky-2/ (accessed 18 August 2016).

Smith, Adam. 1776. *An Inquiry into the Nature and Causes of the Wealth of Nations*. London: Methuen & Co.

—. 1967. 'The Principles Which Lead and Direct Philosophical Enquiries: Illustrated by the History of Astronomy'. In *The Early Writings of Adam Smith*, by Adam Smith, ed. by J. Ralph Lindgren. London: Kelley, 30-109.

Tai, Chaokang. 2017. 'Left Radicalism and the Milky Way: Connecting the Scientific and Socialist Virtues of Anton Pannekoek'. *Historical Studies in the Natural Sciences* 47(2): 200-254.

Voss, Jeronimo. 2011. *Восстание рыбаков* (*Aufstand der Fischer*). Berlin: Galerie Cinzia Friedlaender.

Williams, Alena. 2015. 'Akte der Enthüllung. Das Planetarium und László Moholy-Nagys Kunst der Projektion'. In *Bewegung/mouvement. Über die dynamischen Potenziale der Kunst*. Ed. by Andreas Beyer and Guillaume Cassegrain. Berlin/Munich: Deutscher Kunstverlag, 181-202.

—. 2017. 'The Planetary Cinema of Jeronimo Voss'. In *Jeronimo Voss: Inverted Night Sky*, ed. by GWK-Gesellschaft zur Förderung der Westfälischen Kunstarbeit. Dortmund: Kettler, 18-39.

## About the Author

**Alena J. Williams** is Assistant Professor of Film/Media History and Theory in the Department of Visual Arts at the University of California, San Diego. Her research areas include nineteenth- and twentieth-century technology and visual culture; theories of modernity, and the epistemology of the image. She is currently preparing a book-length study, *The Total Experiment: Cinema and the Modernist Work of Art*, which examines the way in which 'knowledge-making' and 'world-making' practices in astronomy converged with experimental cinema during the long twentieth century.

# Index

For Product Safety Concerns and Information please contact our
EU representative GPSR@taylorandfrancis.com Taylor & Francis
Verlag GmbH, Kaufingerstraße 24, 80331 München, Germany